Agricultural Production: Emerging Plant Growth Regulators

Agricultural Production: Emerging Plant Growth Regulators

Editor: Aila Williams

Callisto Reference
www.callistoreference.com

Callisto Reference,
118-35 Queens Blvd., Suite 400,
Forest Hills, NY 11375, USA

Visit us on the World Wide Web at:
www.callistoreference.com

ISBN: 978-1-64116-815-1 (Hardback)

Cataloging-in-Publication Data

Agricultural production : emerging plant growth regulators / edited by Aila Williams.
p. cm.
Includes bibliographical references and index.
ISBN 978-1-64116-815-1
1. Agricultural productivity. 2. Farm produce. 3. Crops. 4. Growth (Plants).
5. Agriculture. 6. Crop science. I. Williams, Aila.

S494.5.P75 A37 2023
338.16--dc23

Table of Contents

Preface

This book was inspired by the evolution of our times; to answer the curiosity of inquisitive minds. Many developments have occurred across the globe in the recent past which has transformed the progress in the field.

Plants produce hormones naturally, which affects various processes such as gene expression and transcription levels, cellular division, and growth. When these hormones are synthesized artificially, these are called as plant growth regulators (PGRs). They can be used to positively affect plant growth. There are different types of PGRs including auxins, cytokinins and gibberellins. that regulate the growth of a plant. They function as chemical messengers for intercellular communication. These plant regulators play a cardinal role in plant signaling networks as they are involved in the plant developmental process and a wide range of biotic and abiotic stress responses. Plant growth regulators can be extensively used within agricultural production for various purposes such as influencing reproductive development, vegetative growth and aging, along with preservation after harvesting. This book outlines the role of plant regulators in increasing agricultural productivity in detail. It presents researches and studies performed by experts across the globe. The book will prove immensely beneficial to professionals and students involved in this area of agriculture science at various levels.

This book was developed from a mere concept to drafts to chapters and finally compiled together as a complete text to benefit the readers across all nations. To ensure the quality of the content we instilled two significant steps in our procedure. The first was to appoint an editorial team that would verify the data and statistics provided in the book and also select the most appropriate and valuable contributions from the plentiful contributions we received from authors worldwide. The next step was to appoint an expert of the topic as the Editor-in-Chief, who would head the project and finally make the necessary amendments and modifications to make the text reader-friendly. I was then commissioned to examine all the material to present the topics in the most comprehensible and productive format.

I would like to take this opportunity to thank all the contributing authors who were supportive enough to contribute their time and knowledge to this project. I also wish to convey my regards to my family who have been extremely supportive during the entire project.

Editor

1

Application of Plant Growth Regulators on Soft White Winter Wheat under Different Nitrogen Fertilizer Scenarios in Irrigated Fields

Ruijun Qin [1,*], Christos Noulas [2], Don Wysocki [3], Xi Liang [4], Guojie Wang [5] and Scott Lukas [1]

1 Hermiston Agricultural Research and Extension Center, Oregon State University, Hermiston, OR 97838, USA; scott.lukas@oregonstate.edu
2 Institute of Industrial and Forage Crops, Hellenic Agricultural Organization-"Demeter", 41335 Larissa, Greece; noulaschristos@gmail.com
3 Columbia Basin Agricultural Research Center, Oregon State University, Adams, OR 97810, USA; dwysocki@oregonstate.edu
4 Aberdeen Research and Extension Center, University of Idaho, Aberdeen, ID 83210, USA; xliang@uidaho.edu
5 Eastern Oregon Agriculture and Natural Resources Program, La Grande, OR 97850, USA; guojie.wang@oregonstate.edu
* Correspondence: ruijun.qin@oregonstate.edu

Abstract: Lodging in cereal crops can result in yield loss and harvesting difficulties for growers. Application of plant growth regulator (PGR) has been an indispensable management practice to reduce lodging problems that are often exacerbated during high wind growing conditions and/or high nitrogen (N)/water environments, but the data is limited in the Columbia Basin of Oregon. The objective of this research was to evaluate the effect of two PGR products (chlormequat chloride-CC, trinexapac-ethyl-TE) at different rates and application timings on two soft white winter wheat varieties (ORCH-102 and SY Ovation). Crop growth (stem height and thickness), yield-related (spike density as ears m^{-2}, seeds per spike, grain weight) and quality parameters (test weight, protein) were measured for two cropping seasons from October 2017 to July 2019 following the application of the two PGR products at tillering (GS21-26), stem elongation (GS30-32), and/or flag leaf (GS37-39) stages under a high-N fertilizer scenario. In both growing seasons, no lodging problems were recorded for any treatments. The plant height was reduced after PGR application, but the impact on stem thickness was limited. PGR application slightly affected wheat yield, yield components, testing weight, and protein level in both growing seasons. Our results suggested that the effect of PGR application is relatively limited if no lodging problem occurred.

Keywords: chlormequat chloride-CC; trinexapac-ethyl-TE; nitrogen; stem height; wheat yield; lodging

---

1. Introduction

Soft white winter wheat (SWWW) (*Triticum aestivum* L.) is among the most important grain crops in the Columbia Basin of Oregon, and its production viability depends greatly on the adoption of newly released high-yielding varieties. Therefore, there is a need for updated information on how to best integrate these varieties into intensive production systems [1]. Wheat cultivated in the irrigated fields of the Columbia Basin is grown under various rotational schemes with potatoes, onions, corn, seed grasses, grain legumes, etc. Sustainable wheat production depends on high-yielding varieties responsive to inputs and cost-effective management practices, including efficient use of irrigation water, fertilizers, and pesticides.

The most yield-limiting nutrient for wheat is usually nitrogen (N) even though its optimum level of application is difficult to assess and depends on previous crop and soil type. The consideration of N mineralization plus accurate ammonium-N (NH_4^+-N) and nitrate-N (NO_3^--N) in soil tests is important in order to predict the amounts of fertilizer applications in both non-tilled (NT) and conventionally tilled (CT) wheat cropping systems. Horneck et al. [2] reported that the irrigated SWWW requires 308 to 336 kg N ha^{-1} to achieve optimum yield provided that the plant rooting depth is adequate, and irrigation water is sufficient. Under optimum conditions, wheat yields of 12.6 t ha^{-1} can be achieved [2]. Most commonly, N applications may be split as two times, i.e., in fall at planting and the next spring. It has been found that when wheat is sowed from October to early November under CT systems, it usually does not benefit from "starter" fertilizers when phosphorus (P) and potassium (K) in soil tests are adequate because the previous crop (e.g., grass seed, legumes, peppermint, or vegetable crops) most likely has provided a small amount of N [2,3]. In case that the winter wheat is planted after a crop with limited residual N in the soil (e.g., cereal crops) or it is planted later than the schedule, a small amount of N as 22–34 kg N ha^{-1} is needed [2,4]. Spring N application plays a very important role in wheat yield. Generally, the residual NO_3^--N at 60 cm (even 90 cm) and the residual $NH4^+$-N at 30 cm will be factored as soil available N when calculating supplemental N application rates [2]. Other factors include the previous crops and yield target in order to decide fertilizer N application rate.

A chronic worldwide constraint to sustainable wheat production is lodging that often decreases photosynthetic ability and biomass production, deteriorates seed quality, and creates difficulties to harvest operations [5–7]. Wheat and rice are highly prone to lodging during late vegetative growth and at reproductive stages [8,9], exaggerating the need for additional research efforts. Lodging in cereals can be defined as the permanent displacement of stems from the vertical position, which could result from either plastic failure of the stem base (stem lodging) or failure of the belowground anchorage system (root lodging) [7]. Both stem and root lodging can be classified as plant lodging [10,11]. Wheat grain yield reduction due to lodging varies between genotypes and can reach 8–61% [7,12,13]. However, genetics are not the only reason for resistance to lodging [14]; several agronomic management practices can also affect the production of thicker stems and stronger root systems [15,16]. Nitrogen levels, plant density, sowing date, seed type, soil strength, light intensity, soil moisture, diseases (i.e., eyespot infection), as well as weather extremes, such as wind speed and heavy rainfall, are all factors contributing to lodging [8]. Under high plant densities and/or luxuriant N applications, cereal crops substantially elongate stems, which can increase the risk of lodging and a significant reduction in yield [7,8,16].

In field practices, three plant growth stage scales, such as BBCH scales [17], Zadoks scale [18] and Feekes scale [19], are often referred in deciding the PGR application timing. In order to prevent lodging in winter wheat, many of the available plant growth regulators (PGRs) are applied prior to and/or during stem elongation (BBCH GS30-32) with the aim to produce lower internodes with shorter lengths (between 7 and 20 cm depending on variety and PGR applications), thicker stems and stem cell walls. PGRs are synthetic compounds that either mimic plant hormones or interrupt biosynthesis of plant hormones, which may alter the growth and development of the plant [20]. By reducing the overall plant height or increasing stem thickness, applications of PGRs may lessen the lodging potential of plants. Currently, there are five classes of plant hormones (i.e., auxin, gibberellin, cytokinin, ethylene, and abscisic acid) which can affect plant growth individually or interactively [20].

PGRs are effective if applied at the proper crop growth stage. Splitting applications may increase the likelihood of matching application timing with appropriate growing conditions and phonological stage. Selective breeding has produced wheat varieties, which are semi-dwarf, due to the presence of *Rht* genes, which were selected for good stem strength and lodging resistance [3]. When semi-dwarf genes are absent, applications of PGRs could reduce lodging occurrence. Applications of PGRs are also found to increase the number of the seeds produced per unit area, thus becoming a well-adopted practice for cereal productions in Europe (e.g., France, Germany, and the UK), New Zealand, United

States, and South America (i.e., Chile). Harvest index may also be increased by producing a greater grain yield and by reducing biomass produced [9,21].

The role of several main PGRs was found as inhibiting gibberellin biosynthesis at an early stage (e.g., chlormequat chloride (CC) and mepiquat chloride) or at a late stage (e.g., trinexapac-ethyl (TE) and prohexadione-Ca) or releasing ethylene (i.e., Ethephon) [22]. Cereal growers have used Chlormequat chloride for over 60 years [9], as an important group of PGRs for growth retardants. The use of CC to control lodging in wheat is considered the largest application for PGRs [23,24]. Trinexapac-ethyl (4-(cyclopropyl-a-hydroxymethylene)-3, 5-dioxo-cyclohexanecarboxylic acid ethylester) was introduced a few decades ago as an acylcyclohexanedione inhibitor of the 3b-hydroxylation of gibberellic acid [22,25]. It offers more flexibility in the timing of application and is commercially labeled as Palisade EC™ and registered in the United States by Syngenta Crop Protection Inc. According to its label, common application rates for cereal crops range from 841 to 1009 g ha^{-1}. The optimum time for TE application is at the onset of stem elongation (GS30-32) and before the emergence of the flag leaf (GS37-39). TE may interfere with the biosynthesis of gibberellic acid to reduce cell elongation, and as a result, it shortens the internodes and strengthens the stem. A shorter and thicker stem may help cereal crops reduce lodging and improve yields. Applications of TE in combination with N applications as early as pseudo-stem erection (GS30) has gained more interest in the Columbia Basin of Oregon as a means to reduce lodging and potentially improve winter wheat yields, but in-depth research is still limited.

In order to better understand the interaction of N fertilization and PGR applications (CC, TE) on wheat growth and productivity, in-depth research on effects of PGRs on grain yield and quality across different classes of wheat cultivars is needed. It is indispensable to guarantee that farmers in the region have the tools and management options necessary to maximize profitability; therefore, it is very important for them to better understand their agronomic management options when growing irrigated SWWW on fields with high soil N, such as those of Columbia Basin of Oregon. Precise application and kind of PGRs at the optimum growth stage to individual SWWW cultivars may achieve more consistent grain yield gains. From 2017 to 2019, we conducted two field trials with two predominantly grown SWWW cultivars. Through evaluating lodging potential, plant growth, yield, and yield components, the objectives of the study were: (i) To examine how applications of PGRs affect growth and production of SWWW under high N rates as early as stem elongation and/or flag leaf emergence; and (ii) to determine the interactions of N fertilization and PGR application and identify the best combination of N rate and PGR application to reduce lodging and improve grain yield and quality.

2. Materials and Methods

2.1. Experimental Location

Field trials were conducted at the Oregon State University-Hermiston Agricultural Research and Extension Center, Hermiston, OR (Latitude: 45°50′25″ N, Longitude: 119°17′22″ W, and Elevation above sea level: 140 m) from 2017 to of 2019 on an Adkins fine sandy loam (*mesic Fluvaquentic Endoaquepts*) soil. Soil properties can be referred to in Table 1. The location of the study belongs to a semi-arid climate with an annual precipitation of 264 mm, an averaged low temperature of 5 °C, and an averaged high temperature of 18.9 °C [26].

Table 1. Soil chemical properties in the two-year field trials at Hermiston, OR.

Growth Season	NO_3-N (ppm)	NH_4-N (ppm)	P (ppm)	K (ppm)	SO_4-S (ppm)	Zn (ppm)	$pH_{1:1}$	Organic Matter (%)
2017–2018	19.6	1.7	28.1	196	13.2	3.6	6.1	0.9
2018–2019	7.5	3.2	30.0	297	8.5	3.6	6.3	1.2

2.2. Plant Material

Two widely grown SWWW varieties that are regionally well adapted were selected for the present study: ORCF-102 and SY Ovation. The choice of these SWWW varieties was based on reasons related to resistance on disease pressure, on grain yield target and grain quality specifications and seed availability. These varieties are popular among farmers in Columbia Basin of Oregon. ORCF-102 is a variety developed by Oregon State University and the BASF Corporation in cooperation with USDA-ARS. It is an awned, short-statured, semi-dwarf variety with high yields across a wide range of environments in Oregon and Washington and disease resistance (e.g., straw breaker (eyespot) footrot, and current races of stripe rust) [27]. Straw strength of ORCF-102 is good, and lodging has not been observed in any production environment. SY Ovation is a short-statured SWWW variety with medium maturity and released in 2012 by AgriPro (Syngenta Cereals). It has been listed with a consistently high yield potential across the Pacific Northwest, with good straw strength and disease resistance [28]. It is well adapted in moderate to high rainfall regions of western Idaho, eastern Washington, north-central and northeastern Oregon and in irrigated production in Washington and the southern Snake River region of Idaho.

2.3. Planting and Management

Both varieties were sowed on 30 October 2017, and 26 October 2018, under a central pivot irrigation system. The irrigation was scheduled from April to June in each growing season with a water amount of 54.6, 144.8, and 139.7 mm in the 2017–2018 trial and of 50.8, 148.1, 197.1 mm in the 2018–2019 trial, respectively. The precipitation in March, April, May, and June was 11.7, 36.3, 15.5, and 7.4 mm in the 2017–2018 trial and 18.8, 26.2, 29.7, and 3.6 mm in the 2018–1029 trial.

In both growing seasons, sowing density was 300 seeds m^{-2} and sowing depth 1.3 cm. Each variety was sowed with a 9-row cone planter in a pass of 2 m wide. Plot dimensions were 2 by 10 m. For each variety, there were 36 plots in the first trial and 40 plots in the second trial, respectively, arranged in a randomized complete block design with four replicates. In both seasons, a starter fertilizer (a combination of urea (CH_4N_2O) and ammonium sulfate ($(NH_4)_2SO_4$)), including 67 kg N ha^{-1} and 11 kg S ha^{-1} was incorporated into the soil prior to planting to a depth of 20 cm.

In spring, N and S fertilizers were top-dressed to satisfy the requirements of plant growth and grain production according to Oregon State University production guide. Seasonal fertilizers were applied on 30 March 2018, for the first trial. Within the trial, one treatment did not receive any fertilizer (N0), one treatment received 168 kg N ha^{-1}, 22 kg S ha^{-1}, and 0.6 kg Zn ha^{-1} (normal N regime, N1) that was consistent with growers' practices in the region, and seven treatments received 280 kg N ha^{-1}, 22 kg S ha^{-1}, and 0.6 kg Zn ha^{-1} (high N regime, N2), (Table 2). Throughout the trial, the total N rates were 67, 235, and 347 kg ha^{-1} for N0, N1, and N2, respectively.

Table 2. Treatment details for the two-year field trials at Hermiston, OR.

Treatment Number *	Treatment Coding	Seasonal N Application	PGR Application (L ha^{-1} a.i.) at Different Growing Stages		
			GS21-26 (Tillering)	GS30-32 (Onset of Stem Elongation)	GS37-39 (Emergence of Flag Leaf)
1	N0	No N applied	0	0	0
2	N1	Normal N rate	0	0	0
3	N2	High N rate	0	0	0
4	CC-A	High N rate	0	1.004 (Adjust)	0
5	CC-B	High N rate	0	0	1.004 (Adjust)
6	CC-AB	High N rate	0	0.562 (Adjust)	0.442 (Adjust)
7	CC-C	High N rate	1.004 (Adjust)	0	0
8	TE	High N rate	0	0.126 (Palisade)	0
9	TE/CC1	High N rate	0	0.126 (Palisade)	0.442 (Adjust)
10	TE/CC2	High N rate	0	0.126 (Palisade)	1.004 (Adjust)

* Treatment 1 to 3 did not receive plant growth regulator (PGR) treatments, while treatment 4 to 9 received PGR treatments and high N rate; Treatment 7 was only used in the trial of 2018–2019.

In the second trial, one treatment did not receive any fertilizer (N0), while nine treatments received a total amount of 168 kg N ha^{-1}, 22 kg S ha^{-1}, and 0.6 kg Zn ha^{-1} on 3 April 2019, which represented the growers' standard practices (N1). Among these treatments, on 19 April 2019, eight treatments received an additional 168 kg N ha^{-1} to create high N regimes (N2). Based on the observations of the first trial in 2018 where no lodging occurred when a high amount of N fertilizer was applied in one dose, it was decided to split the N fertilization into two doses. The split N fertilizer application strategy was used in order to minimize N losses through leaching (due to the coarse-textured soil types) and improve N use efficiency. Throughout the trial, the total N rates were 67, 235, and 403 kg ha^{-1}, for N0, N1, and N2, respectively. For both trials, the high N regimes were applied in order to promote the occurrence of lodging and to thoroughly test the efficacy of the PGR products.

2.4. *Application of PGRs*

The PGR treatments were applied in April and May in 2018 and 2019, respectively, including all combinations of PGR products, application rates, and timing (Table 2). The PGR products include Palisade (a.i. 12% TE) and Adjust (Pending Registration; a.i. 55% CC). The PGR application timing was at three BBCH plant growth stage, i.e., tillering (GS21-26), stem elongation (GS30-32), and flag leaf (GS37-39).

In the first trial, PGRs were applied to plots of high N regimes, including 1.004 L ha^{-1} of Adjust (a.i.) applied at GS30-32 (CC-A), GS37-39 (CC-B), and at both stages (i.e., 10.562 L ha^{-1} at GS30-32 and 0.442 L ha^{-1} at GS37-39) (CC-AB). Additionally, 0.126 L ha^{-1} of Palisade (a.i.) applied at GS30-32, a combination of 0.126 L ha^{-1} of Palisade (a.i.) applied at GS30-32 and 0.442 or 1.004 L ha^{-1} of Adjust (a.i.) applied at GS37-39, representing as TE, TE/CC1, and TE/CC2, respectively, as well as a non-PGR application control (N2). In the second trial, additional treatment with 1.004 L ha^{-1} of Adjust (a.i.) was applied during the tillering stage (GS21-26), representing as CC-C. Meanwhile, a normal N treatment (N1) and non-fertilizer treatment (N0) were included in the trial. The PGR applications were sprayed on crop foliage using a backpack sprayer. For the first trial, the PGR applications were carried out on 21–22 April at GS31-32 stage and/or on 4 May 2018, at GS37-39. For the second trial, PGRs were applied on 3 April, during GS21-26, 20 April during GS30-32, and/or 22 May 2019, during GS39. It should be noted that herbicide or fungicides were not co-applied as a tank mix with PGR treatments.

2.5. *Plant Growth, Yield Measurement, and Data Analysis*

The lodging resistance score was assessed with visual field observations with a numeric scale from 1–9, representing as the most serious-lodging to no-lodging status. Plant height (cm) and stem thickness (mm) were measured twice: At water ripe when first grains have reached half of their final size (GS 71) on 25 May 2018 and on 3 June 2019, and at physiological maturity (GS92) on 25–26 June 2018 and on 15–17 July 2019, respectively.

Plots were harvested using a small-plot combine on 31 July 2018 and 1 August 2019. Grain yield (t ha^{-1}), grain moisture (%), protein content (%), test weight (kg L^{-1}), and thousand kernel weight (g) were measured. One week before harvesting, plants were cut at soil surface from one middle row in each plot in a length of 0.5 m. Plant samples were dried in an air-forced oven at 65 °C to a constant weight and separated into vegetative tissues (including leaf and stem) and spikes. The number of spikes were counted from each plant sample and converted to spike density (ear m^{-2}). The number of grains per spike was counted from 10 randomly selected spikes that were manually threshed. Grain weight (g) was measured from two random samples of 50 grains from each plant sample and converted to 1000 seed weight (g).

Data were analyzed using the GLIMMIX procedure (SAS 9.4, SAS institute 2008) by separating trials and varieties. In each trial and variety, treatments of N rate and PGR application were treated as fix effects and block as a random effect. Since the statistical analysis indicated that there were no significant treatment × year interactions, data are presented over the years. Means were separated using the LSD test at the 5% level when the F-test was significant. Pearson's correlation coefficients between wheat parameters were determined using PROC CORR.

3. Results and Discussion

3.1. Plant Growth

In both trials, the two SWWW varieties exhibited markedly shorter plant height at the two monitoring stages, GS71 or GS92 under the N0 treatment as compared to the rest treatments that included a combination of N fertilizer and PGRs ($p < 0.05$) (Figure 1). These observations were rather expectable since insufficient N supply may significantly affect crop growth.

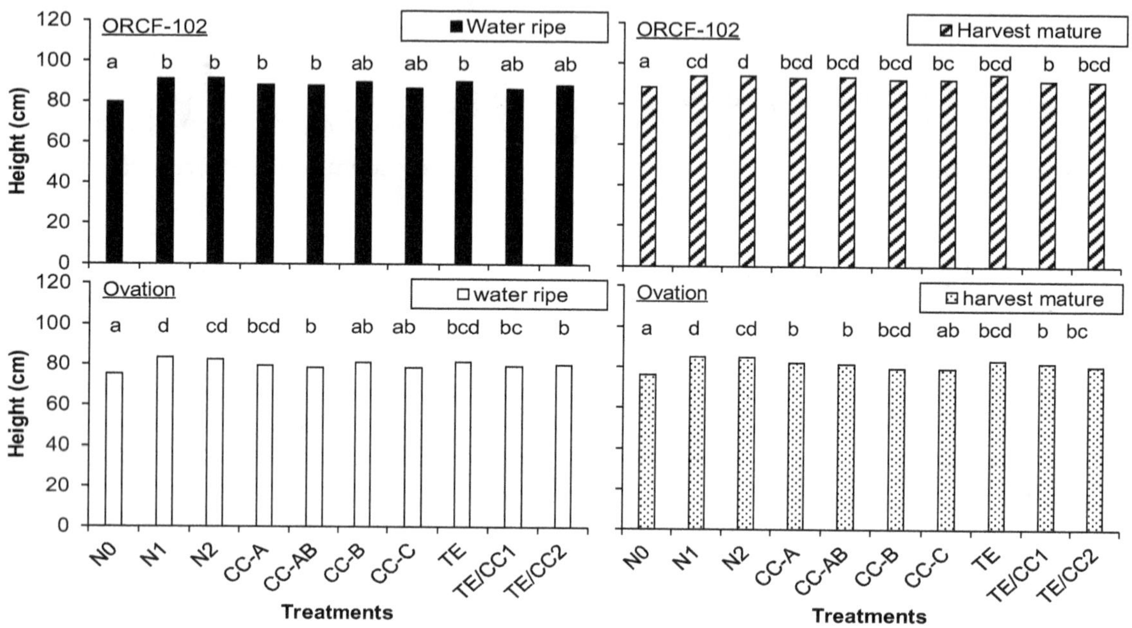

Figure 1. Plant height (cm) at water ripe (GS71) and harvest mature (GS95) for the two soft white winter wheat (SWWW) varieties among PGR-N treatment combinations across two years in Adkins fine sandy loam soil in Oregon. N0 = no fertilization; N1 = normal N rate; N2 = high-N rate without PGR application; CC-A = applying CC at GS30-32; CC-AB = applying CC at GS30-32 and GS37-39; CC-B = applying CC at GS37-39; CC-C = applying CC at GS21-26; TE = applying TE at GS30-32; TE/CC1 = applying TE at GS30-32 and applying CC (0.442 L ha^{-1} a.i.) at GS37-39; TE/CC2 = applying TE at GS30-32 and applying CC (1.004 L ha^{-1} a.i.) at GS37-39. Different letters on the top of each column indicate significant differences among treatments at 5% level according to LSD test.

At GS71, no significant differences were found for the height of ORCF-102 regardless of PGRs application. When the growth stage reached GS92, the height under N2 was the tallest (94.0 cm) among the treatments. The plant height of TE/CC11 and CC-C was shorter than N2 significantly ($p < 0.05$) (Figure 1).

For SY Ovation at GS71, plants under the CC-AB, CC-B, CC-C, TE/CC1, and TE/CC2 treatments were shorter than the N1 treatment ($p < 0.05$). At GS92, wheat plants under CC-A, CC-AB, CC-C, TE/CC1, and TE/CC2 treatments were shorter than the N1 treatments. No significant differences were found for the plant height under the N1 and N2 treatments at either growth stage, suggesting that the seasonal N application rate of 168 kg ha^{-1} is sufficient for wheat growth in the studied field conditions.

Our data indicated that PGRs reduce the plant height of SWWW, especially during the early growing stage (tillering; GS21-26). Among the PGR treatments, the application of CC at tillering (CC-C) produced a shorter plant height compared to the treatments without PGR applications (e.g., N1 and N2), suggesting that an early application of PGR (at tiller stage) might be an effective solution on controlling plant height, which was an earlier application date than the common practices with PGR applied at GS30-32 or GS37-39. Under the conditions of the present experiments when comparing the effect of CC and TE, no conclusive remark can be made, suggesting that CC may play a similar role as TE in winter wheat. Similarly, the application of PGRs on reducing wheat height was reported in other studies (e.g., [29–31]). In a Canadian study, Zhang et al. [32] reported that both TE and CC reduced plant height of six spring wheat cultivars. They also found the synergistic effect on reducing plant height when co-application of TE and CC were administered [32], which was not identified in our studies.

In contrast to stem height, no significant differences were found in stem thickness (or diameter of the stem) among the treatments in this study (data not shown). The stem thickness ranged from 3.5 to 3.8 mm for ORCF-102 and 3.6–3.9 mm for SY Ovation. Different from our findings, the application of PGRs was generally reported to thicken and strengthen stem thickness of cereal crops (e.g., References [32,33]). In general, the treatment without fertilization (N0) produced the smallest stem thickness although no difference was found statistically, because the N supply is not sufficient for promoting the wheat growth in the treatment without N fertilization.

As one of the most important traits, lodging control is commonly reported in studies. For example, the significant effect of TE and CC application on lodging control was reported in a Canadian study [32]. Wiersma et al. [31] reported that a PGR named Ethephon reduced the lodging problem of winter wheat. Webster and Jackson [30] found that PGRs delayed the severity of the lodging problem. However, it should be noted that throughout the field trials, no lodging occurred in our field trials, most possibly because the selected varieties have excellent straw strength. Therefore, the effectiveness of PGRs application on lodging should be evaluated with other varieties that are popular, but have relatively weak straw strength.

3.2. Yield and Protein Content

As expected, no addition of N (N0) resulted in significantly lower grain yields for both SWWW varieties when data were averaged across experimental years (Figure 2), because of the insufficient N supply. No yield benefit was found in the high-N regime (N2) when comparing to the standard N regime (N1), suggesting that 168 Kg N ha^{-1} is sufficient for acceptable wheat production. Grain protein of ORCF-102 and SY Ovation was the lowest in the N0 treatment, followed by N1, while the treatments under the high-N regime (i.e., N2) had greater protein levels ($p < 0.05$, Figure 3).

Our results in grain yield and protein in response to N supplies were consistent with the findings of Yang et al. [34] and Klikocka et al. [35]. Webster and Jackson [30] found that N top-dressing at anthesis increased grain protein content by 1.6 to 7.4% depending on N rate and soil residual N before anthesis.

Generally, it is believed that the application of PGRs increase grain yield (e.g., Reference [31]). Peake et al. [36] reported that PGR application improved grain of most wheat varieties in well-irrigated fields, which has sufficient N supply. However, differently from the above findings, some studies reported that the PGRs application reduced the grain yield [32,37]. The yield data in our field trials showed that no significant differences among any PGR and N2 treatments (Figure 2), suggesting that applications of CC and TE would not decrease grain yield in these studies.

No significant differences were found for protein content among the treatments with the high-N regime for ORCF-102. However, for SY Ovation, among the treatments under the high-N regime, CC-AB has the highest protein level, followed by CC-A, CC-C, N2, TE-CC1, then TE, TE-CC2, and CC-B ($p < 0.05$). The data suggests that the PGR applications would not reduce grain protein. Meanwhile, the effect of PGR on protein level is highly related to wheat cultivars. Corroborating with our result on ORCF-102, McMillan et al. [33] also did not find any effect of PGR on protein levels. No significant differences were found on test weight among treatments (data not shown). For both varieties, the test weight was in a range of 0.76–0.78 kg L^{-1}.

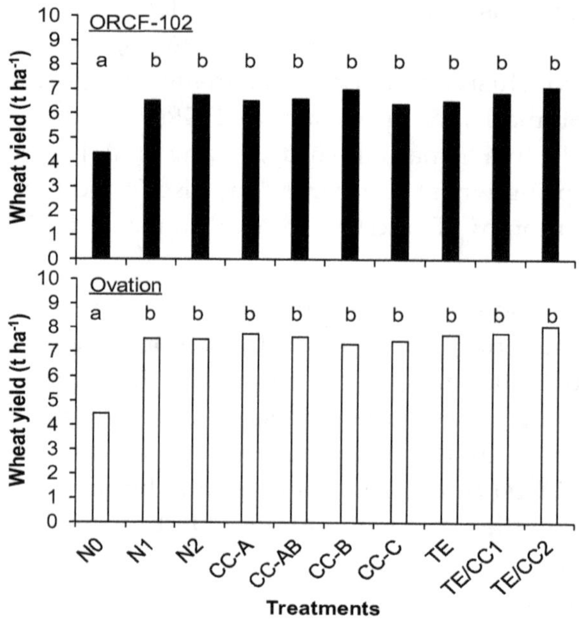

Figure 2. SWWW yield (t ha^{-1}) among PGR—fertilization treatments across two years in Adkins fine sandy loam soil in Oregon. N0 = no fertilization; N1 = normal N rate; N2 = high-N rate without PGR application; CC-A = applying CC at GS30-32; CC-AB = applying CC at GS30-32 and GS37-39; CC-B = applying CC at GS37-39; CC-C = applying CC at GS21-26; TE = applying TE at GS30-32; TE/CC1 = applying TE at GS30-32 and applying CC (0.442 L ha^{-1}a.i.) at GS37-39; TE/CC2 = applying TE at GS30-32 and applying CC (1.004 L ha^{-1} a.i.) at GS37-39. Different letters on the top of each column indicate significant differences among treatments at 5% level according to LSD test.

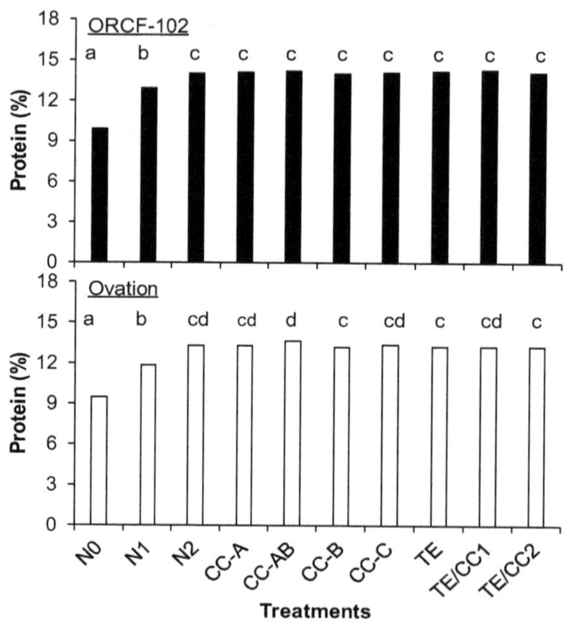

Figure 3. The protein content (%) of two winter wheat varieties among the treatments across the data of two years in Adkins fine sandy loam soil. N0 = no fertilization; N1 = normal N rate; N2 = high-N rate without PGR application; CC-A = applying CC at GS30-32; CC-AB = applying CC at GS30-32 and GS37-39; CC-B = applying CC at GS37-39; CC-C = applying CC at GS21-26; TE = applying TE at GS30-32; TE/CC1 = applying TE at GS30-32 and applying CC (0.442 L ha^{-1} a.i.) at GS37-39; TE/CC2 = applying TE at GS30-32 and applying CC (1.004 L ha^{-1} a.i.) at GS37-39. Different letters on the top of each column indicate significant differences among treatments at 5% level according to LSD test.

3.3. Yield Components

For both SWWW varieties, spike density (ear m^{-2}) was found the lowest in the N0 treatment (Figure 4). The combination of high N regime and PGR application, resulted in no significant difference in spike density (ear m^{-2}) for ORCF-102, while for SY Ovation, significantly lower spike density was found in the TE treatment compared with the TE/CC2 treatment.

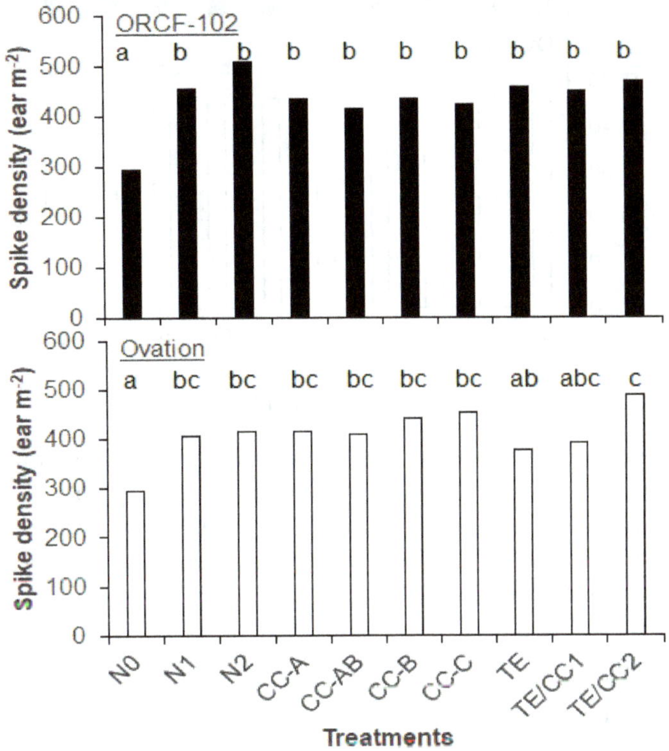

Figure 4. Spike density (ear m^{-2}) of two winter wheat varieties among the treatments across the data of two years in Adkins fine sandy loam soil. N0 = no fertilization; N1 = normal N rate; N2 = high-N rate without PGR application; CC-A = applying CC at GS30-32; CC-AB = applying CC at GS30-32 and GS37-39; CC-B = applying CC at GS37-39; CC-C = applying CC at GS21-26; TE = applying TE at GS30-32; TE/CC1 = applying TE at GS30-32 and applying CC (0.442 L ha^{-1}a.i.) at GS37-39; TE/CC2 = applying TE at GS30-32 and applying CC (1.004 L ha^{-1} a.i.) at GS37-39. Different letters on the top of each column indicate significant differences among treatments at 5% level according to LSD test.

No significant differences were observed on seeds per spike in either variety. For ORCF-102, seeds per spike were in the range of 41–47, and varied from 43 to 49 seeds per spike for SY Ovation (data not shown).

Among the treatments, the largest grain weight for both wheat varieties was found under the N0 and N1 treatments across both experimental years ($p < 0.05$; Figure 5). Among the treatments of the high-N regime in ORCF-102, the treatment CC-C had the lowest grain weight, followed by CC-B that was not different from other treatments of PGR applications. For SY Ovation, the TE/CC1 treatment had the lowest grain weight, followed by CC-C, CC-A, and CC-B, although no significant difference was found among PGR treatments. Similarly, McMillan et al. [33] also found that the PGR application reduced grain weight or kernel size of barley.

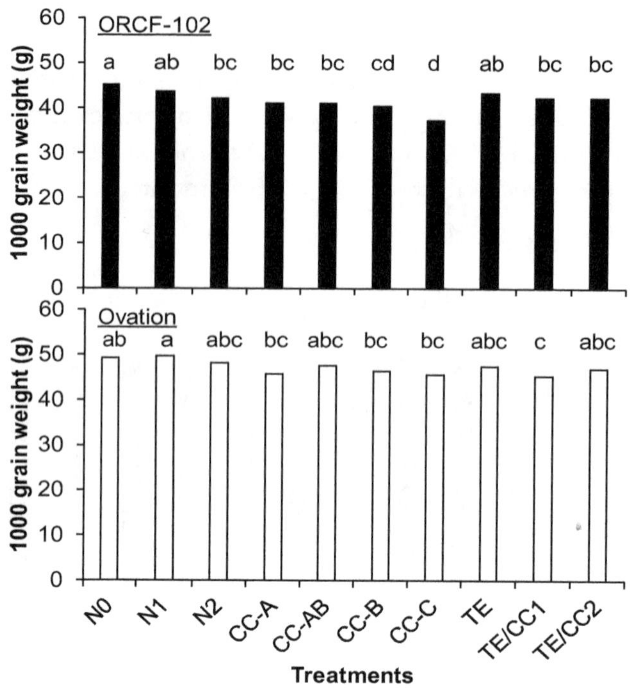

Figure 5. 1000-grain weight (g) of two SWWW varieties among the treatments across the data of two years in Adkins fine sandy loam soil. N0 = no fertilization; N1 = normal N rate; N2 = high-N rate without PGR application; CC-A = applying CC at GS30-32; CC-AB = applying CC at GS30-32 and GS37-39; CC-B = applying CC at GS37-39; CC-C = applying CC at GS21-26; TE = applying TE at GS30-32; TE/CC1 = applying TE at GS30-32 and applying CC (0.442 L ha^{-1} a.i.) at GS37-39; TE/CC2 = applying TE at GS30-32 and applying CC (1.004 L ha^{-1} a.i.) at GS37-39. Different letters on the top of each column indicate significant differences among treatments at 5% level according to LSD test.

3.4. Correlations Among Plant Growth, Yield, and Yield Components

Pearson correlation coefficients data showed that the yield was most positively correlated to grain protein and plant height ($p < 0.01$), as well as to test weight and spike density (ear m^{-2}) ($p < 0.05$) for both SWWW varieties (Tables 3 and 4).

Table 3. Pearson correlation coefficients among yield, protein content, test weight, plant height, stem thickness, spike density (ear m^{-2}), seeds per spike, and grain weight for ORCF-102 across two growing seasons in Columbia Basin of Oregon ($n = 76$).

Parameters	Protein	Test Weight	Plant Height	Stem Thickness	Spike Density	Seeds Per Spike	Grain Weight
Yield	0.59 **	0.39 **	0.53 **	0.00	0.30 *	0.00	−0.45 **
Protein		0.06	0.27 *	−0.05	−0.02	−0.02	−0.50 **
Test weight			0.05	0.16	0.04	−0.06	−0.12
Plant height				0.31 **	0.51 **	0.24 *	−0.04
Stem thickness					0.45 **	0.23	0.06
Spike density						0.24	0.16
Seeds per spike							0.11

* and ** indicate significant correlations at $p < 0.05$ and $p < 0.01$.

A negative correlation was found between yield and grain weight for ORCF-102 ($p < 0.01$), but an insignificant correlation in Ovation. Grain yield primarily consists of grain weight and the number of grains per unit area, but variation in grain weight is usually smaller [38]. Grain yield is, thus, not primarily determined by grain weight, but more dominantly by the number of grains per unit area [39].

Table 4. Pearson correlation coefficients among yield, protein content, test weight, plant height, stem thickness, spike density (ear m^{-2}), seeds per spike, and grain weight for SY-Ovation across two growing seasons in Columbia Basin of Oregon ($n = 76$).

Parameters	Protein	Test Weight	Plant Height	Stem Thickness	Spike Density	Seeds Per Spike	Grain Weight
Yield	0.54 **	0.29 *	0.48 **	0.031	0.42 **	0.11	−0.22
Protein content		0.04	0.28	−0.16	0.11	−0.15	−0.57 **
Test weight			−0.00	−0.02	0.06	0.04	−0.04
Plant height				−0.10	0.18	−0.05	0.03
Stem thickness					0.47 **	0.32 **	0.40 **
Spike density						0.18	0.23
Seeds per spike							0.21

* and ** indicate significant correlations at $p < 0.05$ and $p < 0.01$.

In both varieties, grain protein was negatively correlated to grain weight ($p < 0.01$), while it was positively correlated to plant height only for the ORCF-102. Stem thickness was found to positively correlated with spike density (ear m^{-2}) for both varieties ($p < 0.01$), while it was only correlated to spike density (ear m^{-2}) and grain weight for Ovation ($p < 0.01$). Plant height was correlated to stem thickness, spike density (ear m^{-2}), and grain number per spike for ORCH-102.

4. Conclusions

The application of PGRs was found to reduce the stem height of winter wheat, but their effects were minimal on stem thickness, grain yield, and protein in the absence of lodging conditions. The effect of PGR on spike density (ear m^{-2}) and grain weight varied with the different combinations of applications (e.g., rate, timing, and products) and wheat varieties. However, the application of CC tended to have a lower grain weight. There was no significant difference between the two tested PGRs on crop growth and production—indicating more PGR options for growers to use. Throughout our field trials, no lodging occurred at any treatment. Therefore, the effectiveness of the PGR application on lodging control was not feasible to be evaluated in this study. Nitrogen application is the essential factor limiting crop yield. In the studied region with the fine sandy loam soil, a base N fertilizer rate of 67 kg ha^{-1} and a seasonal N rate of 168 kg ha^{-1} should be sufficient to secure wheat grain yield. Spike density (ear m^{-2}) was the lowest without N application, contrary to grain weight, which was higher. The data implies that spike density (reproductive tillers) is a critical yield component in wheat production. Future research over multiple years and varieties will shed light on the effect of PGRs and N application to sustainable winter wheat production.

Author Contributions: All authors contributed substantially to the work reported in this paper. Conceptualization, R.Q.; methodology, R.Q.; investigation, R.Q.; data curation, R.Q., C.N., and X.L.; writing—original draft preparation, R.Q.; writing—review and editing, C.N., X.L., D.W., G.W., and S.L.; All authors have read and agreed to the published version of the manuscript.

Acknowledgments: Cory Zita, Dan Childs, and Tim Weinke of OSU-HAREC provided technical support for field management; summer interns (Andrew C. Rothe, Yan Yan, Greg Anderson, Avery Treadwell, Schae Borrego, and Nikkole Duitsman) contributed to the field measurement, sampling, and harvesting.

References

1. Schillinger, F.W.; Papendick, I.R. Then and now: 125 years of dryland wheat farming in the Inland Pacific Northwest. *Agron. J.* **2008**, *100*, S-166–S-182. [CrossRef]
2. Horneck, D.A.; Hart, J.M.; Flowers, M.D.; Lutcher, L.K.; Wysocki, D.J.; Corp, M.K.; Bohle, M. *Irrigated Soft White Wheat (Eastern Oregon)*; EM 9015-E; Oregon State University Extension Service: Corvallis, OR, USA, 2010.

3. Hart, J.M.; Flowers, M.D.; Roseburg, R.J.; Christensen, N.W.; Mellbye, M.E. *Nutrient Management Guide for Soft White Winter Wheat (Western Oregon)*; EM 8963-E; Oregon State University Extension Service: Corvallis, OR, USA, 2009.
4. Hart, J.M.; Christensen, N.W.; Jackson, T.L.; Karow, R.; Kronstad, W.E. *Fertilizer Guide. Winter Wheat (Western Oregon—West of Cascades)*; FG 9. OR; Oregon State University Extension Service: Corvallis, OR, USA, 2000.
5. Pinthus, M.J. Lodging in wheat, barley and oats: The phenomenon, its causes and preventative measures. *Adv. Agron.* **1973**, *25*, 210–263.
6. Ma, B.L.; Biswas, D.K.; Zhou, Q.P.; Ren, C.Z. Comparisons between cultivars of wheat, covered and hulless oats: Effects of N fertilization on growth and yield. *Can. J. Plant Sci.* **2012**, *92*, 1213–1222. [CrossRef]
7. Berry, P.M.; Sterling, M.; Spink, J.H.; Baker, C.J.; Sylvester-Bradley, R.; Mooney, S.J.; Tams, A.R.; Ennos, A.R. Understanding and reducing lodging in cereals. *Adv. Agron.* **2004**, *84*, 217–271.
8. Shah, L.; Yahya, M.; Shah, S.; Nadeem, M.; Ali, A.; Wang, J.; Riaz, M.W.; Rehman, S.; Wu, W.; Khan, R.M.; et al. Improving lodging resistance: Using wheat and rice as classical examples. *Int. J. Mol. Sci.* **2019**, *20*, 4211. [CrossRef]
9. Foulkes, M.; John, G.A.; Slafer, W.J.; Davies, P.; Berry, M.; Sylvester-Bradley, R.; Martre, P.; Calderini, D.F.; Griffiths, S.; Reynolds, M.P. Raising yield potential of wheat. III. Optimizing partitioning to grain while maintaining lodging resistance. *J. Exp. Botany* **2011**, *62*, 469–486. [CrossRef]
10. Berry, P.M.; Spink, J.H.; Gay, A.P.; Craigon, J. A comparison of root and stem lodging risks among winter wheat cultivars. *J. Agric. Sci. Camb.* **2003**, *141*, 191–202. [CrossRef]
11. Berry, P.M.; Sylvester-Bradley, R.; Berry, S. Ideotype for lodging resistant wheat. *Euphytica* **2007**, *154*, 165–179. [CrossRef]
12. Berry, P.; Spink, J. Predicting yield losses caused by lodging in wheat. *Field Crop. Res.* **2012**, *137*, 19–26. [CrossRef]
13. Acreche, M.; Slafer, G. Lodging yield penalties as affected by breeding in Mediterranean wheats. *Field Crop. Res.* **2011**, *122*, 40. [CrossRef]
14. Kashiwagi, T.; Ishimaru, K. Identification and functional analysis of a locus for improvement of lodging resistance in rice. *Plant Physiol.* **2004**, *134*, 676–683. [CrossRef]
15. Pham, Q.D.; Akira, A.; Hirano, M.; Sagawa, S.; Kuroda, E. Analysis of lodging resistant characteristic of different rice genotypes grown under the standard and nitrogen–free basal dressing accompanied with sparse planting density practices. *Plant Prod. Sci.* **2004**, *7*, 243–251.
16. Zhang, W.-J.; Wu, L.-M.; Ding, Y.-F.; Weng, F.; Wu, X.-R.; Li, G.-H.; Liu, Z.-H.; Tang, S.; Ding, C.-Q.; Wang, S.-H. Top–dressing nitrogen fertilizer rate contributes to reduce culm physical strength through decreasing in structural carbohydrates contents in japonica rice. *J. Integr. Agr.* **2016**, *15*, 992–1004. [CrossRef]
17. Meier, U. Growth Stages of Mono- and Dicotyledonous Plants. *BBCH Monogr.* **2001**. [CrossRef]
18. Zadoks, J.C.; Chang, T.T.; Konzak, C.F. A decimal code for the growth stages of cereals. *Weed Res.* **1974**, *14*, 415–421. [CrossRef]
19. Feekes, W. *De tarwe en haar milieu*; Hoitsema: Groningen, The Newtherlands, 1941; Volume 17, pp. 523–888.
20. Van Der Zanden, A.M. How Hormones and Growth Regulators Affect Your Plants. OSU Extension Service. 2012. Available online: https://extension.oregonstate.edu/gardening/techniques/how-hormones-growth-regulators-affect-your-plants (accessed on 7 May 2020).
21. Anderson, N.P.; Monks, D.P.; Chastain, T.G.; Rolston, M.P.; Garbacik, C.J.; Ma, C.-H.; Bell, C.W. Trinexapac-ethyl effects on red clover seed crops in diverse production environments. *Agron. J.* **2015**, *107*, 951–956. [CrossRef]
22. Rademacher, W. Growth retardants: Effects of gibberellin biosynthesis and other metabolic pathways. *Annu. Rev. Plant Physiol. Plant Mol. Biol.* **2000**, *51*, 501–531. [CrossRef]
23. Tolbert, N.E. (2-Chloroethyl) trimethylammonium chloride and related compounds as plant growth substances.1. Chemical structure and bioassay. *J. Biol. Chem.* **1960**, *235*, 475–479.
24. Hedden, P.; Sponsel, V. A century of gibberellin research. *J. Plant Growth Regul.* **2015**, *34*, 740–760. [CrossRef]
25. King, R.W.; Junttila, O.; Mander, L.N.; Beck, E.J. Gibberellin structure and function: Biological activity and competitive inhibition of gibberellin 2- and 3-oxidases. *Physiol. Plant.* **2004**, *120*, 287–297. [CrossRef]
26. Climate Hermiston—OR. U.S. Climate Data. 2020. Available online: https://www.usclimatedata.com/climate/hermiston/oregon/united-states/usor0159 (accessed on 7 May 2020).

27. Flowers, M.; Peterson, J.; Hulting, A.; Burns, J.; Kuehner, J. ORCF-I02 CLEARFIELD * Soft White Winter Wheat. 2008. Available online: http://washingtoncrop.com/documents/Wheat/Winter/Soft%20White/ORCF-102.pdf (accessed on 18 May 2020).
28. Washington State Crop Improvement Association. SY Ovation (03PN108-21). 2015. Available online: http://washingtoncrop.com/wp-content/uploads/2015/10/SY-Ovation-SWW.pdf (accessed on 7 May 2020).
29. Spolidorio, F.; Lollato, R. Plant growth regulators to decrease wheat height in high fertility scenarios. *Kans. Agric. Exp. Stn. Res. Rep.* **2019**, *5*. [CrossRef]
30. Webster, J.R.; Jackson, L.F. Management practices to reduce lodging and maximize grain yield and protein content of fall-sown irrigated hard red spring wheat. *Field Crop. Res.* **1993**, *33*, 249–259. [CrossRef]
31. Wiersma, D.W.; Oplinger, E.S.; Guy, S.O. Environment and cultivar effects on winter wheat response to Ethephon plant growth regulator. *Agron. J.* **1986**, *78*, 761–764. [CrossRef]
32. Zhang, Y.; Su, S.; Tabori, M.; Yu, J.; Chabot, D.; Baninasab, B.; Wang, X.; Ma, B.; Li, C.; Khanizadeh, S. Effect of selected plant growth regulators on yield and stem height of spring wheat in Ontario. *J. Agric. Sci.* **2017**, *9*. [CrossRef]
33. McMillan, T.; Tidemann, B.D.; O'Donovan, J.T.; Izydorczyk, M.S. Effects of plant growth regulator application on the malting quality of barley. *J. Sci. Food Agric.* **2019**. [CrossRef] [PubMed]
34. Yang, R.; Liang, X.; Torrion, J.A.; Walsh, O.S.; Brien, K.O.; Liu, Q. The Influence of water and nitrogen availability on the expression of end-use quality parameters of spring wheat. *Agronomy* **2018**, *8*, 257. [CrossRef]
35. Klikocka, H.; Cybulska, M.; Barczak, B.; Narolski, B.; Szostak, B.; Kobiałka, A.; Nowak, A.; Wójcik, E. The effect of sulphur and nitrogen on grain yield and technological quality of spring wheat. *Plant Soil Environ.* **2016**, *62*, 230–236. [CrossRef]
36. Peake, A.S.; Bell, K.L.; Fischer, R.A.; Gardner, M.; Das, B.T.; Poole, N.; Mumford, M. Cultivar × management interaction to reduce lodging and improve grain yield of irrigated spring wheat: Optimising plant growth regulator use, N application timing, row spacing and sowing date. *Front. Plant Sci.* **2020**. [CrossRef] [PubMed]
37. Zhang, G.; Chen, J.; Bull, D.A. The effects of timing of N application and plant growth regulators on morphogenesis and yield formation in wheat. *Plant Growth Regul.* **2001**, *35*, 239–245.
38. Sadras, V.O.; Slafer, G.A. Environmental modulation of yield components in cereals: Heritabilities reveal a hierarchy of phenotypic plasticities. *Field Crop. Res.* **2012**, *127*, 215–224. [CrossRef]
39. Liang, X.; Liu, Y.; Chen, J.; Adams, C. Late-season photosynthetic rate and senescence were associated with grain yield in winter wheat of diverse origins. *J. Agron. Crop Sci.* **2017**, *204*, 1–12. [CrossRef]

2

p-Aminobenzoate Organic Salts as Potential Plant Growth Regulators for Tomatoes

Radu-Liviu Sumalan [1], Lilia Croitor [2,3], Mihaela Petric [2], Isidora Radulov [1], Paulina Bourosh [3], Renata-Maria Sumalan [1] and Manuela Crisan [2,*]

1 Faculty of Horticulture and Forestry, Banat's University of Agriculture Science and Veterinary Medicine "King Michael Ist of Romania" from Timisoara, Calea Aradului nr 119, 300645, Timisoara, Romania; sumalanagro@yahoo.com (R.-L.S.); isidoraradulov@yahoo.com (I.R.); srenata_maria@yahoo.com (R.-M.S.)

2 "Coriolan Dragulescu" Institute of Chemistry, 24 Mihai Viteazul Blvd., 300223, Timisoara, Romania; croitor.lilia@gmail.com (L.C.); mihaelapetric@yahoo.com (M.P.)

3 Institute of Applied Physics, Academiei Street 5, MD2028, Chisinau, Moldova; bourosh.xray@phys.asm.md

* Correspondence: mdorosencu@yahoo.com

Abstract: The discovery of environmentally friendly and inexpensive plant growth regulators (PGRs) for agronomically important crops is a necessity and must be considered a priority worldwide. This study provides the synthesis, structure determination and the biological evaluation of two binary organic salts as potential PGRs. New compounds have dual biological activity and are based on natural metabolite *p*-aminobenzoic acid (*p*ABAH) and different alkanolamines. Studied compounds exhibit hydrogen-bonded 3D supramolecular architectures with different crystal packing due to the formation of one homosynthon and various heterosynthons. The biological profile of new compounds was investigated in laboratory and greenhouse on *Solanum lycopersicum* L., revealing the efficiency in promoting plant rooting and plant productivity. The results may have a positive impact on agricultural economics, developing new sustainable PGRs for tomatoes.

Keywords: *p*-aminobenzoic acid; alkanolammonium salt; biological activity; plant growth regulator; crystal structure; supramolecular assembly

---

1. Introduction

Growing populations have imposed the use of plant growth regulators (PGRs), which have become an integral part of agricultural and horticultural practices, maximizing crop production. To mitigate the harmful effects of conventional synthetic PGRs, such as 2,4-dichlorophenoxyacetic acid-2,4-D, on the environment and plant health [1], extensive efforts have been devoted to discovering sustainable alternatives. Since natural compounds extracted from plants or produced by bacteria rapidly degrade, limiting the applicability of the product [2], the use of PGRs based on synthetic analogues of natural products or active pharmaceutical ingredients commercially available are considered to be more effective [3]. Therefore, developing innovative, environmentally friendly and cost competitive PGRs remains an important task for researchers and a necessity at the global level for nutrition security. Understanding of chemical and crystal structure, inter- and intramolecular interactions as well as the structure–biological activity relationship is essential for investigation of new PGRs. Extensive studies show the importance of the carboxyl group and the planar aromatic ring in the structure of auxin-like PGRs, essential molecules that control almost every aspect of dormancy, seed germination and plant development [4].

Thus, benzoic acids, well-known for their importance as building blocks in drug development, gained increasing attention in plant science, being involved in various physiological processes from the regulation of seed germination [5–7] to disease resistance and stress tolerance in plants [8,9]. Among these compounds, *p*-aminobenzoic acid (*p*ABAH) is a well-known natural metabolite present in plant and animal tissues, widely described in literature as a precursor of folic acid [10] and recently of coenzyme Q [11]. It possesses numerous biological activities in medicine such as antioxidant [12], antibacterial [13], antimutagenic [14], anticoagulant [15], fibrinolytic and immunomodulating agent [16], protective drug against UV-irradiation [17] and also in agriculture as a chemical inducer associated with thermotolerance [18] and pathogens resistance [19] in plants. Moreover, an examination of the Cambridge Structural Database (CSD) has confirmed the *p*ABAH versatility as a fundamental building block in the design of soluble forms of organic multicomponent crystals and coordination compounds with various supramolecular architectures [20,21]. Besides being frequently used as co-formers, such as aliphatic and heterocyclic amines, alkanolamines are very few despite their lower toxicity [22], and their use as intermediates for the production of active pharmaceutical and cosmetic ingredients.

Our previous studies show the huge potential of alkanolamine-substituted benzoic acid systems which can generate supramolecular architectures with different topologies (1D, 2D and 3D) guided by different non-covalent interactions [23–25], have low toxicity [25,26], thermal stability [27] and promising applicability as auxin-like PGRs on the model plant *Arabidopsis thaliana* Col 0 [6]. The aim of this work was to develop new alkanolammonium *p*-aminobenzoates with characteristics of ideal PGRs: low toxicity, easily synthesized in laboratories, non-expensive, soluble in water and to investigate their role in growth and development of the most important vegetable cultivated in Romania and European Union in the last years, *Solanum lycopersicum* L. The discovery of new PGRs for commercial vegetables is essential, because vegetables play a vital role in food front, being the cheapest natural sources. Herein, two new alkanolammonium *p*-aminobenzoates were synthesized, structurally and physicochemically characterized and evaluated for their PGRs' activity on tomatoes (*S. lycopersicum* L).

2. Results and Discussion

The new compounds were prepared by proton exchange reaction of *p*ABAH with different alkanolamines (HEEA-*p*ABA—ethylethanolammonium *p*-aminobenzoate; HDEEA-*p*ABA—diethylethanolammonium *p*-aminobenzoate) and isolated in excellent yields (>95%). The pKa rule [28] predicted the salts formation, because ΔpKa values pKa(alkanolamine)-pKa(*p*ABAH) were greater than three. Physicochemical and structural characterizations of new compounds are presented below.

2.1. Crystallographic Study

The single crystal X-ray diffraction data indicated that HEEA-*p*ABA crystallizes in the monoclinic centrosymmetric $P2_1/c$ space group, while HDEEA-*p*ABA crystallizes in the orthorhombic asymmetric $Pna2_1$ space group (Table 1).

The analysis of both crystal structures revealed the formation of 1:1 organic salts HEEA-*p*ABA and HDEEA-*p*ABA with proton transfer from carboxylic group of *p*ABAH to nitrogen atoms of alkanolamine molecules (Figure 1). The $(p\mathrm{ABA})^-$ anion formed non-planar system in HEEA-*p*ABA and a practically planar system in HDEEA-*p*ABA, since the dihedral angle between the least squares plane of the phenyl ring C_6 and the COO^- group was equal to 20.1° and 5.6°, respectively.

The nitrogen atoms in the amino group in both structures had almost a pyramidal configuration, its valence angles being equal to 108.07, 109.00 and 111.22° in $(\mathrm{HEEA})^+$ of HEEA-*p*ABA, and 108.80, 109.83 and 110.01° in $(\mathrm{HDEEA})^+$ of HDEEA-*p*ABA. The $(\mathrm{HEEA})^+$ and $(\mathrm{HDEEA})^+$ cations adopt the Syn-Clinal conformation with the N(2)CCO(3) torsion angles equal to −66.90 and −66.65°, respectively. The alkanolamine cations and $(p\mathrm{ABA})^-$ anions were held together by two charge-assisted $O\text{-}H\cdots O^-$ and $N^+\text{-}H\cdots O$ hydrogen bonds in HEEA-*p*ABA as distances C-O(1) and C-O(2) are equal to 1.270(2) and 1.254(3) Å and one charge-assisted $N^+\text{-}H\cdots O^-$ and one classic O-H···O hydrogen bonds in HDEEA-*p*ABA with C-O(1) and C-O(2) distances equal 1.254(8) and 1.276(8) Å. As a result, in both compounds, a similar $R_2^2(9)$ graph set was formed. In both structures, the NH_2 group of $(p\mathrm{ABA})^-$ anion participated to charge-assisted N-H···O⁻ and classic N-H···O hydrogen bonds with carboxylic oxygen atoms of adjacent *p*ABA anions (Table 2).

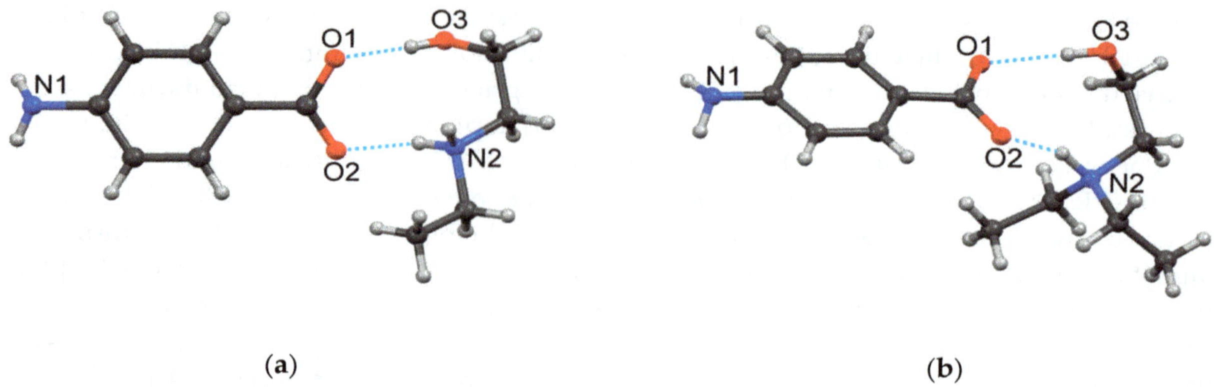

Figure 1. View of the HEEA-*p*ABA (**a**) and the HDEEA-*p*ABA (**b**) with partial atomic labelling and charge-assisted hydrogen bonds.

Table 1. Crystallographic data and structure refinement details for new compounds.

Compound	HEEA-*p*ABA	HDEEA-*p*ABA
Empirical formula	$C_{11}H_{18}N_2O_3$	$C_{13}H_{22}N_2O_3$
Fw	226.27	254.32
T (K)	293(2)	293(2)
Crystal system	Monoclinic	Orthorhombic
Space group	$P2_1/c$	$Pna2_1$
a (Å)	8.3465(4)	17.9492(17)
b (Å)	8.1802(5)	11.1812(9)
c (Å)	18.8676(9)	7.1445(6)
β (°)	102.684(5)	90
V (Å^3)	1256.76(11)	1433.9(2)
Z	4	4
ρ_{calcd} (g cm^{-3})	1.196	1.178
μ (mm^{-1})	0.087	0.084
Crystal size (mm)	0.60 × 0.25 × 0.05	0.400 × 0.200 × 0.040
2Θ range (°)	3.332 to 24.990	3.569 to 24.996
Reflections collected/unique	4344/2199 (R_{int} = 0.0243)	3177/1870 (R_{int} = 0.0417)
Reflections with ($I > 2\ \delta(I)$)	1576	1059
Parameters	148	167
R_1, wR_2 ($I > 2\ \delta(I)$)	0.0528, 0.1382	0.0609, 0.1259
R_1, wR_2 (all data)	0.0774, 0.1564	0.1147, 0.1484
GOF [c]	1.000	1.000
Largest diff. peak/hole (e Å^{-3})	0.332/−0.306	0.178/−0.151

Cations and anions in the HEEA-*p*ABA structure were held together in a supramolecular network (Figure 2a) obtained from $R_2^2(9)$ heterosynthons (Figure 2b) and $R_4^4(22)$ homosynthons formed by four anions (Figure 2c) with the involvement of fine C-H···O hydrogen bond between amine and carboxylic groups (Table 2), in addition to the mentioned N(O)-H···O hydrogen interaction.

Table 2. Hydrogen bond distances (Å) and angles (°) in HEEA-*p*ABA and HDEEA-*p*ABA.

D-H···A	d(D–H)	d(H···A)	d(D···A)	∠(DHA)	Symmetry Transformations for Acceptor
			HEEA-*p*ABA		
O(3)-H(1)···O(1)	0.82	1.83	2.650(2)	178	x, y, z
N(1)-H(1)···O(3)	0.87	2.27	3.077(3)	155	$-x + 1, y + 1/2, -z + 1/2$
N(1)-H(2)···O(2)	0.87	2.34	3.010(3)	135	$-x, y + 1/2, -z + 1/2$
N(2)-H(1)···O(1)	0.89	1.89	2.775(2)	174	$-x + 1, -y + 1, -z + 1$
N(2)-H(2)···O(2)	0.89	1.92	2.787(2)	165	x, y, z
C(8)-H(8)···O(3)	0.97	2.62	3.508(3)	152	$-x + 1, -y, -z + 1$
			HDEEA-*p*ABA		
O(3)-H(3)···O(1)	0.82	1.87	2.676(5)	170	x, y, z
N(1)-H(1)···O(2)	0.87	2.09	2.908(8)	158	$-x + 3/2, y + 1/2, z + 1/2$
N(1)-H(2)···O(1)	0.87	2.29	2.858(9)	123	$-x + 3/2, y + 1/2, z-1/2$
N(2)-H(2)···O(2)	0.98	1.78	2.701(6)	156	x, y, z
C(8)-H(8)···O(3)	0.97	2.54	3.501(9)	172	$-x + 1, -y + 1, z-1/2$
C(13)-H(8)···O(2)	0.96	2.60	3.285(8)	129	x, y, z

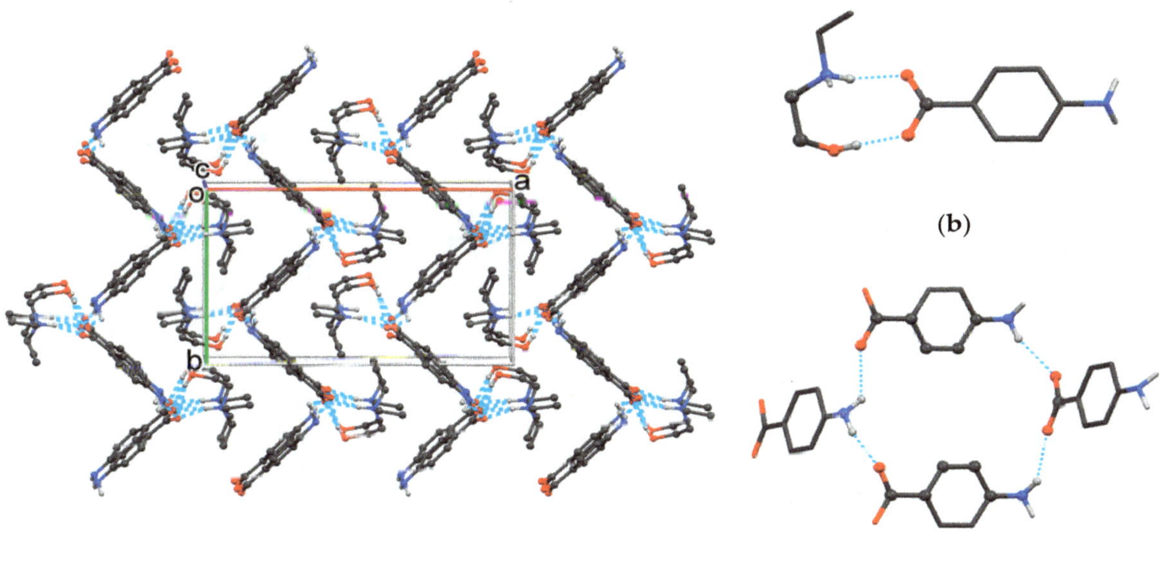

(b)

(a) (c)

Figure 2. The crystal packing of HEEA-*p*ABA (**a**) with representation of $R_2^2(9)$ heterosynthon (**b**) and $R_4^4(22)$ homosynthon in crystal (**c**).

The crystal packing of HDEEA-*p*ABA (Figure 3a) was based on seven heterosynthons between HDEEA cations and *p*ABA anions: bicomponent *R* (9) heterosynthon discussed above, three tetracomponent heterosynthons formed by two cations and two anions (Figure 3b–d) and three hexacomponent heterosynthons assembled by two cations and four anions (Figure 3e–g). The crystalline structure was additionally stabilized by fine C-H···O hydrogen bonds with the involvement of C(8) and C(13) as donors of H (Table 2).

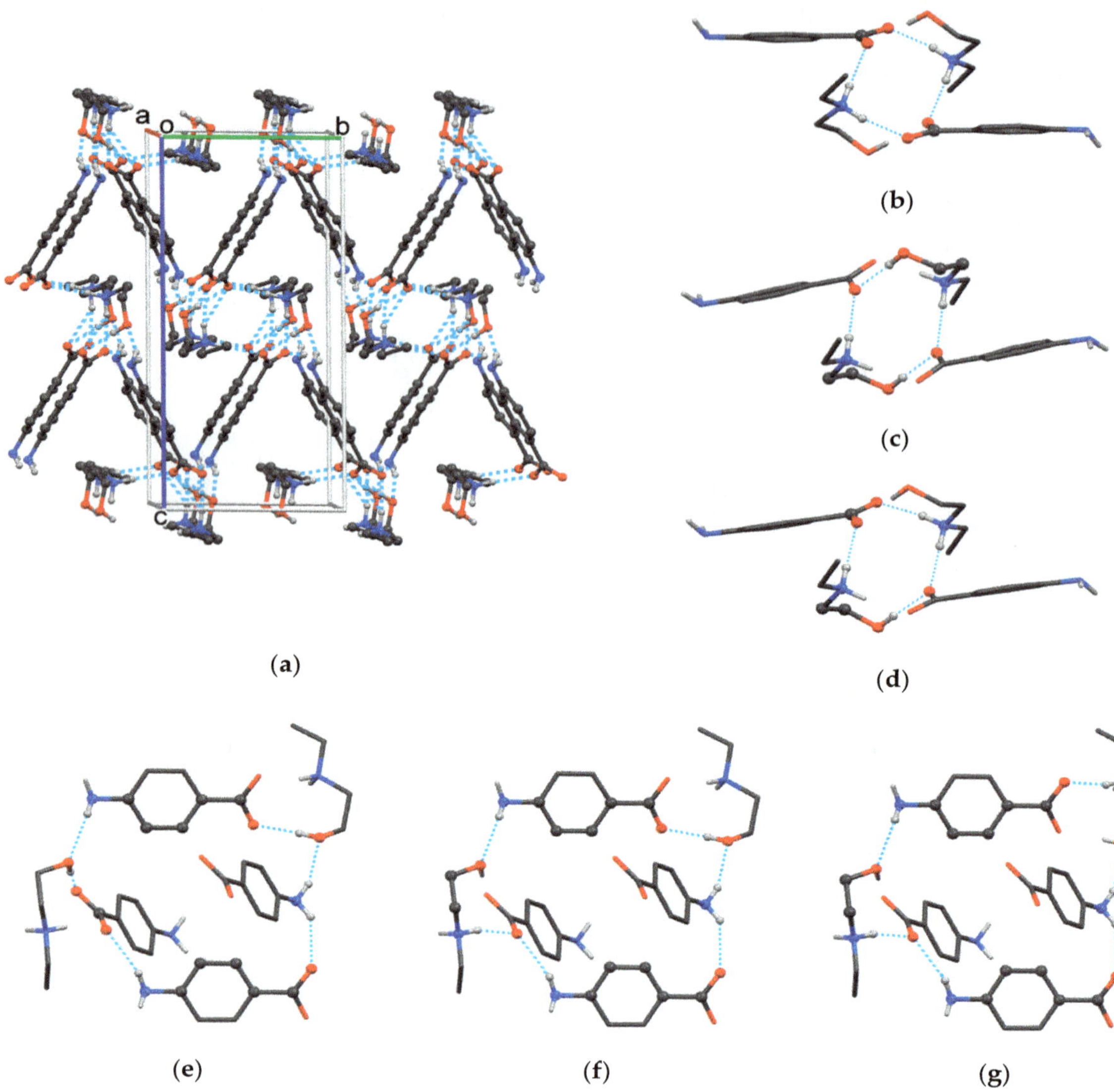

Figure 3. The crystal packing (**a**) and multicomponent heterosynthons found in HDEEA-*p*ABA crystal: $R_4^4(12)$ (**b**), $R_2^4(14)$ (**c**), $R_3^4(13)$ (**d**), $R_6^6(26)$ (**e**), $R_5^6(27)$ (**f**) and $R_5^6(30)$ (**g**).

2.2. FTIR Spectroscopy

The FTIR spectroscopic investigations (Figure 4) provide a supplementary insight into alkanolammonium salts formation by exhibiting the differences in the intensities and wavenumbers of C=O, N-H and O-H stretching modes in new compounds compared to the corresponding acid.

The HEEA-*p*ABA and HDEEA-*p*ABA spectra had certain similarities: the presence of strong bands responsible for the symmetric (1378 cm^{-1}; 1381 cm^{-1}) and asymmetric (1602 cm^{-1}; 1599 cm^{-1}) stretching vibrations of carboxylate group which do not exist in the spectra of *p*ABAH, as well as the absence of νC=O (1671 cm^{-1}), νC-OH (1287 cm^{-1}) and δOH (1417 cm^{-1}) bands characteristic of the carboxylic group [29]. The lower values for symmetric and asymmetric νCOO$^-$ stretching vibrations compared to the free acid indicate that carbonyl was H-bonded in the anion–cation and anion–anion systems. Furthermore, the absence of bands characteristic of the –COOH group confirmed the formation of HEEA-*p*ABA and HDEEA-*p*ABA salts.

The strong and narrow absorption bands (3464 cm^{-1} and 3360 cm^{-1}) assigned for asymmetric and symmetric NH_2 stretching vibrations in *p*ABAH spectrum decreased in intensity, and wavenumbers in the case of two alkanolammonium salts, becoming 3425 cm^{-1} and 3341 cm^{-1} for HEEA-*p*ABA and 3367 cm^{-1} for HDEEA-*p*ABA. These indicate N-H···O$^-$ and N-H···O intermolecular interactions between anion–anion in HEEA-*p*ABA and anion–anion/anion–cation in HDEEA-*p*ABA. The broad bands in the range 3200–2700 cm^{-1} are attributed to O-H stretching vibrations and overlap with NH_2^+ vibration band (3043 cm^{-1}) in the case of secondary alkanolamine. Weak bands of N-H deformation vibrations were observed at 1640 cm^{-1} and 1655 cm^{-1}, respectively, in salt spectra. Supplementary proof of the salt formation was confirmed by the appearance in HEEA-*p*ABA and HDEEA-*p*ABA spectra of C-O stretching vibrations at 1100–1000 cm^{-1}, belonging to alkanolamines. The differences between two new alkanolammonium salts spectra also illustrate the diversity of the synthons in crystals.

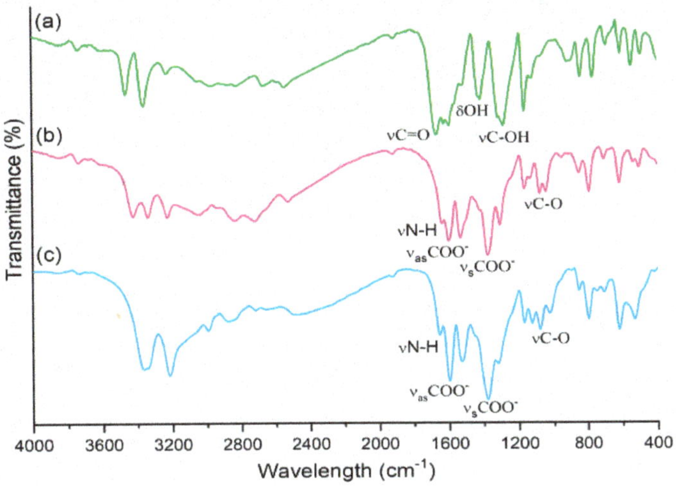

Figure 4. FTIR spectra of *p*ABAH (**a**), HEEA-*p*ABA (**b**) and HDEEA-*p*ABA (**c**).

2.3. Thermal Analysis

The salts were characterized by thermal analysis (thermogravimetric analysis—TGA; derivative thermogravimetry—DTG; and heat flow—HF) in order to determine thermal stability (Figure 5). The TGA and DTG curves show that HEEA-*p*ABA and HDEEA-*p*ABA are thermally stable up to 124 °C and 90 °C, respectively when they begin to decompose into two steps with a total mass loss of over 99.4%.

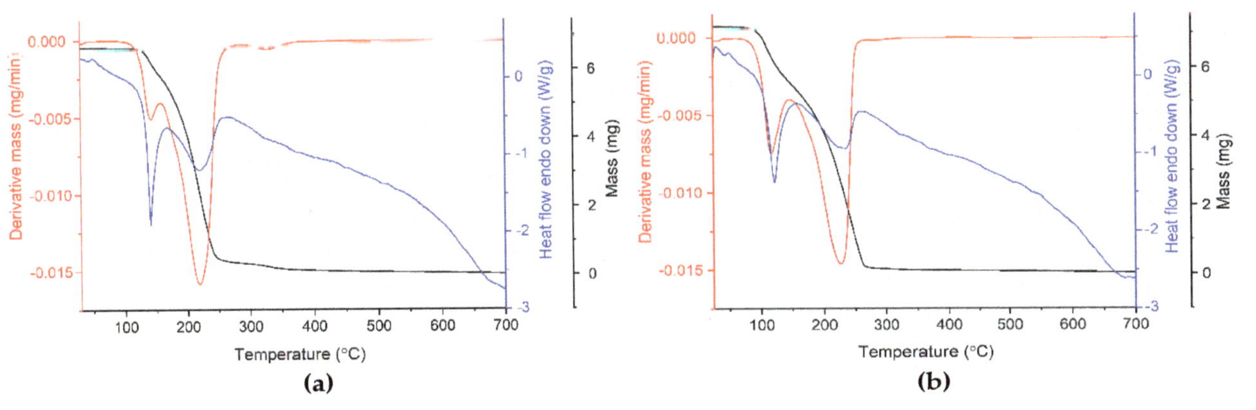

Figure 5. Thermo-analytical curves of HEEA-*p*ABA (**a**) and HDEEA-*p*ABA (**b**).

Therefore, HEEA-*p*ABA had a first mass loss of 13.1% in the range 124.77 °C–160.15 °C, probably due to the removal of the ethyl group from the cation. This decomposition process was accompanied by melting at 138.52 °C. The second step up to 700 °C corresponded to 81.68% mass loss and complete decomposition of the compound. In the case of HDEEA-*p*ABA, the decomposition process, up to 700 °C, began with a 22.59% mass loss due to the elimination of diethyl groups from the cation and continued with the second step which involved a final 76.25% mass loss ascribed to the $C_9H_{12}N_2O_3$ molecule. The DTG curve accompanied the decomposition process by one sharp peak at 115.87 °C and another broad peak at 226.13 °C. HF curve showed a first strong and sharp endothermic peak at

120.85 °C, attributed to the melting point of the compound. Thermogravimetric studies indicate that the studied compounds were anhydrous and HEEA-*p*ABA was more thermally stable than HDEEA-*p*ABA.

2.4. Biological Studies

The auxin-like PGRs' effect on new compounds was investigated in laboratory and greenhouse conditions, on tomato, the main commercial vegetable cultivated in Romania and Europe in the last years. Commercial indole-3-acetic acid (IAA), the most abundant and studied natural auxin in plants, was used as the reference compound. The results of experiment 1 (Table 3) showed that new compounds (HEEA-*p*ABA, HDEEA-*p*ABA) and IAA inhibited germination at all tested concentrations or even blocking it (e.g., IAA, at concentrations above 0.5 mM) especially in the first interval of treatment (after 7 days), compared with the control. The results obtained show that increasing the concentration leads to inhibition of germination. Thus, in the case of HEEA-*p*ABA application, increasing concentration from 0.1 mM to 0.5 mM and then to 1 mM, reduced significantly seed germination by about 20% and 9%, respectively. Instead, HDEEA-*p*ABA decreased seed germination by approximately 24.5% only from 0.5 mM to 1 mM.

Table 3. The influence of treatments on tomato seed germination (tomato cultivar (cv.) Tomtim).

Moment	After 7 Days			After 10 Days		
Treatment	**0.1 mM**	**0.5 mM**	**1 mM**	**0.1 mM**	**0.5 mM**	**1 mM**
Control (MES)	91.11 ^{a}	91.11 ^{a}	91.11 ^{a}	100 ^{a}	100 ^{a}	100 ^{a}
Indole-3-acetic acid (IAA)	^{x} 8.89 ^{c}	^{x} 0 ^{d}	^{x} 0 ^{c}	^{x} 75.56 ^{b}	^{y} 0 ^{b}	^{y} 0 ^{b}
HEEA-*p*ABA	^{x} 46.67 ^{b}	^{y} 26.67 ^{c}	^{y} 17.78 ^{b}	^{x} 88.89 ^{ab}	^{x} 93.33 ^{a}	^{x} 91.11 ^{a}
HDEEA-*p*ABA	^{x} 40.00 ^{b}	^{x} 44.44 ^{b}	^{y} 20.00 ^{b}	^{x} 97.78 ^{a}	^{x} 93.33 ^{a}	^{x} 88.89 ^{a}
	Treatment. $LSD_{5\%}$ = 17.68 (a,b,c,d); Concentration. $LSD_{5\%}$ = 11.21 (x,y)			Treatment. $LSD_{5\%}$ = 16.64 (a,b); Concentration. $LSD_{5\%}$ = 17.11 (x,y)		

The differences among the variants noted with different letters are considered significant ($p < 0.05$).

A tendency towards uniformization of values was revealed ten days after the induction of germination. An increase of germination with approximately 22% was observed only at the concentration of 0.1 mM, for HDEEA-*p*ABA treatment compared to IAA. In the case of 0.5 and 1 mM concentrations, HEEA-*p*ABA and HDEEA-*p*ABA treatments had similar effects; the increase of the concentration did not significantly influence seed germination. Analysis of the effects of the three treatments at the same concentration level shows that the treatments with HEEA-*p*ABA and HDEEA-*p*ABA had significantly higher effects than IAA on tomato seed germination.

Recent studies have characterized auxins as regulators that act positively by maintaining dormancy [30] and negatively by inhibiting seed germination [31]. The hormonal signalling of natural auxin IAA has often been associated with physiological processes manifested during plant growth and development, including seed germination [32]. Some previous studies have shown that exogenous application of IAA inhibits and delays seed germination [33,34]. The mechanism underlying germination blockade is not yet fully elucidated, but it seems to be due to the fact that auxin induces seminal break by enhancing signal transduction of abscisic acid, which associates auxin with maintaining seminal dormancy [30].

Therefore, the results of the tested compounds on tomato seeds germination confirm that IAA shows clear germination blocking effects, and the new compounds, even if they belong to the same category of compounds, have inhibitory effects only in the first part of the interval, the differences from the control being insignificant after a 10 day treatment.

Otherwise there are some studies that prove that besides phytohormones, there are chemical compounds that have the ability to regulate plant growth and development, such as nitrogen oxide and ROS (superoxide; hydrogen peroxide; hydroxyl radical; hydroxyl ion; and nitric oxide), both types being proven to regulate seed dormancy and germination [35–38].

Regarding experiment 2 (Table 4), the chlorophyll content of tomato seedlings was influenced by HEEA-*p*ABA and HDEEA-*p*ABA treatments, showing a significant increase compared to the control and IAA. The effect of new treatments on morphological characters in tomato seedlings indicated that plant height and primary root length presented variations comparatively with control and IAA, while maximum length of secondary roots had similar values to control and significantly smaller effect to IAA. The number of secondary roots (SR) registered the highest values under the effect of IAA treatment compared to the other treatments. The HEEA and HDEEA generated a significant increase (more than 86% of the number of SR) compared to the control.

Table 4. The influence of treatments on tomato seedlings (cv. Tomtim).

Character/ Treatment	Height (cm)	Chlorophyll (SPAD)	Primary Root Length (cm)	No of Secondary Roots	Maximum Length of Secondary Roots (cm)
Control (MES)	5.50 ± 0.50^{b}	31.20 ± 1.49^{c}	11.10 ± 0.37^{ab}	25.00 ± 2.99^{c}	9.27 ± 0.32^{a}
IAA	3.20 ± 0.18^{c}	32.25 ± 0.25^{c}	8.65 ± 0.55^{c}	104.50 ± 18.42^{a}	3.50 ± 0.55^{b}
HEEA-*p*ABA	5.25 ± 0.75^{b}	38.20 ± 0.60^{a}	11.50 ± 0.50^{a}	46.50 ± 8.47^{b}	8.28 ± 0.43^{a}
HDEEA-*p*ABA	6.25 ± 0.95^{a}	36.15 ± 0.25^{b}	10.50 ± 0.50^{b}	51.00 ± 1.00^{b}	8.39 ± 0.37^{a}
$LSD_{5\%}$	0.61	1.51	0.59	15.93	1.28

The differences among the variants noted with different letters are considered significant ($p < 0.05$).

It is known that auxin plays an important role in controlling the roots growth and plant development. The total amount of this major phytohormone is generated by the local biosynthesis and transport. The natural auxin IAA reveals an inhibitory effect on primary root (PR) growth. SR are essential components of the root system which contribute to maximizing the absorption capacity of water and nutrients, facilitating rapid adaptation to environmental changes [39,40]. It is well known that IAA is involved in each stage of SR's formation, the disorders in auxin biosynthesis and transport resulting in a reduced number of SR [41–45].

Therefore, our results attest that HEEA-*p*ABA and HDEEA-*p*ABA treatment solutions stimulate the plants height and the length growth of PR in contrast to IAA, which has inhibitory effects. Also, new compounds promote the formation of a large number of SR and support their growth, compared to control. Contrary, exogenous IAA generates the formation of a large number of SR (104.5 ± 18.42), probably due to the ability of acropetal migration within the root [46], but without ensuring their length growth. These results can also be correlated with the higher capacity of chlorophyll biosynthesis in variants treated with 0.5mM HEEA-*p*ABA and HDEEA-*p*ABA exogenous solutions.

The dynamics of the plants height (Table 5) shows that the foliar treatments with HDEEA-*p*ABA and IAA have a similar and significantly superior influences to the treatments with HEEA-*p*ABA and control, in all the stages of development, from the beginning of flowering until the first stage of fruit emergence.

Table 5. The treatments effect on height (cm) of tomato plants (cv. Tomtim) in different stages of development.

Development Stage/ Treatments	Beginning of Flowering	Flowering	Emergence of First Fruiting Floor
Control (MES)	$^{z}\,42.50 \pm 3.07^{b}$	$^{z}\,66.75 \pm 5.02^{b}$	$^{yz}\,83.75 \pm 7.53^{b}$
IAA	$^{z}\,49.00 \pm 0.58^{a}$	$^{z}\,77.00 \pm 2.89^{a}$	$^{yz}\,96.67 \pm 3.67^{a}$
HEEA-*p*ABA	$^{z}\,43.00 \pm 4.51^{b}$	$^{z}\,64.67 \pm 4.67^{b}$	$^{yz}\,85.67 \pm 7.85^{b}$
HDEEA-*p*ABA	$^{x}\,47.00 \pm 1.00^{a}$	$^{zx}\,79.67 \pm 1.20^{a}$	$^{yz}\,104.33 \pm 3.39^{a}$

Treatments $LSD_{5\%}$ = 4,36 (a,b); developmental stage $LSD_{5\%}$ = 48.87 (x,y,z). The differences among the variants noted with different letters are considered significant ($p < 0.05$).

The three treatments and control did not differ significantly in their effect on the number of leaves at the beginning of flowering (Table 6). Significantly higher number of leaves compared to control has been stimulated by IAA during the flowering period and by HDEEA-*p*ABA treatment at the emergence of first fruiting floor of tomato plants. All treatments presented a significant increase in the number of leaves from the beginning of flowering to full flowering, with differences between 3.33 in HEEA-*p*ABA and 5.33 in HDEEA-*p*ABA. In the period from flowering to the emergence of first fruiting floor, the number of leaves per plant showed little and no significant variations in all cases.

Table 6. The influence of treatments on leaves number of tomato plants (cv. Tomtim) in different stages of development.

Development Stage/ Treatments	Beginning of Flowering	Flowering	Emergence of First Fruiting Floor
Control (MES)	^{y} 6.50 ± 0.50 ^{a}	^{x} 10.50 ± 0.87 ^{b}	^{x} 12.75 ± 0.48 ^{b}
IAA	^{y} 7.33 ± 0.67 ^{a}	^{x} 12.33 ± 0.67 ^{a}	^{x} 13.67 ± 0.88 ^{ab}
HEEA-*p*ABA	^{y} 7.33 ± 0.33 ^{a}	^{x} 10.67 ± 1.45 ^{b}	^{x} 13.33 ± 0.88 ^{ab}
HDEEA-*p*ABA	^{y} 6.67 ± 0.88 ^{a}	^{x} 12.00 ± 1.00 ^{ab}	^{x} 14.33 ± 0.67 ^{a}

Treatments *LSD5%* = 1.54 (a,b); Developmental stage *LSD5%* = 3.24 (x,y). The differences among the variants noted with different letters are considered significant ($p < 0.05$).

It is known that auxins can stimulate plant growth by enhancing photosynthesis due to the fact of chlorophyll increases [47]. The chlorophyll pigments are probably the most relevant natural biomolecules due to the fact of their importance in photosynthesis. Therefore, a direct correlation between their quantity and gross primary productivity has been demonstrated [48]. Our results present a significant increase in chlorophyll content induced by HDEEA-*p*ABA (41.43 ± 0.69) and a significantly lower one for HEEA-*p*ABA (37.07 ± 1.74) compared to IAA treatment and control, at the beginning of flowering (Table 7). In the flowering and emergence of first fruiting floor periods, the highest chlorophyll content was determined in the tomato plants treated with HDEEA-*p*ABA and IAA, respectively, while in the variants HEEA-*p*ABA and control, the chlorophyll content was lower. Control plants recorded a significant increase of chlorophyll content from the beginning of flowering to the emergence of first fruiting floor, in contrast to treated plants, which showed a notable increase only between the first two phenophases.

Table 7. Effect of treatments on chlorophyll content (SPAD) of tomato plants' foliar apparatus (cv. Tomtim) in different stages of development.

Development Stage/ Treatments	Beginning of Flowering	Flowering	Emergence of First Fruiting Floor
Control (MES)	^{u} 39.13 ± 1.43 ^{b}	^{z} 47.00 ± 2.14 ^{b}	^{yz} 48.75 ± 0.70 ^{b}
IAA	^{y} 39.33 ± 1.53 ^{b}	^{x} 51.50 ± 1.45 ^{a}	^{x} 50.73 ± 0.69 ^{a}
HEEA-*p*ABA	^{z} 37.07 ± 1.74 ^{c}	^{y} 47.43 ± 2.38 ^{b}	^{y} 49.30 ± 1.46 ^{b}
HDEEA-*p*ABA	^{y} 41.43 ± 0.69 ^{a}	^{x} 50.37 ± 0.03 ^{a}	^{x} 51.17 ± 0.23 ^{a}

Treatments $LSD_{5\%}$ = 0.64 (a,b,c); developmental stage $LSD_{5\%}$ = 6.73 (u,x,y,z). The differences among the variants noted with different letters are considered significant ($p < 0.05$).

These results support the hypothesis that the initiation of the flowering and fruiting processes, which means the passage of plants from the vegetative to the generative stage, diminishes or stagnates the dynamics of chlorophyll biosynthesis, the plant concentrating the metabolic activity towards the biosynthesis of photosynthetic compounds such as carbohydrates. Therefore, the cellular metabolism was modified by reducing the biosynthesis of functional molecules, with nitrogen, and stimulating the synthesis of structural biomolecules, with carbon [49].

There are studies which show that foliar application of auxinic compounds, at different concentrations, induced increases in plant height, fresh and dried mass, number of shoots and leaves per plant, as well as productivity-related components [50,51]. Other research, on the contrary, concluded that the application of low or moderate doses of exogenous auxins did not generate significant changes on the plant growth parameters and the high doses even had effects of reducing the values of these parameters compared to the untreated variants [52–55].

The foliar apparatus, or the number of leaves per plant, is the essential component of the photosynthetic process. The leaves are considered the primary photosynthetic organs, and their number, along with their surface, determines the amplitude of the assimilation process. Previous studies have shown that phytohormones, especially auxin (IAA) and gibberellins (GA), play a key role in the emergence and development of foliar apparatus [56]. The results prove the stimulatory effect of foliar auxin application in tomatoes plant growth, the best results being obtained by using HDEEA-*p*ABA and IAA, respectively, at a 0.5 mM concentration. Plants with a large number of leaves show an increased photosynthetic potential, which allows the synthesis, transport, and accumulation of larger amounts of valuable bioactive compounds [57,58].

3. Materials and Methods

3.1. Materials and Physical Measurements

Reagents *p*ABAH, EEA (ethylethanolamine) and DEEA (diethylethanolamine) were purchased from Fluka Chemie AG (Buchs SG, Switzerland)) in analytical purity and were used without further purification. Melting points of recrystallized compounds were determined on a Boetius melting point apparatus and were uncorrected. The FTIR spectra (KBr pellet) were recorded on a JASCO-FTIR-4200 spectrometer (Easton, MD, USA), in the range 4000–400 cm^{-1} with a resolution of 4.0 cm^{-1} and a scanning speed of 16 mm s^{-1}. Thermal analysis was performed using a TGA/SDTA 851-LF 1100 Mettler apparatus (Columbus, OH, USA) The sample weight of about 7 mg was used for the test. The measurements were carried out in a dynamic air atmosphere, in the temperature range of 25–700 °C with a heating rate of 10 °C min^{-1}.

3.2. Synthesis Procedure of HEEA-pABA and HDEEA-pABA

New compounds were prepared in a 1:1 molar ratio, adding dropwise EEA (0.71 mL, 7.28 mmol) and DEEA (0.97 mL, 7.3 mmol) respectively into a solution of *p*ABA (1 g, 7.29 mmol) in acetone (20 mL), under constant stirring at room temperature. White microcrystalline precipitates in high yields (>95%) were formed after the addition of alkanolamines. The compounds were collected by filtration, washed with acetone, and dried in air. Colourless crystals suitable for single-crystal X-ray diffraction analysis were obtained after few days by slow evaporation in acetone. Melting points (Boetius) were: HEEA-*p*ABA m.p. = 140–141 °C and HDEEA-*p*ABA m.p. = 122–123 °C.

3.3. Single Crystal X-ray Diffraction Study

Diffraction measurements for compounds HEEA-*p*ABA and HDEEA-*p*ABA were carried out at room temperature on an Xcalibur E diffractometer (Abingdon, Oxfordshire, United Kingdom) equipped with a CCD area detector and a graphite monochromator utilizing MoKα radiation. Final unit cell dimensions were obtained and refined on an entire data set. All calculations to solve the structures and to refine the proposed models were carried out with the SHELXL2014 program package [59]. All non-hydrogen atoms were refined anisotropically. Hydrogen atoms attached to carbon, nitrogen and oxygen atoms were positioned geometrically and treated as riding atoms. The X–ray data and the details of the refinement for both compounds are summarized in Table 1, and selected geometric parameters are given in Table 2. Figures 1–3 were produced using the Mercury program (Cambridge, United Kingdom) [60]. Crystallographic data of the new compounds reported herein were deposited with the Cambridge Crystallographic Data Centre (Cambridge, United Kingdom) and allocated under the deposition numbers CCDC 1989303–1989304 (Supplementary Materials).

3.4. *Biological Assays*

Seeds of tomato cultivar named Tomtim created by BUASVM Timisoara, included in the CPVO (Community Plant Variety Office) test protocols, UPOV (The International Union for the Protection of New Varieties of Plants) and National Tests Guidelines were used in this study.

3.4.1. Germination Tests (Experiment 1)

Tomatoes' mature seeds were surface sterilized by keeping them in a commercial 50% (v/v) sodium hypochlorite solution for 10 min and rinsed three times with distilled water before transferring them to Petri dishes. Twenty-five sterilized seeds were each placed into a 9 cm Petri dish containing two Whatman filter papers moistened with 5 mL solution of MES (2-(N-morpholino)ethanesulfonic acid) 5 mM and CaSO4 0.5 mM (pH 6) for control and different concentrations (0.1, 0.5 and 1 mM) of HEEA-*p*ABA, HDEEA-*p*ABA or IAA solutions for treated samples. Each experiment was repeated three times. Petri dishes of all variants (control and treated) were placed in a growth chamber at 25 ± 2 °C, 70%–75% relative humidity, and 16/8 light/dark conditions. The seeds with emerged radicle were counted at 7 and 10 days and percentage were calculated.

3.4.2. Seedling Growth Tests (Experiment 2)

Germinated seeds of each control and 0.5 mM treated samples were transplanted into pots (200 mL) with peat. Two germinated seeds/pot were placed in 15 replications for each experimental variant as follows: control watered with distilled water and other variants treated with 0.5 mM IAA, HEEA-*p*ABA and HDEEA-*p*ABA solutions. The treatments were performed by applying the solutions in pots, at roots, 25 mL at intervals of 7 days. The determinations were made after 3 applications, after 21 days of growing plants in the greenhouse (25/18 °C, day/night; 65% rh, and 12/12 light/dark conditions). Parameters included seedling height (cm), chlorophyll content (SPAD), main root length (cm), number and maximum length of secondary roots (cm) were determined.

3.4.3. Tests on Tomato Plants (Experiment 3)

After 56 days, 10 seedlings from each experimental variant were planted in the greenhouse, on a fertile soil substrate with fermented manure (5 kg/sqm). The same experimental variants were maintained as in the case of seedling tests, being carried out three treatments with foliar application, in doses correlated with the plant sizes of 8 mL/plant (the first treatment), 12 mL/plant (2nd) and 16 mL/plant (third) at 7 day intervals.

3.4.4. Data and Statistical Analysis

In this study, ten replicates of each moment of measurement were carried out for each sample, regarding the effect of different plant growth regulator substances and concentrations. All data were expressed as means ± standard error of the mean (*SEM*). Means were compared using least significant difference (*LSD*) test. The significant differences between the sample means ($p < 0.05$) were expressed by different letters.

4. Conclusions

Two new binary organic salts of *p*ABA with different alkanolamines were successfully synthesized and characterized, and their auxin-like PGRs' properties investigated in laboratory and greenhouse on *Solanum lycopersicum* L. by comparison with the most abundant and studied natural auxin in plants, IAA. The single crystal structures' interpretation was correlated with the FTIR spectral data showing the intermolecular interactions established within the salts. Both new structures exhibited hydrogen-bonded supramolecular network architectures obtained from: (i) $R_2^2(9)$ heterosynthons and $R_4^4(22)$ homosynthons in HEEA-*p*ABA; and (ii) $R_2^2(9)$, $R_4^4(12)$, $R_2^4(14)$ and $R_3^4(13)$, $R_6^6(26)$, $R_5^6(27)$ and $R_5^6(30)$ heterosynthons in HDEEA-*p*ABA. The presence of $R_2^2(9)$ graph set in both structures indicates the relevance of charge-assisted N^{+}-H···O^{-} hydrogen bond in the formation of salts. The results obtained by investigating the biological profile of new alkanolammonium *p*-aminobenzoates as potential PGRs reveal the possibility of their use in stimulating the rooting and growth processes of tomato seedlings, as well as the positive involvement in plant growth and chlorophyll pigment biosynthesis. Considering the distinct supramolecular synthons in the design of molecular crystals and the promising biological results, this study sheds light on the alkanolammonium *p*-aminobenzoates as potential alternatives in the search of sustainable PGRs for tomatoes.

Author Contributions: M.C. and M.P. carried out synthesis, crystallization and physicochemical characterization of new compounds; L.C. and P.B. performed the diffraction measurements and the structure refinement of single crystals; R.-L.S. and I.R. carried out the biological activity; R.-M.S. performed the statistical analysis. All authors discussed and approved the final manuscript.

Acknowledgments: The authors thank for the support of the "Coriolan Drăgulescu" Institute of Chemistry - Program 2, Project 2.1 and of the Institute of Applied Physics - project ANCD 20.80009.5007.15.

References

1. Song, Y. Insight into the mode of action of 2,4-dichlorophenoxyacetic acid (2,4-D) as an herbicide. *J. Integr. Plant Biol.* **2014**, *56*, 106–113. [CrossRef] [PubMed]
2. Small, C.C.; Degenhardt, D. Plant growth regulators for enhancing revegetation success in reclamation: A review. *Ecol. Eng.* **2018**, *118*, 43–51. [CrossRef]
3. George, E.F.; Hall, M.A.; De Klerk, G. Plant growth regulators I: Introduction; auxins, their analogues and inhibitors. In *Plant Propagation by Tissue Culture*, 3rd ed.; George, E.F., Hall, M.A., Klerk, G.J.D., Eds.; Springer: Dordrecht, The Netherlands, 2008; pp. 175–204. [CrossRef]
4. Paque, S.; Weijers, D. Q&A: Auxin: The plant molecule that influences almost anything. *BMC Biol.* **2016**, *14*, 67–71. [CrossRef] [PubMed]
5. Ng, P.L.L.; Ferrarese, M.L.L.; Huber, D.A.; Ravagnani, A.L.S.; Ferrarese-Filho, O. Canola (Brassica napus L.) seed germination influenced by cinnamic and benzoic acids and derivatives: Effects on peroxidase. *Seed Sci. Technol.* **2003**, *31*, 39–46. [CrossRef]
6. Crisan, M.E.; Bourosh, P.; Maffei, M.E.; Forni, A.; Pieraccini, S.; Sironi, M.; Chumakov, Y.M. Synthesis, crystal structure and biological activity of 2-hydroxyethylammonium salt of p-aminobenzoic acid. *PLoS ONE* **2014**, *9*, e101892. [CrossRef]
7. Grozav, M.; Neamtiu, I.; Dorosencu, M.; Laichici, M.; Mercea, M. The synthesis of some ammonium salts of benzoic acids with etanolamine, possible plant growth stimulators. *Rev. Chim.* **2003**, *54*, 287–288.
8. Dempsey, D.A.; Klessig, D.F. How does the multifaceted plant hormone salicylic acid combat disease in plants and are similar mechanisms utilized in humans? *BMC Biol.* **2017**, *15*, 23–33. [CrossRef]
9. Senaratna, T.; Merritt, D.; Dixon, K.; Bunn, E.; Touchell, D.; Sivasithamparam, K. Benzoic acid may act as the functional group in salicylic acid and derivatives in the induction of multiple stress tolerance in plants. *Plant Growth Regul.* **2003**, *39*, 77–81. [CrossRef]
10. Basset, G.J.C.; Quinlivan, E.P.; Ravanel, S.; Rebeille, F.; Nichols, B.P.; Shinozaki, K.; Seki, M.; Adams-Phillips, L.C.; Giovannoni, J.J.; Gregory, J.F.; et al. Folate synthesis in plants: The p-aminobenzoate branch is initiated by a bifunctional PabA-PabB protein that is targeted to plastids. *Proc. Natl. Acad. Sci. USA* **2004**, *101*, 1496–1501. [CrossRef]
11. Marbois, B.; Xie, L.X.; Choi, S.; Hirano, K.; Hyman, K.; Clarke, C.F. para-Aminobenzoic acid is a precursor in Coenzyme Q6 biosynthesis in Saccharomyces cerevisiae. *J. Biol. Chem.* **2010**, *285*, 27827–27838. [CrossRef]
12. Galbinur, T.; Obolensky, A.; Berenshtein, E.; Vinokur, V.; Chowers, I.; Chevion, M.; Banin, E. Effect of para-aminobenzoic acid on the course of retinal degeneration in the rd10 mouse. *J. Ocul. Pharmacol. Ther.* **2009**, *25*, 475–482. [CrossRef] [PubMed]

13. Richards, R.M.R.; Xing, D.K.L.; King, T.P. Activity of p-aminobenzoic acid compared with other organic acids against selected bacteria. *J. Appl. Bacteriol.* **1995**, *78*, 209–215. [CrossRef] [PubMed]
14. Vasilieva, S. Para-aminobenzoic acid inhibits a set of SOS functions in Escherichia coli K12. *Mutat. Res. Genet. Toxicol. Environ. Mutagen.* **2001**, *496*, 89–95. [CrossRef]
15. Drozd, N.N.; Makarov, V.A.; Miftakhova, N.T.; Kalugin, S.A.; Stroeva, O.G.; Akberova, S.I. Antithrombotic activity of para-aminobenzoic acid. *Eksp. Klin. Farmakol.* **2000**, *63*, 40–44. [PubMed]
16. Andreenko, G.V.; Karabasova, M.A.; Liutova, L.A.; Podorolskaia, L.V.; Serebriakova, T.N.; Sologub, A.A.; Akberova, S.I.; Stroeva, O.G. Effect of para-aminobenzoic acid on the fibrinolytic activity of blood. *Dokl. Akad. Nauk* **1996**, *346*, 268–270. [PubMed]
17. Hanson, K.M.; Gratton, E.; Bardeen, C.J. Sunscreen enhancement of UV-induced reactive oxygen species in the skin. *Free Radic. Biol. Med.* **2006**, *41*, 1205–1212. [CrossRef] [PubMed]
18. Lu, Z.; Kong, X.; Lu, Z.; Xiao, M.; Chen, M.; Zhu, L.; Shen, Y.; Hu, X.; Song, S. para-Aminobenzoic acid (PABA) synthase enhances thermotolerance of mushroom Agaricus bisporus. *PLoS ONE* **2014**, *9*, e91298. [CrossRef]
19. Song, G.C.; Choi, H.K.; Ryu, C.M. The folate precursor para-aminobenzoic acid elicits induced resistance against Cucumber mosaic virus and Xanthomonas axonopodis. *Ann. Bot.* **2013**, *111*, 925–934. [CrossRef]
20. Groom, C.R.; Bruno, I.J.; Lightfoot, M.P.; Ward, S.C. The Cambridge structural database. *Acta Crystallogr. B* **2016**, *72*, 171–179. [CrossRef]
21. Crisan, M.E.; Gorobet, A.; Siminel, A.V.; Bourosh, P.N.; Croitor, L. A new supramolecular isomer of p-aminobenzoate Zn (II) coordination polymer: Structure and photoluminescent property. *Polyhedron* **2019**, *171*, 502–507. [CrossRef]
22. Chicu, S.A.; Herrmann, K.; Berking, S. An approach to calculate the toxicity of simple organic molecules on the basis of QSAR analysis in Hydractinia echinata (Hydrozoa, Cnidaria). *Quant. Struct. Act. Relat.* **2000**, *19*, 227–236. [CrossRef]
23. Croitor, L.; Petric, M.F.; Szerb, E.I.; Vlase, G.; Bourosh, P.N.; Chumakov, Y.; Crisan, M.E. Role of 4-nitrobenzoic acid polymorphs in the crystallization process of the organic acid-base multicomponent system. *CrystEngComm* **2019**, *21*, 6038–6047. [CrossRef]
24. Crisan, M.; Vlase, G.; Plesu, N.; Petric, M.; Croitor, L.; Kravtsov, V.; Chumakov, Y.; Bouros, P.; Vlase, T. Ethylethanolammonium 4-nitrobenzoate: Synthesis, structural characterization, thermal analysis, non-isothermal kinetic investigations and corrosion inhibitor efficiency. *J. Therm. Anal. Calorim.* **2018**, *134*, 343–352. [CrossRef]
25. Crisan, M.; Halip, L.; Bourosh, P.; Chicu, S.A.; Chumakov, Y. Synthesis, structure and toxicity evaluation of ethanolamine nitro/chloronitrobenzoates: A combined experimental and theoretical study. *Chem. Cent. J.* **2017**, *11*, 129–139. [CrossRef] [PubMed]
26. Chicu, S.A.; Grozav, M.; Kurunczi, L.; Crisan, M. SAR for amine salts of carboxylic acids to *Hydractinia echinata*. *Rev. Chim.* **2008**, *59*, 582–587. [CrossRef]
27. Crisan, M.; Vlase, G.; Szerb, E.I.; Vlase, T. Thermal and kinetics studies of primary, secondary and tertiary alkanolammonium salts of 4-nitrobenzoic acid. *J. Therm. Anal. Calorim.* **2018**, *132*, 1409–1418. [CrossRef]
28. Cruz-Cabeza, A.J. Acid–base crystalline complexes and the pKa rule. *CrystEngComm* **2012**, *14*, 6362–6365. [CrossRef]
29. Pavia, D.L.; Lampman, G.M.; Kriz, G.S.; Vyvyan, J.R. *Introduction to Spectroscopy*, 4th ed.; Brooks/Cole, Cengage Learning: Boston, MA, USA; Belmont, CA, USA, 2009; p. 80.
30. Liu, X.; Zhang, H.; Zhao, Y.; Feng, Z.; Li, Q.; Yang, H.Q.; Luan, S.; Li, J.; He, Z.H. Auxin controls seed dormancy through stimulation of abscisic acid signaling by inducing ARF-mediated ABI3 activation in Arabidopsis. *Proc. Natl. Acad. Sci. USA* **2013**, *110*, 15485–15490. [CrossRef]
31. Ye, N.; Zhu, G.; Liu, Y.; Zhang, A.; Li, Y.; Liu, R.; Shi, L.; Jia, L. Ascorbic acid and reactive oxygen species are involved in the inhibition of seed germination by Abscisic acid in rice seeds. *J. Exp. Bot.* **2012**, *63*, 1809–1822. [CrossRef]
32. Miransari, M.; Smith, D.L. Plant hormones and seed germination. *Environ. Exp. Bot.* **2014**, *99*, 110–121. [CrossRef]
33. Shuai, H.; Meng, Y.; Luo, X.; Chen, F.; Zhou, W.; Dai, Y.; Qi, Y.; Du, J.; Yang, F.; Liu, J.; et al. Exogenous auxin represses soybean seed germination through decreasing the gibberellin/abscisic acid (GA/ABA) ratio. *Sci. Rep.* **2017**, *7*, 1–11. [CrossRef] [PubMed]

34. Tsygankova, V.; Andrusevich, Y.; Shtompel, O.; Myroljubov, O.; Hurenko, A.; Solomyanny, R.; Mrug, G.; Frasinyuk, M.; Shablyn, O.; Brovarets, V. Study of auxin, cytokinetin and gibberellins -like activity of heterocyclic compounds derivatives of pyrimidine, pyridine, pyrazole and isoflavones. *Eur. J. Biotechnol. Biosci.* **2016**, *4*, 29–44.
35. Bethke, P.C.; Libourel, I.G.; Jones, R.L. Nitric oxide reduces seed dormancy in Arabidopsis. *J. Exp. Bot.* **2006**, *57*, 517–526. [CrossRef]
36. Oracz, K.; Karpiński, S. Phytohormones signaling pathways and ROS involvement in seed germination. *Front. Plant Sci.* **2016**, *7*, 1–6. [CrossRef] [PubMed]
37. Mittler, R. ROS are good. *Trends Plant Sci.* **2017**, *22*, 11–19. [CrossRef] [PubMed]
38. Noctor, G.; Reichheld, J.P.; Foyer, C.H. ROS related redox regulation and signaling in plants. *Semin. Cell Dev. Biol.* **2017**, *80*, 3–12. [CrossRef]
39. Dash, M.; Yordanov, Y.S.; Georgieva, T.; Tschaplinski, T.J.; Yordanova, E.; Busov, V. Poplar PtabZIP1-like enhances lateral root formation and biomass growth under drought stress. *Plant J.* **2017**, *89*, 692–705. [CrossRef]
40. Qin, H.; Huang, R. Auxin controlled by ethylene steers root development. *Int. J. Mol. Sci.* **2018**, *19*, 3656. [CrossRef]
41. Du, Y.; Scheres, B. Lateral root formation and the multiple roles of auxin. *J. Exp. Bot.* **2018**, *69*, 155–167. [CrossRef]
42. Cai, X.T.; Xu, P.; Zhao, P.X.; Liu, R.; Yu, L.H.; Xiang, C.B. Arabidopsis ERF109 mediates cross-talk between jasmonic acid and auxin biosynthesis during lateral root formation. *Nat. Commun.* **2014**, *5*, 1–13. [CrossRef]
43. Zhou, R.; Yu, X.; Ottosen, C.O.; Rosenqvist, E.; Zhao, L.; Wang, Y.; Yu, W.; Zhao, T.; Wu, Z. Drought stress had a predominant effect over heat stress on three tomato cultivars subjected to combined stress. *BMC Plant Biol.* **2017**, *17*, 1–13. [CrossRef] [PubMed]
44. Zhang, G.; Xu, N.; Chen, H.; Wang, G.; Huang, J. OsMADS25 regulates root system development via auxin signalling in rice. *Plant J.* **2018**, *95*, 1004–1022. [CrossRef] [PubMed]
45. Jing, H.; Yang, X.; Zhang, J.; Liu, X.; Zheng, H.; Dong, G.; Nian, J.; Feng, J.; Xia, B.; Qian, Q.; et al. Peptidyl-prolyl isomerization targets rice Aux/IAAs for proteasomal degradation during auxin signalling. *Nat. Commun.* **2015**, *6*, 1–10. [CrossRef] [PubMed]
46. Reed, R.C.; Brady, S.R.; Muday, G.K. Inhibition of auxin movement from the shoot into the root inhibits lateral root development in Arabidopsis. *Plant Physiol.* **1998**, *118*, 1369–1378. [CrossRef] [PubMed]
47. Dao, G.H.; Wu, G.X.; Wang, X.X.; Zhuang, L.L.; Zhang, T.Y.; Hu, H.Y. Enhanced growth and fatty acid accumulation of microalgae Scenedesmus sp. LX1 by two types of auxin. *Bioresour. Technol.* **2018**, *247*, 561–567. [CrossRef] [PubMed]
48. Li, Y.; He, N.; Hou, J.; Xu, L.; Liu, C.; Zhang, J.; Wang, Q.; Zhang, X.; Wu, X. Factors influencing leaf chlorophyll content in natural forests at the biome scale. *Front. Ecol. Evol.* **2018**, *6*, 1–10. [CrossRef]
49. Beinsan, C.; Camen, D.; Sumalan, R.; Babau, M. Study concerning salt stress effect on leaf area dynamics and chlorophyll content in four bean local landraces from Banat area. In Proceedings of the 44th Croatian and 4th International Symposium on Agriculture, Opatija, Croatia, 16–20 February 2009; pp. 416–419.
50. El-Saeid, H.M.; Abou-Hussein, S.D.; El-Tohamy, W.A. Growth characters, yield and endogenous hormones of cowpea plants in response to IAA application. *Res. J. Agric. Biol. Sci.* **2010**, *6*, 27–31.
51. Finet, C.; Jaillais, Y. Auxology: When auxin meets plant evo-devo. *Dev. Biol.* **2012**, *369*, 19–31. [CrossRef]
52. Khalil, S.; Mandurah, H.M. Growth and metabolic changes of cowpeas plants as affected by water deficiency and indole acetic acid. *J. Agron. Crop Sci.* **1989**, *16*, 160–166. [CrossRef]
53. El-Mergawi, R.A. Sensitivity of faba bean cultivars to low glyphosate doses and the efficiency of IAA as indicator to glyphosate effects. *Egypt J. Hort.* **2003**, *30*, 197–214.
54. Gaspar, T.H.; Kevers, C.; Faivre-Rampant, O.; Crèvecour, M.; Penel, C.L.; Greppin, H.; Dommes, J. Changing concepts in plant hormone action. *Vitro Cell. Dev. Biol. Plant* **2003**, *39*, 85–106. [CrossRef]
55. Alarcón, M.V.; Salguero, J.; Lloret, P.G. Auxin modulated initiation of lateral roots is linked to pericycle cell length in maize. *Front. Plant Sci.* **2019**, *10*, 1–10. [CrossRef] [PubMed]
56. Cui, M.; Jia, B.; Liu, H.; Kan, X.; Zhang, Y.; Zhou, R.; Li, Z.; Yang, L.; Deng, D.; Yin, Z. Genetic mapping of the leaf number above the primary ear and its relationship with plant height and flowering time in maize. *Front. Plant Sci.* **2017**, *8*, 1–13. [CrossRef] [PubMed]

57. Shaver, G.R. Mineral nutrition and leaf longevity in *Ledum palustre*: The role of individual nutrients and the timing of leaf mortality. *Oecologia* **1983**, *56*, 160–165. [CrossRef] [PubMed]
58. Mauro, P.R.; Agnello, M.; Distefano, M.; Sabatino, L.; san Bautista, P.A.; Leonardi, C.; Giuffrida, F. Chlorophyll fluorescence, photosynthesis and growth of tomato plants as affected by long-term oxygen root zone deprivation and grafting. *Agronomy* **2020**, *10*, 137. [CrossRef]
59. Sheldrick, G.M. Crystal structure refinement with SHELXL. *Acta Cryst.* **2015**, *C71*, 3–8. [CrossRef]
60. Macrae, C.F.; Edgington, P.R.; McCabe, P.; Pidcock, E.; Shields, G.P.; Taylor, R.; Towler, M.; Streek, J. Mercury: Visualization and analysis of crystal structures. *J. Appl. Crystallogr.* **2006**, *39*, 453–457. [CrossRef]

3

Rock Phosphate-Enriched Compost in Combination with Rhizobacteria; A Cost-Effective Source for Better Soil Health and Wheat (*Triticum aestivum*) Productivity

Motsim Billah [1], Matiullah Khan [2], Asghari Bano [3,*], Sobia Nisa [4], Ahmad Hussain [5], Khadim Muhammad Dawar [6], Asia Munir [7] and Naeem Khan [8,*]

1 Department of Soil Science, University of Haripur, Haripur 21120, Pakistan; motsimbillah@gmail.com
2 Pakistan Agriculture Research Council, NARC, Islamabad 44000, Pakistan; mukhan65pk@yahoo.co.uk
3 Department of Biosciences, University of Wah, Wah Cantt. 47000, Pakistan
4 Department of Microbiology, University of Haripur, Haripur 21120, Pakistan; sobia@uoh.edu.pk
5 Department of Forestry, University of Haripur, Haripur 21120, Pakistan; ahmad.hussain@uoh.edu.pk
6 Department of Soil and Environmental Science, The University of Agriculture, Peshawar 25000, Pakistan; khadimdawar@yahoo.com
7 Soil and Water Testing Laboratory, Rawalpindi 46000, Pakistan; asiamunir800@gmail.com
8 Department of Agronomy, Institute of Food and Agricultural Sciences, University of Florida, Gainesville, FL 32611, USA
* Correspondence: bano.asghari@gmail.com (A.B.); naeemkhan@ufl.edu (N.K.)

Abstract: Organic materials from various sources have been commonly adopted as soil amendments to improve crop productivity. Phosphorus deficiency and fixation in alkaline calcareous soils drives a reduction in crop production. A two-year field experiment was conducted to evaluate the impact of rock phosphate enriched composts and chemical fertilizers both individually and in combination with plant growth promoting rhizobacteria (PGPR) on wheat productivity and soil chemical and biological and biochemical properties. The present study demonstrates significant increments in crop agronomic and physiological parameters with *Pseudomonas* sp. inoculated $RPEC_1$ (rock phosphate + poultry litter + *Pseudomonas* sp.) over the un-inoculated untreated control. However, among all other treatments i.e., $RPEC_2$ (rock phosphate + poultry litter solubilized with *Proteus* sp.), RPC (rock phosphate + poultry litter), HDP (half dose inorganic P from Single Super Phosphate-SSP 18% P_2O_5) and SPLC (poultry litter only); $RPEC_1$ remained the best by showing increases in soil chemical properties (available phosphorus, nitrate nitrogen, extractable potassium), biochemical properties (alkaline phosphatase activity) and biological properties (microbial biomass carbon and microbial biomass phosphorus). Economic analysis in terms of Value Cost Ratio (VCR) showed that the seed inoculation with *Pseudomonas* sp. in combination with $RPEC_1$ gave maximum VCR (3.23:1) followed by $RPEC_2$ (2.61:1), FDP (2.37:1), HDP (2.05:1) and SPLC (2.03:1). It is concluded that inoculated rock phosphate (RP) enriched compost ($RPEC_1$) can be a substitute to costly chemical fertilizers and seed inoculation with *Pseudomonas* sp. may further increase the efficiency of composts.

Keywords: available phosphorus; enriched compost; PGPR; poultry litter; rock phosphate; wheat

---

1. Introduction

The function of fertilizers for maximum crop production in under-developed countries is customary and well recognized. Nevertheless, the increasing prices of inorganic phosphate fertilizers and the extensive use of chemical fertilizers in agriculture, is also under debate due to environmental concerns and for consumer health reasons [1]. Reduction of agrochemicals for crop production is of great concern for sustainable agriculture [2]. Moreover, inorganic phosphate fertilizers are not totally soluble in soil matrix due to precipitation reactions with ions of Al and Fe in acidic, and Ca in alkaline calcareous soils [3]. Moreover, high dose application of chemical fertilizers creates negative impacts such as changes in soil pH through alkalization and acidification, pollution of water resources through runoff, suppression of microorganisms and friendly insects, fixation of nutrients, degradation of soil structure due to increased decomposition of organic matter [4]. The research workers are required to look for substitutes to inorganic fertilizers [5], which are cost-effective and environmentally friendly. The use of rock phosphate (RP) as an alternative for P fertilizer is gaining attention in sustainable agriculture through microbial solubilization [6] and preparation of RP-enriched compost [7]. The mixing of RP with organic materials such as animal feces, plant residues and inoculation with acid-producing microbes may enhance P solubility from RP because when organic materials decompose, more soluble P is released due to the action of organic acids produced by the microbes [8]. The incorporation of organic residues either singly or in conjunction with a cheap source of mining element as rock phosphate may help to improve soil quality and productivity [9]. Rock phosphate enriched compost which was solubilized by phosphate solubilizing fungi and applied on a mung-bean crop, significantly enhanced yield and P-uptake [10].

Various RP enriched composts and inorganic fertilizers such as diammonium phosphate (DAP) were applied on wheat in a pot experiment. The data revealed that RP enriched composts showed no significant performance in the earlier stages of wheat growth but at maturity, it gave higher grain yield, nutrient uptake and increased fertility status of P and K in the soils [11]. Isolated phosphate-solubilizing fungi from phosphate mines of China were reported to have efficient biofertilizers and P solubilizers with the capacity to enhance the growth of wheat [12]. Colonization of soil by nonindigenous phosphate-solubilizing microorganisms depend both on their interactions with indigenous microorganisms associated with plants and their ability to utilize diverse substrates in soil [13]. The role of phosphate-solubilizing microorganisms in phosphate solubilization has been attributed mainly to their abilities to reduce the pH of the surroundings by the production of organic acids [13]. Preparing the RP-enriched compost with phosphate solubilizing microbes may not only compensate for the higher cost of manufacturing fertilizers, but also provide a sustainable source of available phosphorus to growing plants in alkaline soils [14].

Plant growth promoting rhizobacteria (PGPR) are important inoculants for integrated nutrient management [15] which help in dissolving inorganic P by excreting organic acids and chelation of P cations to release P in soil solution [16]. It was reported that there are several PGPR inoculants currently commercialized that promote growth either by suppression of plant disease, improved nutrient acquisition, or phytohormone production [17]. Generally, phytohormone in plants plays a vital role in cell division, proliferation, and differentiation, vascular tissue alteration, responses to light and gravity, general root and shoot architecture, seed and tuber germination, ethylene synthesis, vegetative growth processes, fruit development [18–20], initiation of lateral and floral organ and organogenesis [21], initiation of rooting, foliation and flowering [22], formation of lateral and adventitious roots [23], and increasing the growth of cambium and size of xylem cells [24]. Bacterial phytohormone production is widely distributed among plant-associated bacteria and is still considered the primary mechanism that enhances the growth and yield of plants [25].

PGPRs influence direct growth promotion of plants by fixing atmospheric nitrogen, solubilizing insoluble phosphates, secreting hormones such as IAA, GAs, and Kinetins besides ACC (1-aminocycloprapane-1-carboxylic acid) deaminase production [26], that helps in the regulation of ethylene. Amongst the majority of influential P solubilizers, bacterial strains from the genera *Pseudomonas*, *Bacillus*, *Rhizobium* and *Enterobacter* are of great importance. Application of phosphate solubilizing microbes in the production of compost can help to increase the interest of farmers to use organic phosphatic fertilizers in alkaline soils [14]. Therefore, this study aimed to evaluate the availability of phosphorus from RP enriched compost with the application of PGPRs and its comparative effectiveness with inorganic fertilizers (Single Super Phosphate) on soil nutrient status, wheat growth and production.

2. Material and Methods

2.1. Experimental Site and Treatments

Two-year field experiments at National Agricultural Research Centre, Islamabad (73°70′ E and 33°39′ N with an altitude 610 masl during growing months Nov, 2010-May, 2011 and Nov, 2011-May, 2012), were conducted on wheat (var. GA-2002). Soil textural class of the experimental site was silty loam. The meteorological data during the growing season (2010–2012) of wheat is given in Table 1. Different composts being prepared during the previous experiments [27] were used in the study for their effectiveness to get better crop production. The treatments included; Control (Untreated un-inoculated); SPLC (Simple poultry litter compost); RP (rock phosphate 18.5% P_2O_5); $RPEC_1$ (rock phosphate + poultry litter solubilized with *Pseudomonas* sp. during composting process); $RPEC_2$ (rock phosphate + poultry litter solubilized with *Proteus* sp. during the composting process); FDP (Full dose inorganic P from Single Super Phosphate-SSP18% P_2O_5); HDP (Half dose inorganic P from Single Super Phosphate-SSP 18% P_2O_5). Treatments were applied at a rate of 100 kg P ha^{-1}, respectively from composts as well as from inorganic fertilizers on a total P basis during seed bed preparation. The nutrient status of different composts, is given in Table 2. The recommended dose of nitrogen at the rate of 100 kg ha^{-1} was equally applied to each plot (4 m × 3 m) either from inorganic fertilizer (Urea-46% N) or compost on a nutrient basis. However, SPLC was applied at the rate of 4.5t ha^{-1}. There were three replications for each treatment. All the fertilizer treatments were applied to respective plots at the same time of sowing.

Table 1. Meteorological data during the growing seasons of wheat crop (2010–2012).

	2010–2011			2011–2012		
Months	**Av. Temp (°C)**	**Rainfall (mm)**	**R.H (%)**	**Av. Temp. (°C)**	**Rainfall (mm)**	**R.H (%)**
Nov.	18.03	4.13	64.2	16.32	7.09	65.52
Dec.	18.18	26.97	64.89	19.23	0	62.74
Jan.	15.5	8.32	71.29	15.55	59.06	68.48
Feb.	11.91	78.73	75.84	14.21	44.12	70.07
March	15.55	53.19	63.02	16.03	15.95	58.65
April	15.27	53.96	61.08	14.87	40.93	54.68
May	17.71	17.29	44.75	17.84	9.47	38.44
Mean	16.02	34.66	63.58	16.29	25.23	59.8

Adopted from CAEWRI, National Agricultural Research Centre, Islamabad. Av. Temp—Average temperature; R.H—relative humidity, mm = millimeter.

Table 2. Nutrient composition of different composts applied as treatments in the experiments.

Compost	Av. P (%)	Total N (%)	TOC (%)	C:N
SPLC	0.35	1.35	19.36	14.34
$RPEC_1$	1.72	1.29	16.3	12.66
$RPEC_2$	1.24	1.28	17.7	13.83

SPLC—Simple poultry litter; $RPEC_1$—Poultry litter + rock phosphate + *Pseudomonas* sp.; $RPEC_2$—Poultry litter + rock phosphate + *Proteus* sp.; Av. P—Available phosphorus; N—Nitrogen; TOC—Total organic carbon; C:N; Carbon–nitrogen ratio.

2.2. Seed Inoculation

The PGPR strains; *Pseudomonas* sp. (Accession no. KF307201) and *Proteus* sp. (Accession no. KF307202) were used at 6 × 10^8 CFU/mL for seed inoculation. Wheat seeds were inoculated with cultures for 4 h and then the seeds were shade dried before sowing. The inoculants were applied individually as well as in combination with organic and inorganic fertilizer treatments.

2.3. *Yield, Physiology and Plant Nutrient Analysis*

Growth and yield parameters; the number of tillers, grain yield and total dry matter yield were recorded at the time of harvesting. However, for the determination of dry matter yield, the aerial part of the plant from each plot was harvested. Then the spikes were separated from harvested plants of respective treatments and the grains of each pot were weighed to calculate grain yield [kg ha^{-1}].

Chlorophyll and phytohormones (IAA, GA) were analyzed in flag leaves of the wheat plants. Chlorophyll was recorded by using SPAD chlorophyll meter [Konica Minolta, Langenhagen, Germany], while leaf IAA and GA were extracted through the method of Kettner and Doerffling [28] and analyzed on HPLC (Agilent 1100, Waldbronn, Germany) using UV detector and C-18 column (39 × 300 mm). Methanol, acetic acid, and water (30:1:70) were used as mobile phase. The wavelength used for the detection of IAA was 280 nm [29] whereas for GA, it was adjusted at 254 nm. These hormones were identified on the basis of retention time and peak area of the standards. Pure IAA and GA_3 (Sigma Chemicals Co. Ltd. St. Louis, Missouri, USA) were used as standard for identification and quantification of plant hormones. The above ground plants were harvested from each plot, dried at 70 °C for 48 h, ground at the grinding mill and samples were stored in Ziploc polyethylene bags at room temperature till nutrient analysis. Total phosphorus in plant samples and in seeds was analyzed through Olsen and Sommers [30]. However, phosphorus concentration in shoot was used for the calculation of plant P uptake (kg ha^{-1}).

2.4. *Soil Analysis*

Initial soil samples were taken for physicochemical properties (Table 3). Soil samples (0–30 cm) were analyzed for the texture [31], organic matter [32], total P [33]. However, soil samples were extracted through Ammonium Bicarbonate Diphenyl Triamine Penta Acetic Acid (AB-DTPA) solution for determination of available P, NO_3-N and extractable K following the method of Soltanpour and Schwab [34] and soil pH (1:5 soil–water) using the method of Mclean [35]. Undisturbed soil samples were collected for soil bulk density (g cm^{-3}) using stainless steel cylinders [36]. Soil phosphatase activity was determined by the method of Tabatabai and Bremner [37], whereas, microbial biomass carbon and microbial biomass phosphorus was determined following the method adopted by Steel and Torriej [38]. For determining post-harvest soil properties, soil samples were collected after 6 days of wheat crop harvesting.

Table 3. Physicochemical, biological and biochemical properties of soil.

Properties	2010–2011	2011–2012
Texture	Silty Loam	Silty Loam
Sand	18%	18%
Silt	52%	50%
Clay	30%	32%
pH	7.48	7.5
Ec (dSm^{-1})	0.45	0.46
Bulk density (g cm^{-3})	1.43	1.42
NO_3-N (mg kg^{-1})	3.23	3.4
Total Phosphorus (mg kg^{-1})	500	482
Available Phosphorus (mg kg^{-1})	2.7	2.8
Extractable potassium (mg kg^{-1})	96	92

Table 3. *Cont.*

Properties	2010–2011	2011–2012
Organic matter (%)	0.86	0.84
Cu (µg g^{-1})	1.1	0.78
Fe (µg g^{-1})	54.38	58.98
Zn (µg g^{-1})	1.7	1.64
Mn (µg g^{-1})	1	1.32
Microbial Biomass carbon (mg kg^{-1})	83	84
Microbial Biomass phosphorus (mg kg^{-1})	7	8
Alkaline phosphatase activity (µg g^{-1})	110	112

2.5. *Statistical Analysis*

The experiment was laid down following the randomized complete block design (RCBD) with split plot design. Different soil amendments (composts and inorganic fertilizers) were assigned with the main plot while PGPRs were placed in sub-plots of the field. Analysis of variance (ANOVA) was conducted with the General Linear Models and means were compared according to the Tukey HSD test with Statistix 8.1 [39]. Two years of data were pooled because there were not interactions between the two years and year was included as a random effect in statistical model.

3. Results

Means of two-year data (2010–2011 and 2011–2012) are provided here due to the result similarity trend from both the years.

3.1. *Yield and Yield Components*

The data for the number of tillers showed 36%, 34%, 30%, 24% and 21% increases with un-inoculated $RPEC_1$, FDP, $RPEC_2$, SPLC and HDP, respectively, over un-inoculated untreated control (Table 4). The treatment RP did not show any significant increase over un-inoculated untreated control. Seed inoculation with PGPRs without any fertilizer treatment did not show any difference with un-inoculated control. However, seed inoculation with *Pseudomonas* sp. in combination with $RPEC_1$ treatment showed a maximum 5% increase in the number of tillers over un-inoculated $RPEC_1$ and FDP.

C—Control (un-inoculated untreated), SPLC—Simple poultry compost, $RPEC_1$—Rock phosphate enriched compost inoculated with *Pseudomonas* species, $RPEC_2$—Rock phosphate enriched compost inoculated with *Proteus* species, RP—Rock phosphate, HDP—Half dose inorganic P fertilizer, FDP—Full dose inorganic P fertilizer.

The data presented in Table 4, showed that there was a significant ($p \leq 0.05$) effect of PGPR on grain yield of wheat crop. A maximum (18%) increase in grain yield was recorded in plants inoculated with *Pseudomonas* sp. which was 4% higher than inoculation with *Proteus* sp. Without inoculation, maximum (67%) increase in grain yield was recorded with the application of $RPEC_1$ which was 4%, 9% and 16% higher than FDP, $RPEC_2$ and SPLC, respectively over un-inoculated untreated control. However, $RPEC_2$ showed a 52% increase in grain yield over control. The interactive effect of fertilizers × PGPR, was highly significant ($p \leq 0.05$) for grain yield. *Pseudomonas* sp. inoculated $RPEC_1$ and FDP gave maximum (10%) increase over un-inoculated $RPEC_1$ and FDP treatments, respectively. The *Proteus* sp. in combination with $RPEC_1$ also showed 3% increase over un-inoculated $RPEC_1$ treatment, whereas the treatment RP produced minimum grain yield showing 14% increase over un-inoculated RP treatment.

The data in Table 4 show that the treatment $RPEC_1$ resulted in a maximum increase in dry matter yield which was 3.8%, 16%, 27% higher than FDP, $RPEC_2$ and SPLC respectively, over un-inoculated untreated control. The stimulatory effect of PGPR was recorded on dry matter yield. However, the interactive effect of PGPR and fertilizer treatments was significant for dry matter yield of wheat crop. The combination of *Pseudomonas* sp. with $RPEC_1$ gave the maximum increase (62%) similar to

FDP (60%) while with *Proteus* sp. in combination with $RPEC_1$ showed 56% increase over un-inoculated untreated control. RP inoculation with *Proteus* sp. showed nonsignificant difference with un-inoculated untreated control.

Table 4. Effects of plant growth promoting rhizobacteria (PGPR), P-enriched compost and inorganic fertilizers on yield and yield components on wheat.

Treatments	C	SPLC	$RPEC_1$	$RPEC_2$	RP	HDP	FDP
Number of Tillers (m^{-2})							
Without inoculation	260 ^{e} (±2.22)	323 ^{c} (±1.42)	354 ^{a} (±2.83)	339 ^{b} (±3.53)	261 ^{e} (±3.12)	315 ^{c,d} (±3.9)	347 ^{ab} (±2.88)
Proteus sp.	263 ^{e} (±2.98)	331 ^{c} (±2.17)	361 ^{a} (±2.67)	347 ^{b} (±4.94)	264 ^{e} (± 4.24)	321 ^{d} (±3.87)	352 ^{ab} (±4.72)
Pseudomonas sp.	267 ^{f} (±3.61)	343 ^{d} (±3.14)	372 ^{a} (±4.88)	356 ^{b,c} (±5.01)	269 ^{f} (±3.9)	326 ^{e} (±2.84)	364 ^{ab} (±2.92)
Grain yield (kg ha^{-1})							
Without inoculation	2177 ^{f,g} (±29.22)	3120 ^{d} (±36.09)	3629 ^{a} (±210.74)	3317 ^{c} (±24.36)	2195 ^{f} (±29.032)	3020 ^{d,e} (±25.78)	3495 ^{b} (±31.16)
Proteus sp.	2474 ^{f,g} (±20)	3319 ^{e} (±25.45)	3731 ^{a} (±33.22)	3524 ^{c} (±24.67)	2498 ^{f} (±36.103)	3511 ^{c,d} (±23.24)	3620 ^{b} (±29.2)
Pseudomonas sp.	2571 ^{f,j} (±32.89)	3433 ^{d} (±34.36)	3987 ^{a} (±39.84)	3639 ^{c} (±23.48)	2592 ^{f} (±22.86)	3364 ^{d,e} (±38.96)	3848 ^{b} (±29.07)
Dry matter yield (kg ha^{-1})							
Without inoculation	8955 ^{f,g} (±38.43)	10,804 ^{d} (±31.9)	13,714 ^{a} (±26.7)	11,773 ^{c} (±32.33)	8969 ^{f} (±24.41)	10,746 ^{d,e} (±35.9)	13,208 ^{b} (±24.47)
Proteus sp.	8978 ^{e} (±28.97)	10,930 ^{c} (±39.93)	13,926 ^{a} (±35.55)	12,077 ^{c} (±26.23)	8987 ^{e} (±34.48)	10,806 ^{c,d} (±37.13)	13,358 ^{b} (±27.2)
Pseudomonas sp.	9005 ^{f,g} (±34.02)	11,138 ^{d} (±0.67)	14,503 ^{a} (±37.14)	12,651 ^{c} (±36.29)	9050 ^{f} (±45)	10,891 ^{e} (±34.65	14,382 ^{a,b} (±25.37)

All the treatments sharing common letter are similar otherwise they differ significantly at $p \leq 0.05$.

3.2. Leaf Chlorophyll, IAA and GA Contents

Mean data recorded for chlorophyll contents in flag leaves of wheat crop showed that there was a significant ($p \leq 0.05$) difference for the treatments (Figure 1). Among the un-inoculated treatments, $RPEC_1$ showed the highest (28%) increase which was 2%, 6%, 12% and 25% higher than FDP, $RPEC_2$, SPLC and HDP, respectively. Seed inoculation with *Pseudomonas* sp. resulted in an increase (4%) in chlorophyll content over un-inoculated treatments. However, the interactive effect of treatments (PGPRs × fertilizer) showed 29% increase followed by FDP (27%) over un-inoculated untreated control. While the treatment RP showed a nonsignificant difference when applied in combination with *Pseudomonas* as well as *Proteus* sp.

Data in Figure 2 show that $RPEC_1$ resulted maximum (12%) increase in IAA content, having a similar effect as with FDP, followed by $RPEC_2$ showing a 9% increase, while HDP and SPLC showed a similar effect (i.e., 7% increase) in IAA content over un-inoculated untreated control. The inoculation of seeds with *Pseudomonas* sp. gave higher values of IAA by showing 6% increase over un-inoculated treatments. The interactive effect of PGPR × Fertilizer was nonsignificant except for *Pseudomonas* sp. which showed a 20% increase, when used in combination with $RPEC_1$ and FDP treatments.

The treatment $RPEC_1$ resulted in a 13% increase in GA content, followed by FDP (11%), $RPEC_2$ (9%), SPLC (6%), while HDP resulted only a 4% increase over un-inoculated untreated control (Figure 3). However, seed inoculation with *Pseudomonas* sp. showed a maximum (5%) increase over un-inoculated $RPEC_2$ treatment. The data showed that the treatments $RPEC_1$ and FDP in combination with *Pseudomonas* sp. showed a maximum (16%) increase in GA contents of flag leaves. PGPR inoculation with RP and HDP showed a nonsignificant difference among respective un-inoculated treatments.

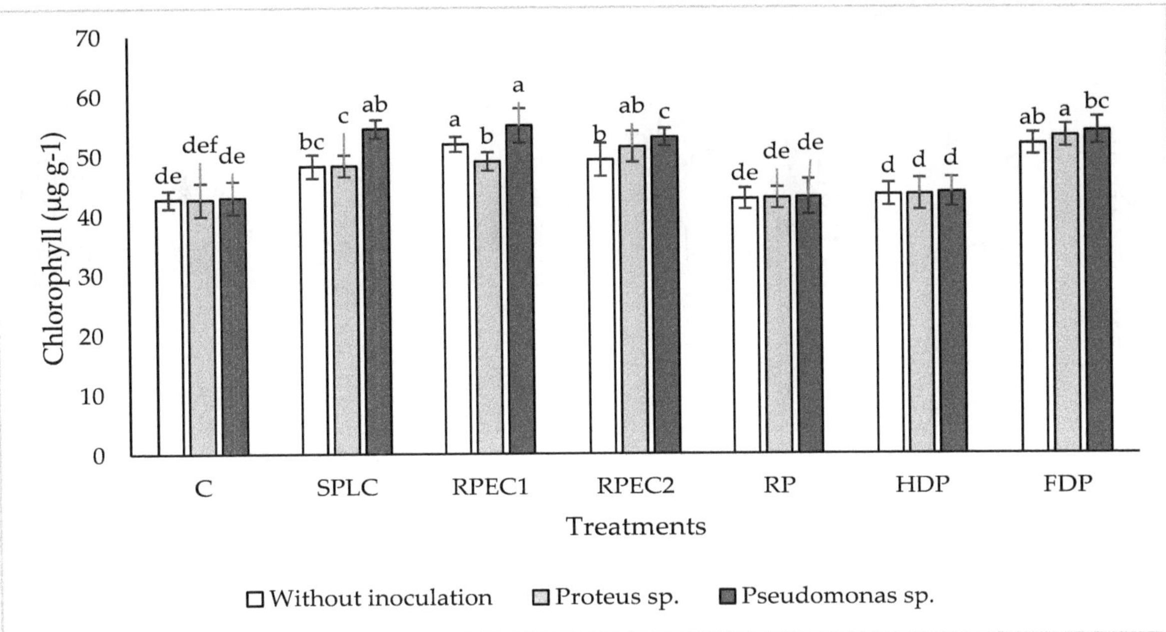

Figure 1. Effects of PGPR, P-enriched compost and inorganic fertilizers on leaf chlorophyll concentration (µg g^{-1}). C—Control; SPLC—Poultry litter only; RPEC1—Rock phosphate + poultry litter solubilized with *Pseudomonas* sp. during the composting process; $RPEC_2$—Rock phosphate + poultry litter solubilized with *Proteus* sp. during composting process), RP—Rock phosphate + poultry litter; HDP—Half dose inorganic P from Single Super Phosphate-SSP 18% P_2O_5; FDP—Chemical fertilizer (Single Super Phosphate). All the treatments sharing a common letter are similar, otherwise they differ significantly at $p \leq 0.05$.

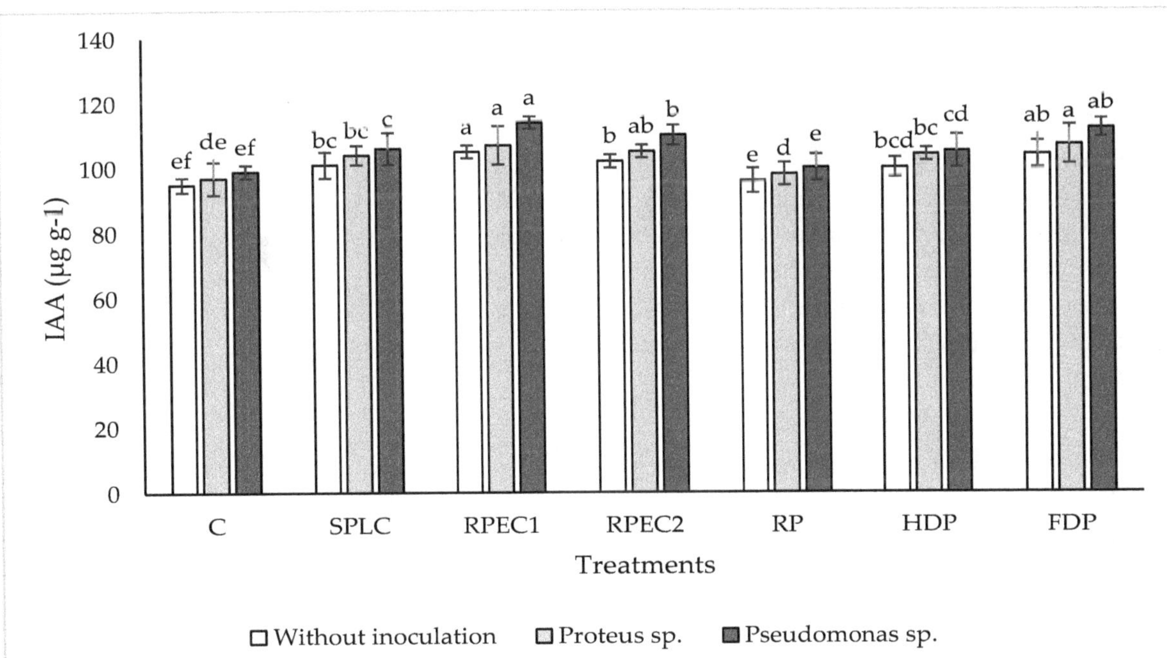

Figure 2. Effects of PGPR, P-enriched compost and inorganic fertilizers on leaf IAA concentration (µg g^{-1}) in wheat. C—Control; SPLC—Poultry litter only; RPEC1—Rock phosphate + poultry litter solubilized with *Pseudomonas* sp. during the composting process; $RPEC_2$—Rock phosphate + poultry litter solubilized with *Proteus* sp. during composting process), RP—Rock phosphate + poultry litter; HDP—Half dose inorganic P from Single Super Phosphate—SSP 18% P_2O_5; FDP- Chemical fertilizer (Single Super Phosphate). All the treatments sharing common letter are similar otherwise they differ significantly at $p \leq 0.05$.

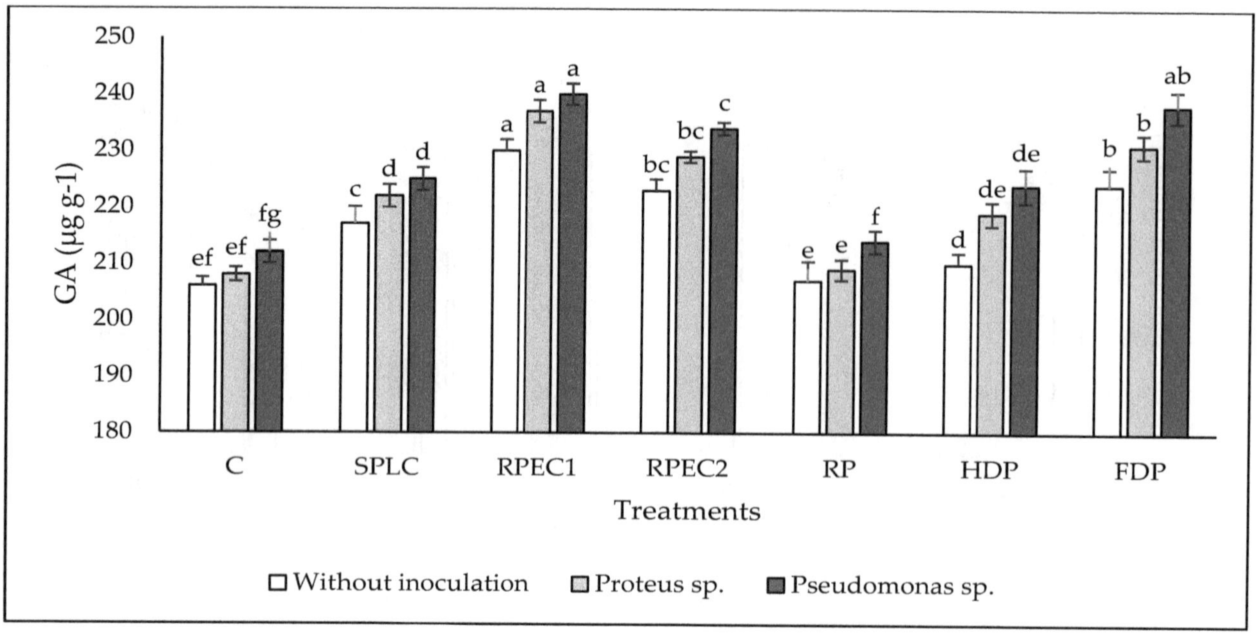

Figure 3. Effects of PGPR, P-enriched compost and inorganic fertilizers on leaf GA concentration (µg g^{-1}) in wheat. C—Control; SPLC—Poultry litter only; RPEC1—Rock phosphate + poultry litter solubilized with *Pseudomonas* sp. during the composting process; $RPEC_2$—Rock phosphate + poultry litter solubilized with *Proteus* sp. during composting process), RP—Rock phosphate + poultry litter; HDP—Half dose inorganic P from Single Super Phosphate—SSP 18% P_2O_5; FDP—Chemical fertilizer (Single Super Phosphate). All the treatments sharing common letter are similar otherwise they differ significantly at $p \leq 0.05$.

3.3. Plant Phosphorus Uptake and Seed Phosphorus

The data presented in Figure 4 show that the phosphorus uptake was maximum (70%) due to the application of $RPEC_1$ followed by $RPEC_2$ (63%) and FDP (60%), while RP treatment showed no significant difference compared to un-inoculated untreated control. Seed inoculation with *Pseudomonas* sp. resulted in a maximum (7%) increase in P-uptake over un-inoculated treatments. The interaction of fertilizer treatments and PGPRs showed that $RPEC_1$ in combination with *Pseudomonas* sp. showed maximum increase (88%) in P-uptake followed by *Proteus* sp. inoculated $RPEC_1$ (79%) over untreated un-inoculated control.

The seed phosphorus content showed a 61% increase following application of $RPEC_1$ over un-inoculated untreated control, which was 12%, 17%, 33% and 41% higher than FDP, $RPEC_2$, SPLC and HDP, respectively (Figure 5). The application of *Pseudomonas* sp. alone also resulted in an increase (3.5%) in seed P contents over un-inoculated untreated control whereas the interactive effect of PGPR × Fertilizer was nonsignificant.

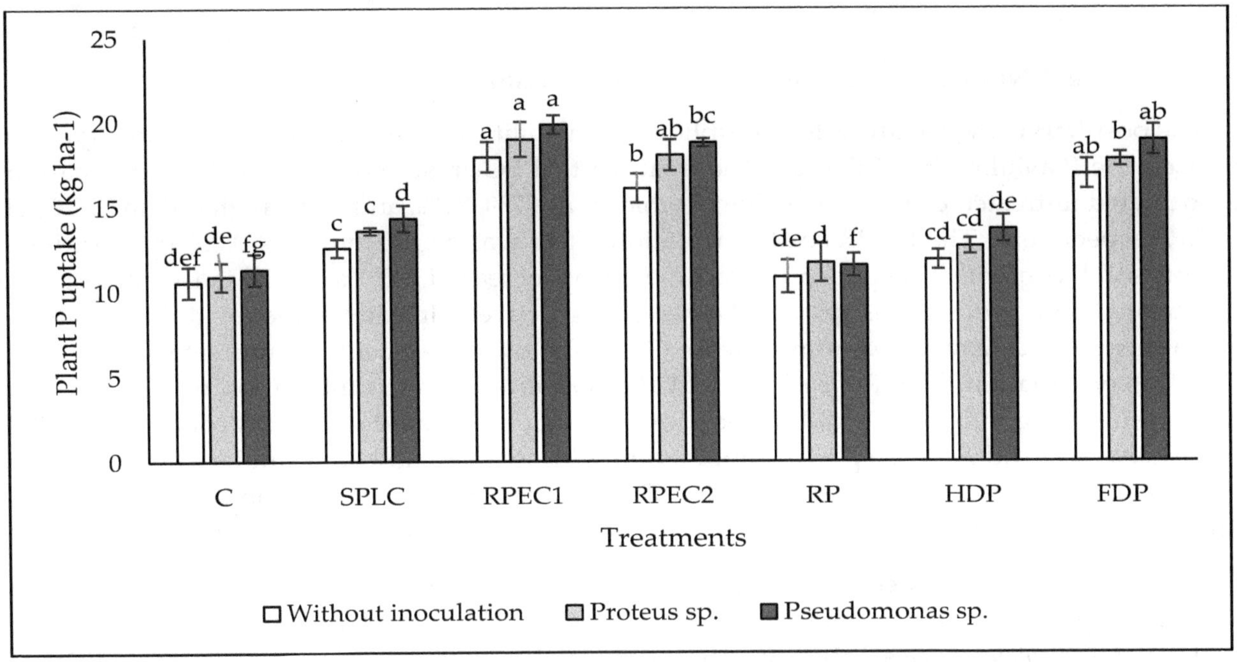

Figure 4. Effects of PGPR, P-enriched compost and inorganic fertilizers on plant P uptake (kg ha^{-1}) in wheat. C—Control; SPLC—Poultry litter only; RPEC1—Rock phosphate + poultry litter solubilized with *Pseudomonas* sp. during the composting process; RPEC2—Rock phosphate + poultry litter solubilized with *Proteus* sp. during composting process), RP—Rock phosphate + poultry litter; HDP— Half dose inorganic P from Single Super Phosphate—SSP 18% P_2O_5; FDP—Chemical fertilizer (Single Super Phosphate). All the treatments sharing common letter are similar otherwise they differ significantly at $p \leq 0.05$.

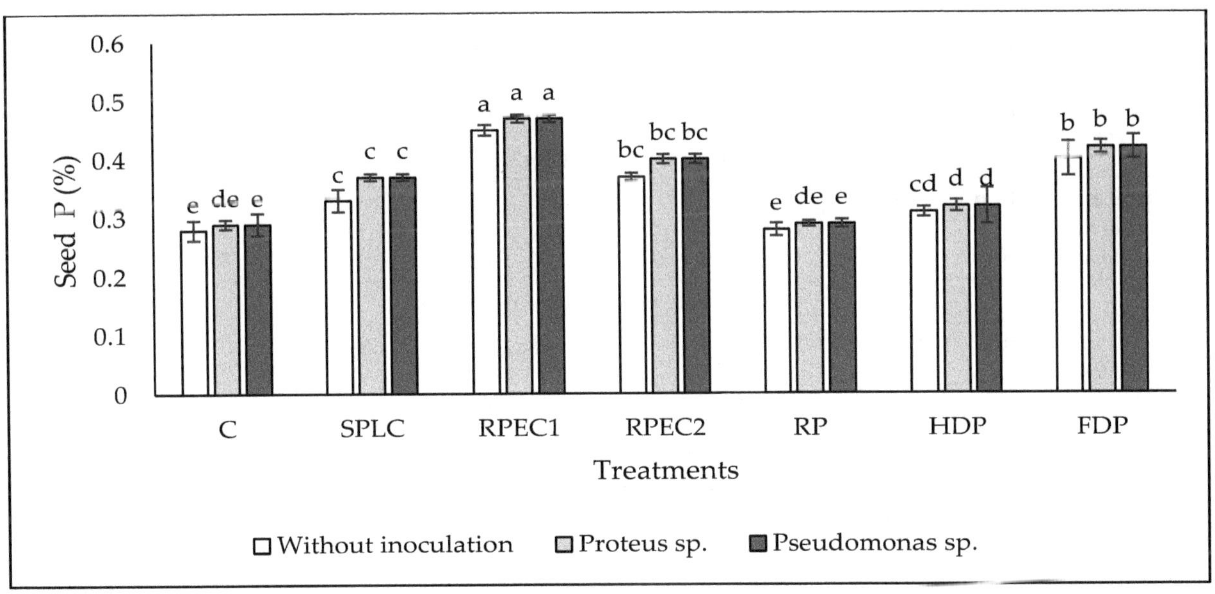

Figure 5. Effects of PGPR, P-enriched compost and inorganic fertilizers on plant P uptake (kg ha^{-1}) in wheat. C—Control; SPLC—Poultry litter only; RPEC1—Rock phosphate + poultry litter solubilized with *Pseudomonas* sp. during the composting process; RPEC2—Rock phosphate + poultry litter solubilized with *Proteus* sp. during composting process), RP—Rock phosphate + poultry litter; HDP— Half dose inorganic P from Single Super Phosphate—SSP 18% P_2O_5; FDP—Chemical fertilizer (Single Super Phosphate). All the treatments sharing common letter are similar otherwise they differ significantly at $p \leq 0.05$.

3.4. Soil Properties

3.4.1. Available P, Nitrate Nitrogen and Extractable Potassium

The post-harvest soil analysis for phosphorus availability showed that the treatments significantly increased the P availability (Table 5). The treatment $RPEC_1$ resulted in a significant increase over un-inoculated untreated control, the value of which was 37%, 82%, and 130% higher than SPLC, HDP and RP, respectively. The PGPR seed inoculation effect was significant ($p \leq 0.05$) for post-harvest available soil P contents. The phosphorus content was increased (20% and 9%) in the rhizosphere of plants treated with *Pseudomonas* sp. and *Proteus* sp. respectively. In combination with *Pseudomonas* sp., the treatment $RPEC_1$ gave maximum increase (3.43-fold) over un-inoculated untreated control, while $RPEC_1$ in combination with *Proteus* sp. and *Pseudomonas* sp. in combination with FDP showed similar results by giving a 3.17-fold increase over the un-inoculated untreated control. The PGPRs (*Pseudomonas* sp. and *Proteus* sp.) in combination with $RPEC_2$ showed a similar effect for increase [8%] in available P over un-inoculated $RPEC_2$. The treatment RP in combination with *Pseudomonas* sp. resulted in 30% increase over un-inoculated RP, which was 16.5% higher than *Proteus* sp. inoculated RP.

C—Control (un-inoculated untreated), SPLC—Simple poultry compost, $RPEC_1$—Rock phosphate enriched compost inoculated with *Pseudomonas* species, $RPEC_2$—Rock phosphate enriched compost inoculated with *Proteus* species, RP—Rock phosphate, HDP—Half dose inorganic P fertilizer, FDP—Full dose inorganic P fertilizer. Soil samples for nutrient and biological analyses were collected two days after wheat harvesting.

Table 5. Effects of PGPR, P-enriched compost and inorganic fertilizers on post-harvest soil of wheat in field experiments.

			Available P (mg kg^{-1})				
Treatments	**C**	**SPLC**	**$RPEC_1$**	**$RPEC_2$**	**RP**	**HDP**	**FDP**
Without inoculation	2.8 ^{g}	6.0 ^{e}	10.9 ^{a}	9.7 ^{b}	4.4 ^{f}	7.8 ^{d}	8.76 ^{c}
Proteus sp.	2.9 ^{g}	6.3 ^{e}	11.6 ^{a}	10.1 ^{b}	4.9 ^{f}	8.4 ^{c,d}	8.90 ^{c}
Pseudomonas sp.	3.2 ^{g}	6.9 ^{e}	14.4 ^{a}	10.48 ^{b}	5.7 ^{f}	9.2 ^{c,d}	9.74 ^{b,c}
			Nitrate nitrogen (mg kg^{-1})				
Without inoculation	3.07 ^{f}	3.87 ^{a,b,c}	3.98 ^{a}	3.92 ^{a,b}	3.07 ^{f}	3.39 ^{e}	3.78 ^{b,c,d}
Proteus sp.	3.08 ^{e}	3.88 ^{a,b}	4.04 ^{a}	4.07 ^{a}	3.07 ^{e}	3.42 ^{d}	3.81 ^{b,c}
Pseudomonas sp.	3.0 ^{f}	3.89 ^{b,c}	4.20 ^{a}	4.14 ^{a,b}	3.08 ^{f}	3.44 ^{e}	3.85 ^{c,d}
			Extractable potassium (mg kg^{-1})				
Without inoculation	95.50 ^{c,d}	106 ^{a,b}	109.5 ^{a}	106 ^{a,b}	96.5 ^{c,d}	101 ^{a,b,c}	102 ^{a,b,c}
Proteus sp.	96.33 ^{d}	108 ^{a,b}	110 ^{a}	107 ^{a,b}	97 ^{c,d}	102 ^{a,b,c}	103 ^{a,b,c}
Pseudomonas sp.	97.2 ^{d}	108 ^{a,b}	112 ^{a}	109 ^{a,b}	98 ^{c,d}	104 ^{b,c}	104 ^{b,c}
			Alkaline phosphatase (μg PNP g^{-1} hr^{-1})				
Without inoculation	117 ^{d}	136 ^{b}	151 ^{a}	143 ^{a,b}	118 ^{d}	127 ^{c}	133 ^{b,c}
Proteus sp.	118 ^{f,g}	142 ^{b,c}	157 ^{a}	147 ^{b}	121 ^{f}	132 ^{d,e}	137.5 ^{c,d}
Pseudomonas sp.	120 ^{e}	155.5 ^{b}	167 ^{a}	156 ^{b}	122 ^{e}	136 ^{c,d}	143.5 ^{c}
			Microbial biomass carbon (μg g^{-1})				
Without inoculation	84 ^{f,g}	112 ^{c}	136 ^{a}	123 ^{b}	87 ^{e,f}	93 ^{e}	106 ^{c,d}
Proteus sp.	89 ^{f,g}	119 ^{c}	144 ^{a}	137 ^{a,b}	92 ^{f}	103 ^{e}	114 ^{c,d}
Pseudomonas sp.	93 ^{f,g}	127 ^{c}	159 ^{a}	146 ^{b}	96 ^{f}	114 ^{d,e}	120 ^{c,d}
			Microbial biomass phosphorus (μg g^{-1})				
Without inoculation	7 ^{d}	14 ^{b}	19 ^{a}	18 ^{a}	7 ^{d}	10 ^{b,c}	11 ^{b,c}
Proteus sp.	8 ^{d}	16 ^{b,c}	21 ^{a}	19 ^{a,b}	8 ^{d}	11 ^{c,d}	12 ^{c,d}
Pseudomonas sp.	8 ^{e}	18 ^{c}	26 ^{a}	24 ^{a,b}	8 ^{e}	13 ^{d}	13 ^{d}

All the treatments sharing a common letter are similar otherwise they differ significantly at $p \leq 0.05$.

Mean data for post-harvest soil nitrate-nitrogen showed significant differences with the application of fertilizer treatments (Table 5). The treatments $RPEC_1$ and $RPEC_2$ resulted a 36% increase followed by SPLC, FDP and HDP showing 29%, 26% and 14% increase over un-inoculated untreated control, respectively. The treatment RP showed nonsignificant difference with the control. There was a nonsignificant effect of seed inoculation on NO_3-N over un-inoculated control. However, the maximum increase was recorded by the application of *Pseudomonas* sp. which was 4% higher over un-inoculated control treatments. The interactive effect of PGPR and fertilizer treatments was nonsignificant with SPLC, RP, HDP and FDP, while the treatments $RPEC_1$ and $RPEC_2$ showed 36% and 35% increases over un-inoculated untreated control. The mean data showed that all the treatments increased extractable potassium except RP (Table 5). A maximum increase (15%) in the content of extractable K was recorded following the treatment $RPEC_1$ followed by SPLC and $RPEC_2$ which were significantly similar in their effect by showing 11% and 12% increase over un-inoculated untreated control, respectively. The treatments FDP and HDP also showed similar effects and increased extractable K by 7% over control. Seed inoculation with *Pseudomonas* sp. showed 2% increase in seed phosphorus contents over un-inoculated untreated control. The interactive effect of PGPR × Fertilizers, was nonsignificant, whereas *Pseudomonas* sp. inoculation in combination with $RPEC_1$ showed maximum (17%) phosphorus contents over un-inoculated untreated control.

3.4.2. Alkaline Phosphatase and Microbial Biomass

Alkaline phosphatase activity was significantly ($p \leq 0.05$) increased as a result of different treatments (Table 5). The treatment $RPEC_1$ resulted a maximum increase (29%) over un-inoculated untreated control, which was 5.6%, 11%, 13.5% and 19% higher than $RPEC_2$, SPLC, FDP and HDP respectively. *Pseudomonas* sp. inoculation showed 8% increase over un-inoculated treatments, which was 2.5% higher than *Proteus* sp. inoculation. Significant ($p \leq 0.05$) increase in alkaline phosphatase activity was recorded due to the combine effects of PGPRs with different fertilizer treatments. The inoculation of *Pseudomonas* sp. in combination with the treatment $RPEC_1$ showed maximum (43%) increase over untreated un-inoculated control. The treatments $RPEC_2$ and SPLC in combination with *Pseudomonas* sp. and $RPEC_1$ in combination with *Proteus* sp. showed similar effect and increased the alkaline phosphatase activity by 34% over un-inoculated untreated control. The treatment FDP in combination with *Pseudomonas* sp. showed a 23% increase over untreated un-inoculated control; the effect of which was significantly similar to un-inoculated $RPEC_2$ treatment. The PGPRs (*Pseudomonas* sp. and *Proteus* sp.) in combination with HDP showed a similar effect but significantly lower percentage increase than un-inoculated HDP treatment.

Mean data showed that the fertilizer treatments significantly ($p \leq 0.05$) improved the microbial biomass carbon contents (Table 5). Significant increase (65%) was recorded in microbial biomass carbon contents in $RPEC_1$ treatment over control; the increase for $RPEC_1$ was also 9%, 22%, 30%, 43%, and 60% higher than that of $RPEC_2$, SPLC, FDP, HDP and RP, respectively. Inoculation of seeds with *Pseudomonas* sp. showed maximum (16%) increase in microbial biomass carbon over un-inoculated treatments, the values of which were 8% higher than *Proteus* sp. inoculated treatments. The interactive effect of PGPR inoculation to seeds and fertilizer treatments was also significant with microbial biomass carbon contents. Among the *Pseudomonas* sp. inoculated treatments, $RPEC_1$ showed 89% increase in microbial biomass carbon over un-inoculated untreated control, while the treatment $RPEC_2$ showed 74% increase, which was significantly similar with *Proteus* sp. inoculated $RPEC_1$ treatment. However, *Proteus* sp. inoculated $RPEC_2$ increased microbial biomass carbon by 63% and showed a nonsignificant difference with un-inoculated $RPEC_1$ treatment. *Pseudomonas* sp. in combination with FDP showed a nonsignificant difference with un-inoculated $RPEC_2$ but was 13% higher than the un-inoculated FDP treatment. The treatment RP in combination with *Pseudomonas* sp. increased microbial biomass by 10% over untreated un-inoculated control.

The data in Table 5 show that microbial biomass phosphorus (MBP) increased significantly ($p \leq 0.05$) with the application of fertilizer treatments than the un-inoculated untreated control. The maximum increase (1.75-fold) in MBP was recorded from the treatment $RPEC_1$ which was 7%, 37% and 83%

higher than $RPEC_2$, SPLC and FDP, respectively while FDP showed nonsignificant difference with HDP. The treatment RP showed a nonsignificant difference with un-inoculated untreated control. There was also a significant effect of PGPR on microbial biomass phosphorus. *Pseudomonas* sp. inoculation increased microbial biomass P by 33% over un-inoculated treatments, the values of which were 14% higher than *Proteus* sp. inoculated treatments. *Pseudomonas* sp. inoculation with $RPEC_1$ showed a maximum increase (2.7-fold) in MBP followed by $RPEC_2$ (2.43-fold) over un-inoculated untreated control. The treatment $RPEC_1$ in combination with *Proteus* sp. showed an (61%) increase over un-inoculated untreated control which was at par with *Pseudomonas* sp. inoculated SPLC. The treatments FDP, HDP increased (46%) microbial biomass P showing nonsignificant difference with each other and RP showed 12% increase in microbial biomass P with PGPR inoculation over un-inoculated untreated control.

3.5. *Economic Analysis*

The economic analysis of applied treatments (Table 6) in terms of value cost ratio (VCR) showed that $RPEC_1$ performed best with and without seed inoculation. Among the un-inoculated treatments, $RPEC_1$ showed maximum VCR (2.72) followed by $RPEC_2$ (2.14), FDP (1.94) while the minimum (0.06) VCR was received from RP. Seed inoculation with *Pseudomonas* sp. in combination with $RPEC_1$ superseded all of the treatments resulting in maximum VCR (3.23). Hence, rock phosphate enriched compost alone or more so in combination with phosphate solubilizing bacteria (PSB), can perform better than chemical fertilizers. The economic analysis revealed that RPEC could be an economically feasible substitute to costly chemical fertilizers for sustainable crop production.

Table 6. Economic Analysis of the applied products presented as value cost ratio (VCR).

Treatments	Grain Yield	Increase in Yield	Increased Yield Value	Cost of Inputs	Net Return	VCR
	kg ha^{-1}		Rs. ha^{-1}			
Control	**2177**	-	-	-	-	-
SPLC	3120	943	28,290	17,760	10,530	1.59:1
$RPEC_1$	3629	1452	43,560	16,010	27,550	2.72:1
$RPEC_2$	3317	1140	34,200	16,010	18,190	2.14:1
RP	2195	18	540	9285	−8745	0.06:1
HDP	3020	843	25,290	13,435	11,855	1.88:1
FDP	3495	1318	39,540	20,385	19,155	1.94:1
Proteus sp. (1S)	2474	297	8910	7310	1600	1.22:1
SPLC + 1S	3319	1142	34,260	18,560	15,700	1.85:1
$RPEC_1$+1S	3731	1554	46,620	16,810	29,810	2.77:1
$RPEC_2$+1S	3524	1347	40,410	16,810	23,600	2.40:1
RP+1S	2498	321	9630	10,085	−455	0.95:1
HDP+1S	3511	1334	40,020	14,235	25,785	2.01:1
FDP+1S	3620	1443	43,290	21,185	22,105	2.04:1
Pseudomonas (2S)	2571	394	11,820	7310	4510	1.62:1
SPLC + 2S	3433	1256	37,680	18,560	19,120	2.03:1
$RPEC_1$+ 2S	3987	1810	54,300	16,810	37,490	3.23:1
$RPEC_2$+ 2S	3639	1462	43,860	16,810	27,050	2.61:1
RP+ 2S	2592	415	12,450	10,085	2365	1.23:1
HDP+ 2S	3364	1187	35,610	14,235	21,375	2.05:1
FDP+ 2S	3848	1671	50,130	21,185	28,945	2.37:1

Increase in yield = Yield of treatment − Yield of control, Increased yield value = Grain price × increase in yield, Net return = Increased yield value − cost of inputs, Value cost ratio (VCR) = Increased yield value/cost of inputs, Poultry litter = Rs. 1.5 kg^{-1}, Rock phosphate = Rs. 5 kg^{-1}, Single super phosphate (SSP) = Rs. 25 kg^{-1}, Urea = Rs. 30 kg^{-1}, Labor charges for compost preparation = Rs. 5250, Seed inoculant = Rs. 200 L^{-1}, Wheat grain price = Rs. 30 kg^{-1}, SPLC—Simple Poultry litter, $RPEC_1$—Rock Phosphate Enriched Compost solubilized with *Pseudomonas* sp., $RPEC_2$—Rock phosphate enriched compost solubilized with *Proteus* sp. 1S-Seed inoculation with *Proteus* sp., 2S—Seed inoculation with *Pseudomonas* sp., RP—Rock phosphate, HDP—Half dose of inorganic fertilizer, FDP—Full dose of inorganic fertilizer. Rs—Refer to national currency (Rupees).

4. Discussion

The use of plant growth promoting rhizobacteria (PGPR) in combination with organic (composts, rock phosphate) and inorganic (chemical fertilizers) phosphorus sources significantly increased the number of tillers per plant and yield components of wheat crop. The results are in conformity to the findings of Akhtar et al. [40] who recorded increase in plant height, the number of tillers, grain yield and 1000 grain weight of wheat with the use of compost and PGPR inoculation.

Maximum grain yield was obtained by the application of $RPEC_1$ which was higher than the full dose of inorganic P fertilizers (FDP), irrespective of the PGPR seed inoculants. The observed yield increase from $RPEC_1$ was indicative of the high P availability and greater photosynthesis as observed by an increase in chlorophyll content and dry matter production, which was maximum in $RPEC_1$ over other treatments. Plant P availability as the key factor for maximum plant growth and higher crop production [41]. Although a full dose of P (FDP) as inorganic fertilizer (SSP) is a source of readily available phosphorus necessary for early growth of the plants, at the site of SSP application, production of the least soluble Ca-P compounds due to surface adsorption and precipitation, reduce P availability [10]. The organic acids produced due to compost might have reduced P exchange sites through chelation and released more soluble forms of plant available P [42] compared to SSP which could help increase growth and yield of wheat. Seed inoculation with *Pseudomonas* sp. increased the grain yield with fertilizer treatments, however the maximum increase was recorded from *Pseudomonas* sp. inoculation with $RPEC_1$ followed by the inoculated FDP treatment. Microbial community in the root rhizosphere might have taken part to release fixed phosphorus through organic acids production which ultimately increased the yield of wheat. Afzal and Bano [43] reported that seed inoculation with PGPR in combination with P fertilizer increased the grain yield of wheat which was 30–40% higher than the un-inoculated P fertilizer. It was reported that organic manures and bio-fertilizers have a high impact on nutrient uptake, physiological process of wheat, and also on water holding capacity of the soil which ultimately increase grain yield of the crop [44]. Amujoyegbe et al. [45] recorded higher grain yield of maize due to the application of chicken manure in combination with microbes compared to chemical fertilizer and chicken manure alone. An association of agronomic traits with grain yield and a positive correlation of 1000 grain weight with grain yield was previously demonstrated in PGPR + manure treated plants of wheat [44].

Increase in the dry matter yield due to the application of $RPEC_1$ compared to FDP may be due to higher vegetative growth, chlorophyll content and the maximum number of tillers during the crop growth, while mobilization of phosphorus due to dissolution of rock phosphate from $RPEC_1$ might have taken part in the physiological processes leading to maximum biomass yield. Higher yields of mung bean were recorded due to bio-inoculated RP enriched compost having higher citrate soluble, water soluble P and organic P, maximum microbial biomass carbon and acid phosphatase activity compared to un-inoculated composts [7]. Similarly, Nishanth and Biswas [11] prepared enriched composts with *Aspergillus awamori* inoculation and tested these on the wheat crop, which gave maximum biomass production in comparison to composts prepared without inoculants. Hossain et al. [46] reported an increase in grain and straw yield of wheat crop with the application of phosphate solubilizing bacteria (PSB) along with different levels of phosphorus. In concurrence with the present results, an increase in dry matter and grain yield of agronomic crops due to phosphate solubilizing microorganisms in combination with different P fertilizers were reported earlier by different workers [47–49].

Phosphorus plays an important role in chlorophyll production and regulation. It has been reported that the partitioning of photosynthates between leaves and reproductive organs is regulated by the availability of phosphorus to the plants [50]. Maximum increase in chlorophyll contents in flag leaves were recorded due to the application of $RPEC_1$ followed by FDP and SPLC. Zafar et al. [51] reported an increase in chlorophyll contents by 10–89% over control in leaves of maize crop following application of P fertilizers in the form of compost and inorganic fertilizers. The PGPR in combination with compost was recorded to be stimulatory for chlorophyll production; this was confirmed for *Pseudomonas* sp. in combination with $RPEC_1$. Seed inoculation with *Pseudomonas* sp. alone or in combination with P fertilizers, was more efficient for improving chlorophyll contents in flag leaves of wheat plants. Naseem and Bano [52] reported that the seed inoculation with *Pseudomonas* sp.

and *Bacillus cereus* increased chlorophyll contents by 8–13% in leaves of wheat crop. An increase in chlorophyll contents with the application of organic manure was also recorded [53].

PGPR alone or in combination with fertilizers showed a significant effect on IAA and GA contents of wheat flag leaves, however, maximum increase was recorded as a result of $RPEC_1$ application followed by FDP and $RPEC_2$. Among the PGPRs, *Pseudomonas* sp. performed better than the *Proteus* sp. Indole Acetic Acid synthesis by bacteria may have various regulatory effects in plant–bacterial interactions and significant effect on plant growth promotion [54]. Generally, phytohormones in plants plays an important role in cell division, proliferation, and differentiation, vascular tissue alteration, responses to light and gravity, general root and shoot architecture, seed and tuber germination, organ differentiation, peak predominance, ethylene synthesis, vegetative growth processes, fruit development and aging. These results are in accordance with the findings of Saharan and Nehra [55], who reported that the phytohormone production through PGPR (*Pseudomonas, Azotobacter, Azospirillum*) may contribute to growth and yield of the crop. IAA acts as a signal molecule for cell expansion, division and differentiation. Higher counts of genus *Pseudomonas* were recorded [56] in winter wheat cultivars and described the developmental phase of wheat crops as a key factor in higher population of the microbes. The GA and IAA were reported to be produced by bacterial strains such as *Bacillus* and *Pseudomonas* [57] and inoculation of wheat with *Pseudomonas* sp. gave maximum increase in growth and yield [58]. Khan et al. [59] found an increase in IAA and GA contents in leaves of wheat inoculated with *Pseudomonas* and *Bacillus* strains. Sivasankari et al. [60] isolated bacterial strains from black gram (*Vigna mungo*) rhizosphere soil and reported maximum IAA production from *Pseudomonas* sp. than *Proteus* sp.

Phosphors uptake increased with the application of RP enriched compost ($RPEC_1$) which would be due to phosphorus in the soluble form. Higher concentration of macronutrients due to the decomposition of organic materials in the soil were recorded [61]. Incorporation of organic materials can enhance phosphorus availability in the soil solution by decreasing P sorption/fixation through chelation [62]. Phosphorus also plays an efficient role in plant photosynthesis, respiration, formation of cell membrane, glycolysis and enzymes activities [63] showing that the growth and development of all crops are dependent upon P availability [64]. The presence of P as an integral part of nucleotides, phospholipids, phosphoproteins, and coenzymes shows its importance for life [65]. An increase in P-uptake due to enriched compost in the present study was due to the maximum available P as well as total organic and readily available carbon. Sharma et al. [66] reported increased N uptake (18–38 kg ha^{-1}), P uptake (2.7–6.6 kg ha^{-1}), and K uptake by (16–41 kg ha^{-1}) in the rice–wheat system when inoculated with *Pseudomonas striata*. It was reported by Nishanth and Biswas [11] that RP enriched compost inoculated with *Aspergillus awamori* can significantly enhance P uptake in wheat crop, which was recorded as 78% more efficient compared to DAP. Ghaderi et al. [67] reported 51%, 29% and 62% release of phosphorus from iron hydroxides by the application of *Pseudomonas putida, Pseudomonas fluorescens, and Pseudomonas fluorescens,* respectively. Shrivastava [10] reported that inoculation of microbes with P enriched manure show maximum P uptake in mung-bean crop compared to SSP fertilizer. The P-enriched compost in combination with effective microbes (EM) can enhance N and P uptakes of the cowpea crop [14].

Crop growth is regulated by the nutrient supply from organic or chemical fertilizer sources. Organic materials are considered to be the best source for nutrient supply to plants but with slow release until the crop maturity, which may create a delay in crop maturity or cause high nutrients content in the produce [68]. Maximum P concentration in the wheat seeds with the application of $RPEC_1$ might be due to a slow release process resulting in P accumulation in the seeds due to mobility of the phosphorus from soil to plant process. The integrated management of P fertilizers at the root zone can increase the mobility of P from plant roots through physiological adaptive mechanisms [69]. Seed inoculation with *Pseudomonas* sp. showed an increase in seed phosphorus. According to Son et al. [70] soybean seed P content increased with inoculation of phosphate solubilizing microorganisms.

Organic and inorganic amendments have a great impact on soil properties [71]; however, while the application of fertilizer increases P availability at all crop growth stages compared to control treatment, the RP compost showed maximum P availability at later stages of wheat crop growth [11]. The increase in post-harvest soil P availability with the application of RP enriched compost may be due to mineralization of both RPEC and soil organic P, and chelation of P through ligand exchange reactions to reduce P fixation throughout the crop growth stages. The ligand exchange reactions can increase P mobilization through organic and phosphate anions adsorption with Fe and Al sites [72]. Slow release of P through mineralization of organic P fraction from enriched compost was reported previously [73]. Organic acids produced by phosphate solubilizing microorganisms are sources of H^+ ions which help mineralize tri-calcium phosphate of RP to mono-calcium phosphate; the available form of phosphorus for better plant growth [74].

The application of compost treatments showed significantly higher nitrate-nitrogen contents in post-harvest soil compared to control. Higher nitrate nitrogen content from compost treated plots would be due to reduced nitrate leaching from the soil [75]. Sommers and Giordano [76] stated that all inorganic N in soil amended with Municipal Solid Waste (MSW) compost was available for plant uptake, but 5 to 75% of the organic N will be mineralized within 1 year after application. The findings are in accordance with the results of Baziramakenga et al. [77] who reported an increase in inorganic nitrogen (NO_3-N) contents of snap-bean post-harvest soil with the application of compost of de-inking paper residues and poultry manure. The reason for higher NO_3-N contents due to the application of composts is attributed to the formation of phospho-protein due to the interaction with rock phosphate from enriched compost, which is less susceptible to volatilization. The proteins are decomposed by soil bacteria and change into ammonium that is further nitrified by nitrifying bacteria. This form of nitrogen from compost is slowly available to plants having fewer chances of loss through volatilization. The escape of ammonia from soil decreases if the nitrogen source is compost, organic manure or green manure [78]. The presence of phosphate preserves the nitrogen resulting in a decrease in the number of denitrifying bacteria [79]. The slow release process of nutrients from enriched compost might be another reason for higher nitrate-nitrogen contents than inorganic fertilizers (FDP) in post-harvest soil. Adeli et al. [80] reported higher residual soil NO_3-N contents after cotton crops with the application of poultry manure compared to inorganic fertilizers. Seed inoculation with PGPR (*Pseudomonas* sp. and *Proteus* sp.) showed an increase in post-harvest nitrate-nitrogen contents. Canbolat et al. [81] also reported increase in soil post-harvest nitrogen contents with application of *Pseudomonas putida* compared to the control on barley crop.

Extractable potassium contents increased in post-harvest soil with the application of compost compared to inorganic P fertilizer (FDP). A significantly higher concentration of K with the application of enriched compost compared to FDP and control might be due to the higher water-soluble potassium present in the enriched composts. Stratoon et al. [82] reported that K in composts remains in water soluble forms and thus does not need to be mineralized before becoming available to plants. The increase in soil extractable K by rock phosphate enriched compost (RPEC) may be related to the direct addition to the available K pool of the soils, and to the reduction of K fixation and increase the release of K from the soil solid phase due to the interaction of organic matter and/or soil microorganisms with K-bearing minerals [77]. It was revealed that potassium in manure and compost is highly plant-available and can be used similar to K fertilizer application [83].

Soil enzymes (alkaline phosphatase and acid phosphatase) play a vital role in conversion of fixed soil phosphorus to plant available form [7]. The increase in alkaline phosphatase activities with the application of RP enriched compost in the present study may be due to the availability of organic C which consequently increased the soil phosphatase activity [84] and the compost might have provided considerable carbon and nitrogen for maximum growth of microbes. It has been emphasized that C and N are interlinked with P mineralization by microbes [85] and Shrivastava et al. [10] concluded that the availability of metabolizable C plays a significant role to increase soil phosphatase activity with the application of P enriched manure on mungbean crop. Soil enzymes such as acid and alkaline phosphatases help to increase mineralization of P_0 to P_i by creating a strong relation between bio-available and unavailable P in the soil [86]. Some researchers [87,88] believe that there is an inverse relationship between available P and phosphatases due to negative feedback of phosphate ions on PHO genes, that suppress phosphatase synthesis by microbes [89]. However, phosphatase activity was not affected by the use of rock phosphate as a phosphate source in RP enriched compost [9] showing long persistence and least biodegradation of enzymes with the application of compost. But Pascual et al. [90] endorsed the decrease in phosphatase activities with time span due to exhaustion of biodegradable substrates by microbial activity.

Microbial biomass is an important factor assessing soil quality and its ability to provide energy for nutrient recycling and transformation in the soils [91]. Kiani et al. [92] documented the microbial biomass responses to different land management systems including fertilizer addition and organic amendment application and identified suitable soil quality indicators. The microbial biomass carbon (MBC) acts as substrate supplying entity for microbial communities in soil [93]. In the present study, maximum MBC with the application of RP enriched compost compared to poultry litter compost is due to higher percentage of existing microbial biomass carbon, mineralizable nitrogen and water-soluble carbon in the former compost. The results are in conformity to the findings of Meena et al. [94] who reported an increase in soil microbial biomass carbon with the application of enriched compost compared to ordinary compost as well as inorganic fertilizer. Previously Ayed et al. [95] found an increase in microbial biomass carbon with the application of compost compared to inorganic chemical fertilizer and control in wheat crop.

The maximum microbial biomass phosphorus (MBP) with the application of RPEC and inorganic P fertilizer could be due to the transformation of labile and nonlabile inorganic phosphorus to the organic pool through microbial activity of compost. As Leytem et al. [96] reported the assimilation of various fractions of P into microbial biomass which ultimately provides available P for plants since most organic phosphorus in microbial cells is hydrolysable. Microbial biomass phosphorus may also help in calcareous soils by providing plant available P with application of manure [97] by the mechanism in which P is immobilized and transformed to labile P, which is safe from fixation and transmitted to available P [98]. The results in the present study showed less microbial biomass P with the addition of a full recommended dose of inorganic P fertilizer, indicating that soil existing organic carbon was limited to support microbial growth and activity. Minimum microbial biomass was recorded with the addition of high P inorganic fertilizer to the soil in an incubation study.

5. Conclusions

The present research revealed that enrichment of rock phosphate with poultry litter and PGPR during the process of composting improves nutrient availability and biological properties of the compost. Application of RP enriched compost in field experiment increased yield and yield components of the wheat crop compared to the full recommended dose of inorganic fertilizer and control. Moreover, seed inoculation with PGPR showed significant results to improve the agronomic effectiveness of RP enriched compost. Chemical (availability of phosphorus) and biological (microbial biomass C & P, alkaline and acid phosphatase activities) properties of post-harvest soil improved with the application of RP enriched compost. It can be concluded that RP enriched compost may be an alternative to chemical fertilizer to improve the growth and yield of the crop.

Author Contributions: Conceptualization, M.B., A.B. and N.K.; Methodology, M.B., M.K., K.M.D., and N.K.; Formal analysis, M.B., M.K., S.N. and N.K.; Software, M.B., K.M.D. and N.K.; Data curation, S.N., K.M.D. and A.M.; Validation, M.B., A.B., A.H., A.M. and N.K., Investigation, M.B., A.B., S.B., A.H. and N.K., Resources, M.B., M.K., A.B., S.N., A.M. and N.K., writing—original draft preparation, M.B., M.K., A.B. and N.K.; writing—review and editing, M.B., M.K., A.B., A.H., A.S. and N.K.; supervision, M.B., A.B. and N.K. All authors have read and agreed to the published version of the manuscript.

Acknowledgments: The author is thankful to Higher Education Commission Islamabad Pakistan and Pakistan Agriculture Research Council, Islamabad, Pakistan for the assistance to conduct this research.

References

1. Elkoca, E.; Turan, M.; Donmez, M.F. Effects of single, dual and triple inoculations with Bacillus subtilis, Bacillus megaterium and Rhizobium leguminosarum bv. Phaseoli on nodulation, nutrient uptake, yield and yield parameters of common bean [Phaseolus vulgaris l. cv.'elkoca-05']. *J. Plant Nutr.* **2010**, *33*, 2104–2119. [CrossRef]
2. Aydin, E.; Šimanský, V.; Horák, J.; Igaz, D. Potential of Biochar to Alternate Soil Properties and Crop Yields 3 and 4 Years after the Application. *Agronomy* **2020**, *10*, 889. [CrossRef]
3. Gyaneshwar, P.; Kumar, G.N.; Parekh, L.J.; Poole, P.S. Role of soil microorganisms in improving P nutrition of plants. *Plant Soil* **2002**, *245*, 83–93. [CrossRef]
4. Ali, S.; Xie, L. Plant Growth Promoting and Stress mitigating abilities of Soil Born Microorganisms. *Recent Pat. Food Nutr. Agric.* **2019**, *15*. [CrossRef] [PubMed]
5. Khan, N.; Bano, A. Exopolysaccharide producing rhizobacteria and their impact on growth and drought tolerance of wheat grown under rainfed conditions. *PLoS ONE* **2019**, *14*, e0222302. [CrossRef] [PubMed]
6. Reddy, M.S.; Kumar, S.; Babita, K. Bio-solubilization of poorly soluble rock phosphates by Aspergillus tubingensis and Aspergillus niger. *Bioresour. Technol.* **2002**, *84*, 187–189. [CrossRef]
7. Achal, V.; Savant, V.V.; Reddy, M.S. Phosphate solubilization by a wild type strain and UV-induced mutants of Aspergillus tubingensis. *Soil Biol. Biochem.* **2007**, *39*, 695–699. [CrossRef]
8. Puente, M.E.; Bashan, Y.; Li, C.Y.; Lebsky, V.K. Microbial populations and activities in the rhizoplane of rock-weathering desert plants. I. Root colonization and weathering of igneous rocks. *Plant Biol.* **2004**, *6*, 629–642. [CrossRef]
9. Gaind, S.; Pandey, A.K.; Lata. Microbial biomass, P-nutrition, and enzymatic activities of wheat soil in response to phosphorus enriched organic and inorganic manures. *J. Environ. Sci. Health Part B* **2006**, *41*, 177–187. [CrossRef]
10. Shrivastava, M.; Kale, S.P.; D'Souza, S.F. Rock phosphate enriched post-methanation bio-sludge from kitchen waste based biogas plant as P source for mungbean and its effect on rhizosphere phosphatase activity. *Eur. J. Soil Biol.* **2011**, *47*, 205–212. [CrossRef]
11. Nishanth, D.; Biswas, D.R. Kinetics of phosphorus and potassium release from rock phosphate and waste mica enriched compost and their effect on yield and nutrient uptake by wheat (Triticum aestivum). *Bioresour. Technol.* **2007**, *99*, 3342–3353. [CrossRef] [PubMed]
12. Naseem, H.; Ahsan, M.; Shahid, M.A.; Khan, N. Exopolysaccharides producing rhizobacteria and their role in plant growth and drought tolerance. *J. Basic Microbiol.* **2018**, *58*, 1009–1022. [CrossRef] [PubMed]
13. Reyes, I.; Valery, A.; Valduz, Z. Phosphate-solubilizing microorganisms isolated from rhizospheric and bulk soils of colonizer plants at an abandoned rock phosphate mine. In *First International Meeting on Microbial Phosphate Solubilization*; Springer: Dordrecht, Germany, 2007; pp. 69–75.
14. Chen, Y.P.; Rekha, P.D.; Arun, A.B.; Shen, F.T.; Lai, W.A.; Young, C.C. Phosphate solubilizing bacteria from subtropical soil and their tricalcium phosphate solubilizing abilities. *Appl. Soil Ecol.* **2006**, *34*, 33–41. [CrossRef]
15. Zayed, G.; Abdel-Motaal, H. Bio-active composts from rice straw enriched with rock phosphate and their effect on the phosphorous nutrition and microbial community in rhizosphere of cowpea. *Bioresour. Technol.* **2005**, *96*, 929–935. [CrossRef]
16. Adesemoye, A.O.; Obini, M.; Ugoji, E. Comparison of plant growth-promotion with Pseudomonas aeruginosa and Bacillus subtilis in three vegetables. *Braz. J. Microbiol.* **2008**, *39*, 423–426. [CrossRef] [PubMed]
17. Khan, N.; Bano, A. Modulation of phytoremediation and plant growth by the treatment with PGPR, Ag nanoparticle and untreated municipal wastewater. *Int. J. Phytoremediat.* **2016**, *18*, 1258–1269. [CrossRef]
18. Duca, D.; Lorv, J.; Patten, C.L.; Rose, D.; Glick, B.R. Indole-3-acetic acid in plant–microbe interactions. *Antonie Van Leeuwenhoek* **2014**, *106*, 85–125. [CrossRef]
19. Davies, P.J. The plant hormones: Their nature, occurrence and functions. In *Plant Hormones*; Davies, P.J., Ed.; Springer: Dordrecht, Germany, 2010; pp. 1–15.
20. Ahemad, M.; Kibret, M. Mechanisms and applications of plant growth promoting rhizobacteria: Current perspective. *J. King Saud Univ. Sci.* **2014**, *26*, 1–20. [CrossRef]

21. Chandler, J.W. The hormonal regulation of flower development. *J. Plant. Growth Regul.* **2011**, *30*, 242–254. [CrossRef]
22. Phillips, K.A.; Skirpan, A.L.; Liu, X.; Christensen, A.; Slewinski, T.L.; Hudson, C.; Barazesh, S.; Cohen, J.D.; Malcomber, S.; McSteen, P. Vanishing tassel2 encodes a grass-specifc tryptophan aminotransferase required for vegetative and reproductive development in maize. *Plant Cell* **2011**, *23*, 550–566. [CrossRef]
23. Bellini, C.; Pacurar, D.; Perrone, L. Adventitious roots and lateral roots: Similarities and differences. *Annu. Rev. Plant Biol.* **2014**, *65*, 639–666. [CrossRef] [PubMed]
24. Uggla, C.; Moritz, T.; Sandberg, G.; Sundberg, B. Auxin as a positional signal in pattern formation in plants. *Proc. Natl. Acad. Sci. USA* **1996**, *93*, 9282–9286. [CrossRef]
25. Chen, T.; Chen, Z.; Ma, G.H.; Du, B.H.; Shen, B.; Ding, Y.Q.; Xu, K. Diversity and potential application of endophytic bacteria in ginger. *Genet. Mol. Res.* **2014**, *13*, 4918–4931. [CrossRef] [PubMed]
26. Glick, B.R. The enhancement of plant growth by free-living bacteria. *Can. J. Microbiol.* **1995**, *41*, 109–117. [CrossRef]
27. Saharan, B.S.; Nehra, V. Plant growth promoting rhizobacteria: A critical review. *Life Sci. Med. Res.* **2011**, *21*, 30.
28. Billah, M.; Bano, A. Role of plant growth promoting rhizobacteria in modulating the efficiency of poultry litter composting with rock phosphate and its effect on growth and yield of wheat. *Waste Manag. Res.* **2015**, *33*, 63–72. [CrossRef]
29. Whitelaw, M.A. Growth promotion of plants inoculated with phosphate-solubilizing fungi. In *Advances in Agronomy*; Academic Press: Cambridge, MA, USA, 1999; Volume 69, pp. 99–151.
30. Kettner, J.; Dörffling, K. Biosynthesis and metabolism of abscisic acid in tomato leaves infected with Botrytis cinerea. *Planta* **1995**, *196*, 627–634. [CrossRef]
31. Olsen, S.; Sommers, L.E. Phosphorus. In *Methods of Soil Analysis: Part 2 Chemical and Microbiological Properties*; Page, A.L., Ed.; American Society of Agronomy: Madison, WI, USA; Soil Science Society of America: Madison, WI, USA, 1982; pp. 403–430.
32. Sarwar, M.; Arshad, M.; Martens, D.A.; Frankenberger, W.T. Tryptophan-dependent biosynthesis of auxins in soil. *Plant Soil* **1992**, *147*, 207–215. [CrossRef]
33. Blake, G.R.; Hartge, K.H. Bulk density. In *Methods of Soil Analysis: Part 1 Physical and Mineralogical Methods*; American Society of Agronomy: Madison, WI, USA; Soil Science Society of America: Madison, WI, USA, 1986; Volume 5.1, pp. 363–375.
34. Soltanpour, P.N.; Schwab, A.P. A new soil test for simultaneous extraction of macro-and micro-nutrients in alkaline soils. *Commun. Soil Sci. Plant Anal.* **1977**, *8*, 195–207. [CrossRef]
35. McLean, E.O. Soil pH and lime requirement. In *Methods of Soil Analysis: Part 2 Chemical and Microbiological Properties*; American Society of Agronomy: Madison, WI, USA; Soil Science Society of America: Madison, WI, USA, 1983; Volume 9.2.2, pp. 199–224.
36. Brookes, P.C.; Powlson, D.S.; Jenkinson, D.S. Measurement of microbial biomass phosphorus in soil. *Soil Biol. Biochem.* **1981**, *14*, 319–329. [CrossRef]
37. Tabatabai, M.A.; Bremner, J.M. Use of p-nitrophenyl phosphate for assay of soil phosphatase activity. *Soil Biol. Biochem.* **1969**, *1*, 301–307. [CrossRef]
38. Steel, R.G.D.; Torrie, J.H. *Principles and Procedures of Statistics*; HcGraw-Hill: New York, NY, USA, 1960.
39. Khan, N.; Bano, A. Effects of exogenously applied salicylic acid and putrescine alone and in combination with rhizobacteria on the phytoremediation of heavy metals and chickpea growth in sandy soil. *Int. J. Phytoremediat.* **2018**, *20*, 405–414. [CrossRef] [PubMed]
40. Akhtar, M.J.; Asghar, H.N.; Shahzad, K.; Arshad, M. Role of plant growth promoting rhizobacteria applied in combination with compost and mineral fertilizers to improve growth and yield of wheat (Triticum aestivum L.). *Pak. J. Bot.* **2009**, *41*, 381–390.
41. Grant, C.A.; Flaten, D.N.; Tomasiewicz, D.J.; Sheppard, S.C. The importance of early season phosphorus nutrition. *Can. J. Plant Sci.* **2001**, *81*, 211–224. [CrossRef]
42. Illmer, P.; Schinner, F. Solubilization of inorganic calcium phosphates—Solubilization mechanisms. *Soil Biol. Biochem.* **1995**, *27*, 257–263. [CrossRef]
43. Afzal, A.; Bano, A. Rhizobium and phosphate solubilizing bacteria improve the yield and phosphorus uptake in wheat (Triticum aestivum). *Int. J. Agric. Biol.* **2008**, *10*, 85–88.
44. Gangwar, K.S.; Singh, K.K.; Sharma, S.K.; Tomar, O.K. Alternative tillage and crop residue management in wheat after rice in sandy loam soils of Indo-Gangetic plains. *Soil Tillage Res.* **2006**, *88*, 242–252. [CrossRef]

45. Amujoyegbe, B.J.; Opabode, J.T.; Olayinka, A. Effect of organic and inorganic fertilizer on yield and chlorophyll content of maize (*Zea mays* L.) and sorghum *Sorghum bicolour* (L.) Moench. *Afr. J. Biotechnol.* **2007**, *6*. [CrossRef]
46. Hossain, M.B.; Sattar, M.A.; Islam, M.Z. Characterization of P and characterization of phosphate solubilizing bacteria on wheat. *J. Agric. Res.* **2004**, *42*, 295–303.
47. Iftikhar, R.; Khaliq, I.; Ijaz, M.; Rashid, M.A.R. Association analysis of grain yield and its components in spring wheat (Triticum aestivum L.). *Am. Eurasian J. Agric. Environ. Sci.* **2015**, *12*, 389–392.
48. Khan, M.S.; Zaidi, A. Synergistic effects of the inoculation with plant growth-promoting rhizobacteria and an arbuscular mycorrhizal fungus on the performance of wheat. *Turk. J. Agric. For.* **2007**, *31*, 355–362.
49. Sharif, M.; Khan, M.; Khan, M.A.; Wahid, F.; Marwat, K.B.; Khattak, A.M.; Naseer, M. Effect of rock phosphate and farmyard manure applied with effective microorganisms on the yield and nutrient uptake of wheat and sunflower crops. *Pak. J. Bot.* **2015**, *47*, 219–226.
50. Dell, B.; Huang, L. Physiological response of plants to low boron. *Plant Soil* **1997**, *193*, 103–120. [CrossRef]
51. Zafar, M.; Abbasi, M.K.; Arjumend, T.; Jabran, K. Impact of compost, inorganic phosphorus fertilizers and their combination on maize growth, yield, nutrient uptake and soil properties. *J. Anim. Plant Sci.* **2012**, *22*, 1036–1041.
52. Naseem, H.; Bano, A. Role of plant growth-promoting rhizobacteria and their exopolysaccharide in drought tolerance of maize. *J. Plant Interact.* **2014**, *9*, 689–701. [CrossRef]
53. Jala-Abadi, A.L.; Siadat, S.A.; Bakhsandeh, A.M.; Fathi, G.; Saied, K.A. Effect of organic and inorganic fertilizers on yield and yield components in wheat (T. aestivum and T. durum) genotypes. *Adv. Environ. Biol.* **2012**, *6*, 756–763.
54. Khan, N.; Zandi, P.; Ali, S.; Mehmood, A.; Adnan Shahid, M.; Yang, J. Impact of salicylic acid and PGPR on the drought tolerance and phytoremediation potential of Helianthus annus. *Front. Microbiol.* **2018**, *9*, 2507. [CrossRef]
55. Khan, N.; Bano, A. Growth and Yield of Field Crops Grown Under Drought Stress Condition Is Influenced by the Application of PGPR. In *Field Crops: Sustainable Management by PGPR*; Springer: Cham, Switzerland, 2019; pp. 337–349.
56. Osborne, D.J.; Jackson, M.B. *Cell Separation in Plants: Physiology, Biochemistry and Molecular Biology*; Springer Science & Business Media: Berlin/Heidelberg, Germany, 2013; Volume 35.
57. Wachowska, U.; Okorski, A.; Głowacka, K. Population structure of microorganisms colonizing the soil environment of winter wheat. *Plant Soil Environ.* **2006**, *52*, 39–44.
58. Gutiérrez-Mañero, F.J.; Ramos-Solano, B.; Probanza, A.N.; Mehouachi, J.R.; Tadeo, F.; Talon, M. The plant-growth-promoting rhizobacteria Bacillus pumilus and Bacillus licheniformis produce high amounts of physiologically active gibberellins. *Physiol. Plant.* **2012**, *111*, 206–211. [CrossRef]
59. Khan, N.; Bano, A.; Curá, J.A. Role of Beneficial Microorganisms and Salicylic Acid in Improving Rainfed Agriculture and Future Food Safety. *Microorganisms* **2020**, *8*, 1018. [CrossRef]
60. Sivasankari, B.; Kavikumar, N.; Anandharaj, M. Indole-3-acetic acid production and enhanced plant growth promotion by indigenous bacterial species. *J. Curr. Res. Sci.* **2013**, *1*, 331–335.
61. Walley, F.L.; Germida, J.J. Response of spring wheat [Triticum aestivum] to interactions between Pseudomonas species and Glomus clarum NT4. *Biol. Fertil. Soils* **1997**, *24*, 365–371. [CrossRef]
62. Ghosh, P.K.; Bandyopadhyay, K.K.; Manna, M.C.; Mandal, K.G.; Misra, A.K.; Hati, K.M. Comparative effectiveness of cattle manure, poultry manure, phosphocompost and fertilizer-NPK on three cropping systems in vertisols of semi-arid tropics. II. Dry matter yield, nodulation, chlorophyll content and enzyme activity. *Bioresour. Technol.* **2004**, *95*, 85–93. [CrossRef] [PubMed]
63. Brady, N.C.; Weil, R.R.; Weil, R.R. *The Nature and Properties of Soils*; Prentice Hall: Upper Saddle River, NJ, USA, 2008; Volume 13, pp. 662–710.
64. Wu, C.; Wei, X.; Sun, H.L.; Wang, Z.Q. Phosphate availability alters lateral root anatomy and root architecture of Fraxinus mandshurica Rupr. seedlings. *J. Integr. Plant Biol.* **2005**, *47*, 292–301. [CrossRef]
65. Wyngaard, N.; Cabrera, M.L.; Jarosch, K.A.; Bünemann, E.K. Phosphorus in the coarse soil fraction is related to soil organic phosphorus mineralization measured by isotopic dilution. *Soil Biol. Biochem.* **2007**, *96*, 107–118. [CrossRef]
66. Sharma, S.N.; Prasad, R.; Davari, M.; Ram, M.; Dwivedi, M.K. Effect of phosphorus management on production and phosphorus balance in a rice (*Oriza sativa* L.)-wheat (*Triticum aestivum*) cropping system. *Arch. Agron. Soil Sci.* **2011**, *57*, 655–667. [CrossRef]

67. Ghaderi, A.; Aliasgharzad, N.; Oustan, S.; Olsson, P.A. Efficiency of three Pseudomonas isolates in releasing phosphate from an artificial variable-charge mineral [iron III hydroxide]. *Soil Environ.* **2008**, *27*, 71–76.
68. Halvin, J.L.; Beaton, J.D.; Tisdale, S.L.; Nelson, W.L. *Soil Fertility and Fertilizers: An Introduction to Nutrient Management*; Prentice Hall: Upper Saddle River, NJ, USA, 2005.
69. Muhammad, A.; Khan, M.A.; Awan, I.U.; Khan, E.A.; Khan, A.A.; Ghulam, J. Effect of single and combined use of various organic amendments on wheat grown over green manured soil: I. Growth and yield attributes. *Pak. J. Nutr.* **2011**, *10*, 640–646.
70. Son, T.T.N.; Diep, C.N.; Giang, T.T.M. Effect of Bradyrhizobia and phosphate solubilizing bacteria application on soybean in rotational system in the Mekong delta. *Omonrice* **2006**, *14*, 48–57.
71. Belay, A.; Claassens, A.; Wehner, F.C. Effect of direct nitrogen and potassium and residual phosphorus fertilizers on soil chemical properties, microbial components and maize yield under long-term crop rotation. *Biol. Fertil. Soils* **2002**, *35*, 420–427.
72. Trolove, S.N.; Hedley, M.J.; Kirk, G.J.D.; Bolan, N.S.; Loganathan, P. Progress in selected areas of rhizosphere research on P acquisition. *Soil Res.* **2003**, *41*, 471–499. [CrossRef]
73. Biswas, D.R.; Narayanasamy, G. Research note: Mobilization of phosphorus from rock phosphate through composting using crop residue. *Fertil. News* **2002**, *47*, 53–56.
74. Khan, N.; Bano, A.; Zandi, P. Effects of exogenously applied plant growth regulators in combination with PGPR on the physiology and root growth of chickpea (Cicer arietinum) and their role in drought tolerance. *J. Plant Interact.* **2018**, *1*, 239–247. [CrossRef]
75. Shiralipour, A.; McConnell, D.B.; Smith, W.H. Physical and chemical properties of soils as affected by municipal solid waste compost application. *Biomass Bioenergy* **1992**, *3*, 261–266. [CrossRef]
76. Sommers, L.E.; Giordano, P.M. Use of nitrogen from agricultural, industrial, and municipal wastes. In *Nitrogen in Crop Production*; American Society of Agronomy, Inc.: Madison, WI, USA; Crop Science Society of America, Inc.: Madison, WI, USA; Soil Science Society of America, Inc.: Madison, WI, USA, 1984; pp. 207–220.
77. Baziramakenga, R.; Lalande, R.R.; Lalande, R. Effect of de-inking paper sludge compost application on soil chemical and biological properties. *Can. J. Soil Sci.* **2001**, *81*, 561–575. [CrossRef]
78. Shen, Y.; Lin, H.; Gao, W.; Li, M. The effects of humic acid urea and polyaspartic acid urea on reducing nitrogen loss compared with urea. *J. Sci. Food Agric.* **2020**. [CrossRef]
79. Iqbal, M.K.; Shafiq, T.; Hussain, A.; Ahmed, K. Effect of enrichment on chemical properties of MSW compost. *Bioresour. Technol.* **2010**, *10*, 5969–5977. [CrossRef]
80. Adeli, A.; Tewolde, H.; Sistani, K.; Rowe, D. Comparison of broiler litter and commercial fertilizer at equivalent N rates on soil properties. *Commun. Soil Sci. Plant Anal.* **2010**, *41*, 2432–2447. [CrossRef]
81. Canbolat, M.Y.; Barik, K.; Çakmakçi, R.; Sahin, F. Effects of mineral and biofertilizers on barley growth on compacted soil. *Acta Agric. Scand. Sec. B Soil Plant Sci.* **2006**, *56*, 324–332. [CrossRef]
82. Stratton, M.L.; Rechcigl, J.E. Agronomic benefits of agricultural, municipal, and industrial by-products and their co-utilization: An overview. In *Beneficial Co-Utilization of Agricultural, Municipal and Industrial By-Products*; Springer: Dordrecht, Germany, 1998; pp. 9–34.
83. Khan, N.; Bano, A. Rhizobacteria and Abiotic Stress Management. In *Plant Growth Promoting Rhizobacteria for Sustainable Stress Management*; Springer: Singapore, 2019; pp. 65–80.
84. Medina, A.; Vassileva, M.; Barea, J.M.; Azcón, R. The growth-enhancement of clover by Aspergillus-treated sugar beet waste and Glomus mosseae inoculation in Zn contaminated soil. *Appl. Soil. Ecol.* **2006**, *33*, 87–98. [CrossRef]
85. Allison, S.D.; Vitousek, P.M. Responses of extracellular enzymes to simple and complex nutrient inputs. *Soil Biol. Biochem.* **2005**, *37*, 937–944. [CrossRef]
86. Olander, L.P.; Vitousek, P.M. Regulation of soil phosphatase and chitinase activityby N and P availability. *Biogeochemistry* **2000**, *49*, 175–191. [CrossRef]
87. Moscatelli, M.C.; Lagomarsino, A.; De Angelis, P.; Grego, S. Seasonality of soil biological properties in a poplar plantation growing under elevated atmospheric CO_2. *Appl. Soil Ecol.* **2005**, *30*, 162–173. [CrossRef]
88. Criquet, S.; Braud, A.; Nèble, S. Short-term effects of sewage sludge application on phosphatase activities and available P fractions in Mediterranean soils. *Soil Biol. Biochem.* **2007**, *39*, 921–929. [CrossRef]
89. Khan, N.; Bano, A. Role of plant growth promoting rhizobacteria and Ag-nano particle in the bioremediation of heavy metals and maize growth under municipal wastewater irrigation. *Int. J. Phytoremediat.* **2016**, *18*, 211–221. [CrossRef]

90. Pascual, J.A.; Moreno, J.L.; Hernández, T.; García, C. Persistence of immobilised and total urease and phosphatase activities in a soil amended with organic wastes. *Bioresour. Technol.* **2002**, *82*, 73–78. [CrossRef]
91. Wang, F.E.; Chen, Y.X.; Tian, G.M.; Kumar, S.; He, Y.F.; Fu, Q.L.; Lin, Q. Microbial biomass carbon, nitrogen and phosphorus in the soil profiles of different vegetation covers established for soil rehabilitation in a red soil region of southeastern China. *Nutr. Cycl. Agroecosyst.* **2004**, *68*, 181–189. [CrossRef]
92. Kiani, M.; Hernandez-Ramirez, G.; Quideau, S.; Smith, E.; Janzen, H.; Larney, F.J.; Puurveen, D. Quantifying sensitive soil quality indicators across contrasting long-term land management systems: Crop rotations and nutrient regimes. *Agric. Ecosyst. Environ.* **2017**, *248*, 123–135. [CrossRef]
93. Mohammadi, K. Soil microbial activity and biomass as influenced by tillage and fertilization in wheat production. *Am. Eurasian J. Agric. Environ. Sci.* **2011**, *10*, 330–337.
94. Meena, M.D.; Biswas, D.R. Phosphorus and potassium transformations in soil amended with enriched compost and chemical fertilizers in a wheat–soybean cropping system. *Commun. Soil Sci. Plant. Anal.* **2014**, *45*, 624–652. [CrossRef]
95. Ayed, L.B.; Hassen, A.; Jedidi, N.; Saidi, N.; Bouzaiane, O.; Murano, F. Microbial C and N dynamics during composting process of urban solid waste. *Waste Manag. Res.* **2007**, *25*, 24–29. [CrossRef] [PubMed]
96. Leytem, A.B.; Turner, B.L.; Raboy, V.; Peterson, K.L. Linking manure properties to phosphorus solubility in calcareous soils: Importance of the manure carbon to phosphorus ratio. *Soil Sci. Soc. Am. J.* **2005**, *69*, 1516–1524. [CrossRef]
97. Khan, N.; Bano, A. Role of PGPR in the Phytoremediation of Heavy Metals and Crop Growth under Municipal Wastewater Irrigation. In *Phytoremediation*; Springer: Cham, Switzerland, 2018; pp. 135–149.
98. Ayaga, G.; Todd, A.; Brookes, P.C. Enhanced biological cycling of phosphorus increases its availability to crops in low-input sub-Saharan farming systems. *Soil. Biol. Biochem.* **2006**, *38*, 81–90. [CrossRef]

4

Paecilomyces variotii as a Plant-Growth Promoter in Horticulture

Alejandro Moreno-Gavíra, Fernando Diánez, Brenda Sánchez-Montesinos and Mila Santos *

Departamento de Agronomía, Escuela Superior de Ingeniería, Universidad de Almería, 04120 Almería, Spain; alejanmoga@gmail.com (A.M.-G.); fdianez@ual.es (F.D.); brensam@hotmail.com (B.S.-M.)
* Correspondence: msantos@ual.es.

Abstract: In the present study, *P. variotii*, an endophytic fungus isolated from plant roots from the Cabo de Gata Natural Park (Parque Nacional Cabo de Gata—Spain), was tested to determine the effect on the growth promotion of tomato and pepper seeds and seedlings. For these purposes, germination trials in a laboratory and two experiments in a commercial nursery and greenhouse conditions were performed. The *P. variotii* isolate has shown a high ability to produce siderophores and IAA, but low ability to solubilize P. High values for germination percentage, seedling vigor, root and shoot length were obtained by *P. variotii* on tomato and pepper against control. *P. variotii* applications resulted in improved most of the growth parameters evaluated, for both horticultural crops, with the best results in the development of pepper seedlings. The application of a higher dose of *P. variotii* improved most of the morphological parameters and the Dickson quality index (DQI) value in tomato in seedlings and plants. The establishment of the endophytic fungus at the root enabled its biostimulant effects to persist after transplantation without any additional application. Few studies have analyzed this species as a biostimulant. The positive results from the tests showed its high potential. The application of this isolate can be of enormous benefit to horticultural crops for its high reproductive and establishment capacity.

Keywords: *Paecilomyces*; PGPF; tomato; pepper; plant probiotic microorganisms

---

1. Introduction

The genus *Paecilomyces* includes more than 100 species known for their multiple activities and habitat heterogeneity [1]. Among them, *Byssochlamys spectabilis* (Udagawa and Shoji Suzuki) Houbraken and Samson, formerly known as *Paecilomyces variotii* Bainier, is an ascomycete characterized by its ability to produce secondary metabolites, which belong to different chemical groups with wide biological activity [2–5]. This species has been described as a biological control agent (BCA) against nematodes [6,7], trematode eggs [8] and phytopathogenic fungi, such as *Biscogniauxia mediterranea*, *Fusarium moniliforme* and *Phytophthora cinnamomi* [9], *Pyricularia oryzae* [10], *Fusarium graminearum* [11] and *Magnaphorte oryzae* [12], among others, that function through their raw extracts, secondary bioactive metabolites or their mycelia. *P. variotii* produces metabolites with herbicidal [13] and insecticidal [14] activity and has been reported to control infections caused by pathogenic bacteria in fish [15] and humans [9]. In turn, this fungal species has even been shown to degrade aromatic compounds [16,17], in addition to removing ammonium from synthetic media and reducing ammonia emissions from chicken manure. [18]. However, it is also associated with many types of human infections in immunosuppressed patients [19]. Nevertheless, only studies related to the possible activity of these metabolites as a hormone-like substance or a promoter of phytohormone production by plant hosts have been published to date [20]. Very few references describe the application of *P. variotii* as a plant-growth promoter. The biocontrol agent ZhiNengCong (ZNC), which is an extract of *P. variotii,* is used in China [21]. ZNC is a highly effective plant elicitor that promotes plant growth by inducing auxin accumulation in root tips with low concentrations [21].

The use of plant probiotic microorganisms (PPMs) is an effective alternative to the use of chemical fertilizers [22–24]. The most studied PPMs are plant growth-promoting bacteria (PGPB), although there are numerous examples of plant growth-promoting fungi (PGPF), which increase crop yield [25–27]. Thus, the most relevant are those that establish endosymbiotic relationships, such as arbuscular mycorrhizal fungi, which solubilize nutrients, such as phosphorous, and micronutrients absorbed by plants [28–30]. *Trichoderma* is one of the most studied genera as PGPF [31], although many others fungi have demonstrated their potential growth-promoting capacity, such as *Penicillium oxalicum* [32], *Penicillium simplicissimum* [33], *Fusarium oxysporum* [34], *Fusarium equiseti* [35], *Alternaria* sp. [36], *Aspergillus* spp. [37] and *Phoma* [38], among others.

In the present study, *P. variotii*, an endophytic fungus isolated from plant roots from the Cabo de Gata Natural Park (Parque Nacional Cabo de Gata—Spain), was tested to evaluate: (a) the effects of seed priming with a fungus suspension on root colonization and tomato and pepper plant vigor; (b) the promotion of growth and quality of pepper and tomato seedlings under a conventional production system and (c) the effects of applying different doses to tomato seedlings and their subsequent transplantation in a greenhouse.

2. Materials and Methods

2.1. Isolation of *P. variotii* from Plant Roots

Twenty roots of different species of autochthonous plants from the Cabo de Gata Natural Park (CGNP; Almería, Spain) were collected for the isolation of fungal organisms in 2017. Collected samples were cleaned under running tap water to remove debris before use, air dried and processed for isolation of endophytic fungi. To remove epiphytic and surface-adhering microbes, the roots were cut into small, 2–3 cm long, pieces, were surface-sterilized with 2% sodium hypochlorite for 3 min and washed three times with sterile distilled water. The surface-sterilized samples were allowed to dry on sterile paper towels. Ten fragments from each root were placed onto potato dextrose agar (PDA, Difco) supplemented with 50 μg mL^{-1} chloramphenicol to suppress bacterial growth. After incubation at 25 °C for 7 d, individual hyphal tips of the developing fungal colonies were removed, placed on PDA medium and incubated for 5–7 d.

Colony morphology of the pure cultured isolates on PDA and conidiophore morphology were examined and identified by light microscopy, and all selected isolates were stored for further studies. Only one isolate, whose identification under a microscope was consistent with the genus *Paecilomyces* (Figure 1), was selected for this study (*P. variotii* CDG33). Molecular identification of the selected fungi was conducted following the procedure described by Diánez et al. [26]. The sequence was analyzed using a BLAST search in the GenBank database of the National Centre for Biotechnology Information (NCBI, http://blast.ncbi.nlm.nih.gov/Blast.cgi) and aligned to the nearest neighbors. The sequence has not been deposited in the GenBank database because the isolate is subject to patent.

The culture of *P. variotii* has been deposited in the CECT (Spanish Type Culture Collection, Valencia, Spain) with the collection number CECT 20957. This strain was selected for the experiments based on the results of a preliminary assay (data not shown).

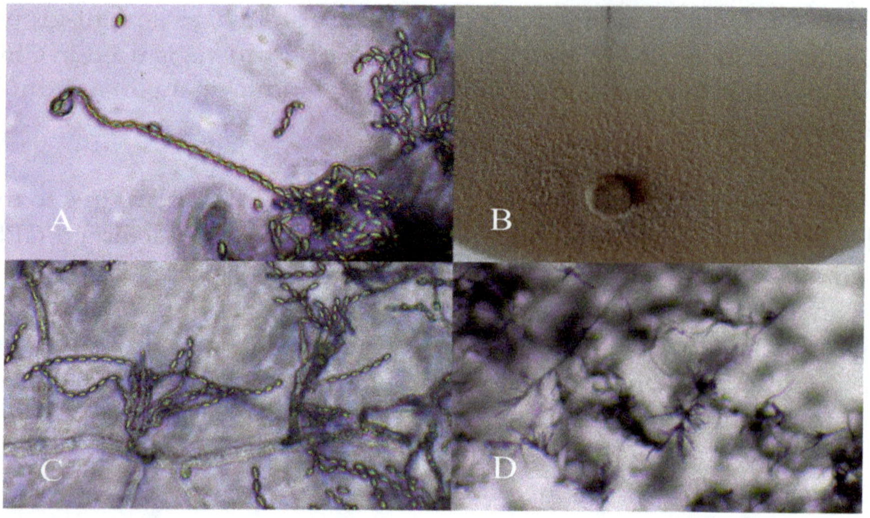

Figure 1. Conidiophore of *Paecilomyces variotii* (**A**: 400×; **C**: 200×; **D**: 100×) and an aspect of colony morphology (**B**) in the potato dextrose agar (PDA) medium.

2.2. Analysis of Plant Growth-Promoting Attributes

Siderophore production was determined on the chrome-azurol S (CAS) medium following the method of Schwyn and Neilands [39] and Louden et al. [40]. Fungal mycelial discs (5 mm) of active culture were transferred to CAS medium and orange halos around the colonies on blue were indicative of siderophore production. The diameter of the orange halo was measured at 24, 48 and 72 h.

Indole-3-acetic acid (IAA) production was estimated according to the procedure described by Diánez et al. [26]. *P. variotii* was grown in 50 mL of glucose peptone broth (GPB) amended with or without L-tryptophan (Sigma-Aldrich) at a concentration of 100 mg L^{-1}. The flasks were inoculated and incubated on an orbital shaker at 150 rpm at 25 °C in the dark for 7 d. After incubation, the suspension from each flask was centrifuged for 30 min at 12,000× *g*. The supernatant was filtered through sterile Millipore membranes (pore size 0.22 µm) and collected in sterile tubes. The culture supernatants (3 mL) were pipetted into test tubes, and 2 mL Salkowski reagent (2 mL of 0.5 mol L^{-1} $FeCl_3$ + 98 mL of 35% $HClO_4$) was added to it. The tubes containing the mixture were left for 30 min for red color development. The intensity of the color was determined by measuring the optical density at 530 nm using a scanning spectrophotometer. The quantity of IAA was determined by comparison with a standard curve for IAA. Five independent replicates of *P. variotii* were analyzed.

The qualitative evaluation of the phosphorus solubilized by *P. variotii* was performed using NBRIP and PVK media supplemented with 2% agar (Difco Laboratories, Detroit, MI, USA). Phosphate solubilization was detected by the formation of transparent zones surrounding fungal colonies in both media [41]. For the quantitative estimation of phosphate solubilization, a modified version of the procedure by Lima-Rivera [42] was followed. Flasks (250 mL capacity) containing 50 mL of NBRIP broth were inoculated with two disks of agar (5 mm diameter) that had been taken from pure cultures of *P. variotii*. Uninoculated flasks were used as a control (three replicates). Incubation was conducted at 26 °C at a shaking speed of 100 rpm for 3, 5, 7, 10 and 15 d. Supernatants of each culture were analyzed for pH and phosphate concentration. Phosphates in culture supernatants were estimated using the Fiske and Subbarow method [43] and expressed as equivalent phosphate (µg mL^{-1}). The experiments were conducted in triplicate and values were expressed as the mean. The total P in flasks was 10 mg mL^{-1}.

2.3. Mass Production of *P. variotii* on Solid Substrates

A mixture of two kinds of substrates, buckwheat husk (BH) and oat (O), were tested for the mass multiplication of *P. variotii*. Different percentages (90–10%, 80–20% and 70–30% *v/v* BH-O) of both substrates were submerged in different percentages of water (10%, 20% and 30% *v/v*) for 24 h. Each mixture was sterilized for 1 h at 125 °C twice on consecutive days. Each mixture was placed on a tray and aseptically inoculated by spraying with 5 mL of spore suspension containing 4×10^6 spores mL^{-1} of *P. variotii*. The trays were kept at 25 °C in the dark for 10 d. In total, three samples (2 g) of the fungus-colonized substrate were removed from the trays in each treatment. The samples were successively diluted in sterile distilled water + 0.01% Tween 20® and the number of conidia g^{-1} of the solid substrate was quantified for each replicate using a Neubauer hemocytometer. There were three replications per treatment. The collected spores were used in the different experiments conducted in this study.

2.4. Analysis of Effects of P. variotii on Seed Germination under Laboratory Conditions

Seeds of tomatoes (*Solanum lycopersicum* 'Red Cherry') and pepper (*Capsicum annuum* 'Largo de Reus') were used in this study. The trial used a random block experimental design with two treatments (control and *P. variotii*) and four repetitions. Each repetition included 50 seeds that were germinated in Petri dishes (150 mm diameter) containing two sheets of Whatman No. 1 filter paper that were moistened with sterile distilled water. The seeds were surface-sterilized sterilized with 1.5% sodium hypochlorite (NaOCl) for 5 min, rinsed twice with sterile distilled water and dried under laminar airflow on sterile paper [26]. Treatments were performed by pipetting 50 μL of *P. variotii* spore suspension (1×10^5 spores mL^{-1}) or 50 μL of water (control) on each seed; all boxes were placed in an incubator (25 ± 1 °C in the dark). Root length (mm) was measured from the tip of the primary root to the base of the hypocotyl. After 7 and 10 d, for tomatoes and peppers, respectively, percent germination, root length and shoot length were recorded and a seed vigor index (SVI) was calculated as follows: SVI (length) = seed germination% (mean root length + mean shoot length) [44].

2.5. Analysis of Promoter Effects of P. variotii on Pepper and Tomato Seedlings: Experiment 1

This experiment was performed in nursery polystyrene planting trays, each with 96 cells (70 mL volume), at a commercial nursery (Almería Province, Spain). Pepper and tomato seeds red cherry and Largo de Reus, respectively, were sown into commercial peat mix and covered with vermiculite. After 2 d (tomato) and 4 d (pepper) in a germination room (relative humidity (RH) = 95%; 25 °C), trays were located in a greenhouse and rinsed with water (control) or a 5 mL spore suspension per cell at 10^5 spores per plant. Seedlings were grown using standard nursery culture conditions (18–28 °C; 75.4% ± 6.7% RH) and four trays were used for each treatment. At 45 d after sowing, 20 plants per treatment and control were randomly selected from the four replications and measured for different growth parameters: number of leaves, stem length, stem base diameter, total leaf area and aerial and root dry weights. Leaf area was measured using the WINDIAS 3.1. (Delta-T Devices Ltd., Cambridge, UK, 2009) leaf area processing software. The Dickson quality index [45] was determined using the formula Dickson quality index (DQI) = TDW/((LS/D) + SDW/RDW)), where TDW is the total dry weight (g), LS is the stem length (cm), D is the stem diameter (mm) and SDW and RDM are the stem and root dry weight (g), respectively. The experiments were conducted in autumn using a completely randomized design.

2.6. Analysis of Effects of Applying Different Doses of P. variotii to Tomatoes: Experiment 2

For this experiment, the procedure described in experiment 1 was followed. Three doses of *P. variotii* conidia (PaeD1: 10^4 spores mL^{-1}, PaeD2: 10^5 spores mL^{-1} and PaeD3: 10^6 spores mL^{-1}) were applied to tomato seedlings growing under commercial plant nursery conditions with irrigation of the substrate, by adding 5 mL of spore suspension to each plant. The test was conducted in winter and four replicates were performed with 96 plants per replicate. The seedlings were harvested 30 d after sowing. Twenty plants per treatment and control were randomly selected from the four replicates and measured for the same parameters as described above. Another 25 plants from each treatment were transplanted into sandy soil in mid-February and analyzed in mid-May. The experiment was performed under greenhouse conditions (Figure 2). Water requirements were established according to climatic conditions and crop needs. Plants were fertilized with a commercial complex nutrient fertilizer.

Figure 2. Tomato plants grown under commercial seedlings (**A**) and field assay, 30 (**B**) and 60 (**C**) days after transplanting.

In all tests (experiments 1 and 2), tomato and pepper roots inoculated with *P. variotii* were collected at the end of the tests. Roots were surface-sterilized in 0.1% sodium hypochlorite and washed with sterilized water. Lastly, 2 cm root fragments were placed in PDA medium to determine root colonization by the fungal isolate.

2.7. Statistical Analysis

Statistical analysis was conducted using Statgraphics Centurion XVIII Software. The data were tested by a one-way analysis of variance (ANOVA) or Student's t-test with significance defined as p values less than 0.05 ($p < 0.05$). The experimental results are presented as the means and standard error (±SE) for the different replicates. Mean separation was performed using Fisher's least significant difference (LSD) test.

3. Results

A total of 42 fungal isolates were obtained from the analyses of 20 root samples taken from plants native to the CGNP. All fungal isolates were obtained in pure cultures using standard techniques. The isolates were identified as filamentous fungi belonging to the phylum Ascomycota. The fungal isolates were identified to the genus level. *Aspergillus*, *Trichoderma* sp., *Rustroemia* sp. and *Penicillium* sp., were the most characteristic genera. They differed in color, type of conidiophore or presence of microsclerotia-like structures. Therefore were considered to be different. The presence of the genus *Paecilomyces* was especially relevant, and this genus was thus selected for this study (isolate CDG33). The amplified sequences of isolate of *Paecilomyces* were compared with available DNA sequences using BLAST, having 99% homology with *Paecilomyces variotii* accession number JX282326.1.

3.1. Mass Production of P. variotii on Solid Substrates

The fungus multiplied well in all mixtures tested. Among the different treatments, whose composition varied in water content and the percentage of oats or buckwheat husk that were tested for mass multiplication of *P. variotii*, the proportion of 70% + 30% for buckwheat husk and oats, respectively, with 20% water content, resulted in significantly higher spore production (Figure 3), followed by 90% + 10% with 10% water content (Table 1). The lowest spore production rate was observed at a percentage of 80% + 20% with 10% water (Figure 3).

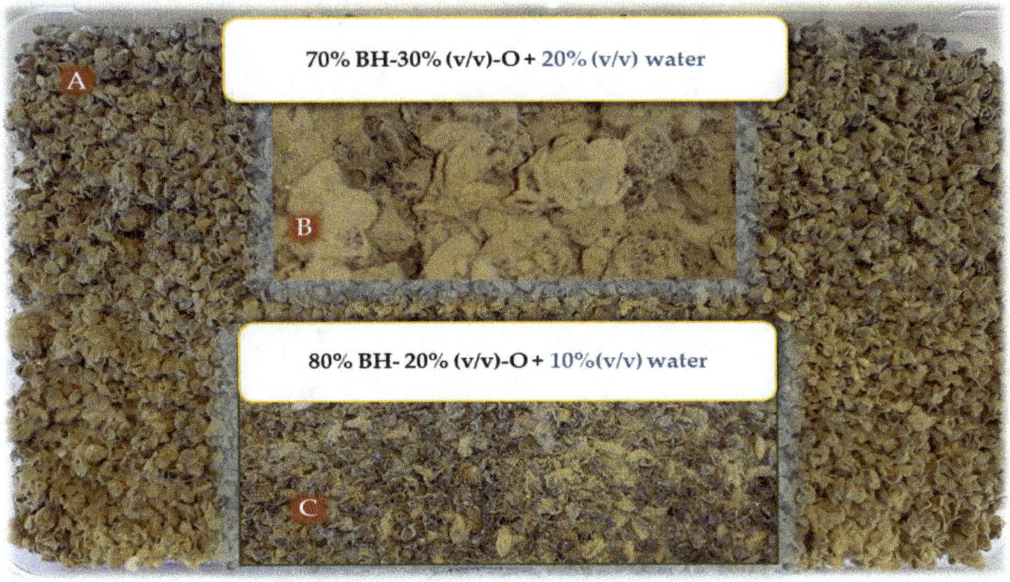

Figure 3. Mass production of *P. variotii* spores at a proportion of 70% + 30% buckwheat husk and oats, respectively, +20% water (**A**,**B**) and at 80% + 20% with 10% water (**C**).

Table 1. Effects of solid media on mass multiplication of *P. variotii* at varying substrates percentages (*v*/*v*) and water contents (*v*/*v*).

Treatments					
% (*v*/*v*) water					
10		20		30	
BH-O%	$CFU \cdot g^{-1}$	BH-O%	$CFU\ g^{-1}$	BH-O%	$CFU\ g^{-1}$
90-10	$1.13 \times 10^8 \pm 1.2 \times 10^7$ b	90-10	$8.5 \times 10^7 \pm 7.09 \times 10^6$ ef	90-10	$9.14 \times 10^7 \pm 9.73 \times 10^6$ de
80-20	$7.0 \times 10^7 \pm 1.32 \times 10^7$ g	80-20	$1.05 \times 10^8 \pm 1.09 \times 10^7$ bc	80-20	$1.0 \times 10^8 \pm 5.11 \times 10^6$ d
70-30	$8.04 \times 10^7 \pm 1.14 \times 10^7$ f	70-30	$1.24 \times 10^8 \pm 1.06 \times 10^7$ a	70-30	$1.07 \times 10^8 \pm 9.54 \times 10^6$ bc

BH: buckwheat husk; O: oats. Data were analyzed by ANOVA and treatment means were compared according to Fisher's least significant difference (LSD) statistical procedure (*F*-test at $p < 0.05$).

3.2. Plant Growth-Promoting Characteristics of P. variotii: Siderophore Production, IAA and P Solubilization

The formation of an orange-colored zone around the fungal colonies was observed, which indicated siderophore production by *P. variotii* (Figure 4A). The diameter of the halo (mm) was estimated at 3.88 ± 0.33, 5.55 ± 0.22 and 8.83 ± 1.29 for 24, 48 and 72 h, respectively.

IAA was produced by *P. variotii* in medium supplemented with 100 mg L^{-1} tryptophan during a 7 d period, and the final concentration was 0.049 ± 0.001 mg mL^{-1}. The final concentration of IAA was 0.03 ± 0.001 mg mL^{-1} in medium supplemented without tryptophan.

No halo of P solubilization by *P. variotii* was detected in any of the media used (NBRIP and PVK media supplemented with 2% agar). The effect of *P. variotii* on the soluble phosphate concentration is shown in Figure 4. The initial concentration of P in the medium was used to quantify the concentration of P solubilized by *P. variotii*. As shown in Figure 4B, no P solubilization was detected during up to 15 d of incubation, assessing a soluble P of 2.01 ± 0.68 g L^{-1} versus 0.74 ± 0.25 in the control ($p = 0.0243$). In turn, no change in the pH of the medium was detected, which remained at approximately 7.

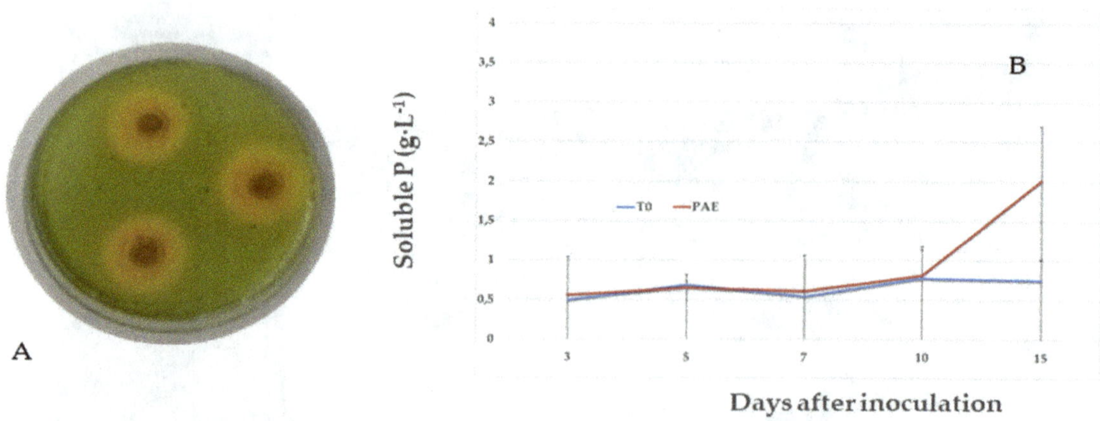

Figure 4. Formation of orange-colored halos owing to production of siderophores by *P. variotii* (**A**). Effects of *P. variotii* on phosphate solubilization (**B**) in NBRIP broth containing tribasic calcium phosphate (10 g). Mean standard deviation is expressed in error bar. The results are shown as the average of the three replicates, in g L^{-1}.

3.3. *P. variotii* Inoculation Effects on Tomato and Pepper Seed Germination

Table 2 outlines the results of the in vitro application of *P. variotii* to tomato and pepper seeds. Direct inoculation of seeds by *P. variotii* isolate spores had a significant effect ($p < 0.05$) on the percent seed germination, root and shoot length, and SVI in tomatoes. The increase in root and shoot length was 18.23% and 17.85%, respectively. However, pepper seeds treated with *P. variotii* showed no improvement in SVI ($p > 0.05$).

Table 2. Effects of *P. variotii* isolate on tomato and pepper seed germination 7 and 10 d after treatment, respectively. pp

Treatment	% Germination	Root Length (cm)	Shoot Length (cm)	Seed Vigor Index
		Pepper		
***p* -value**	*0.5369*	*0.6162*	*0.0010*	*0.9975*
P. variotii	80 ± 9.79	2.37 ± 1.30	1.66 ± 0.55	323.00 ± 127.07
T0	76 ± 7.30	2.26 ± 1.24	1.98 ± 0.56	322.93 ± 119.36
		Tomato		
***p*-value**	*0.0020*	*0.0013*	*0.0036*	*0.0000*
P.variotii	89 ± 3.82	4.41 ± 1.33	3.30 ± 1.15	686.70 ± 210.40
T0	75 ± 3.82	3.73 ± 1.32	2.80 ± 1.02	498.70 ± 164.28

T0: control without *P. variotii*. Data were analyzed using Student's *t*-test ($p < 0.05$).

3.4. *Promoter Effects of P. variotii Isolates on Tomato and Pepper Seedlings: Experiment 1*

The effects of *P. variotii* application on morphological parameters and DQI are shown in Table 3. Most of the growth parameters evaluated in tomato and pepper were improved by *P. variotii*, when compared with those of the experimental control, with the best results in the development of pepper seedlings.

These increases were statistically significant for most parameters. The increases assessed in pepper and tomato seedlings (P/T) were 9.7%/6.9% for stem length, 4.9%/0.8% for stem diameter, 10.6%/6.0% for leaf number, 18.2–6.7% for root dry weight, 16.7%/10.7% for aerial dry weight and 10.1%/7.5% for leaf area. In tomatoes, *P. variotii* applications resulted in a decrease in root dry weight, albeit without significant differences from that of the control. *P. variotii* application improved plant quality, albeit without significant differences for tomato plants ($p = 0.2059$). In all tests (experiments 1 and 2), *P. variotii* was observed in tomato and pepper roots analyzed in PDA medium.

Table 3. Morphological parameters and quality index of pepper and tomato seedlings treated with *P. variotii* isolate at 45 d after sowing.

Treatment	Length of Stem (cm)	Diameter (mm)	Number of Leaves	Areal Dry Weight (g)	Root Dry Weight (g)	Leaf Area mm^2	DQI
				Pepper			
p-value	*0.0000*	*0.0008*	*0.0000*	*0.0000*	*0.0022*	*0.0020*	*0.0054*
P. variotii	29.63 ± 1.75	3.83 ± 0.21	7.39 ± 0.54	0.42 ± 0.05	0.13 ± 0.02	83.58 ± 11.27	0.051 ± 0.00
T0	27.01 ± 2.07	3.65 ± 0.23	6.68 ± 0.70	0.36 ± 0.04	0.11 ± 0.03	75.88 ± 11.15	0.046 ± 0.00
				Tomato			
p-value	*0.0002*	*0.0039*	*0.0330*	*0.0042*	*0.1443*	*0.0397*	*0.2059*
P. variotii	27.37 ± 2.06	3.81 ± 0.27	4.40 ± 0.54	0.62 ± 0.07	0.14 ± 0.02	74.45 ± 11.85	0.067 ± 0.01
T0	25.61 ± 2.07	3.78 ± 0.22	4.15 ± 0.48	0.56 ± 0.10	0.15 ± 0.02	69.28 ± 10.17	0.063 ± 0.01

T0: control without *P. variotii*. Data were analyzed using Student's *t*-test ($p < 0.05$).

3.5. Effects of Dose of Application of P. variotii Isolates on Tomato Seedlings and Transplanted Plants: Experiment 2

The effects of applying three doses on the development of tomato seedlings under commercial plant nursery conditions and subsequent transplantation into soil under greenhouse conditions are shown in Table 4. As shown in Table 4, applying different doses of *P. variotii* promoted seedling shoot development, significantly decreasing, in some cases, root development. The best results were assessed for dose 3, which resulted in a 30.2% increase in stem length, 15.7% in stem diameter, 19.5% in leaf number and 46.2% in aerial dry weight. However, root dry weight decreased 25.0%. The decrease in root length can cause stress when transplanting tomato seedlings, thereby decreasing plant quality. The DQI [45] expresses the global aptitude of a plant to successfully overcome the transplantation phase, based on overall plant development, while considering the balance between plant shoots and roots. Higher values of this index indicate higher seeding quality. In this case, despite the decrease in root length, a higher value of this index was observed for dose 3, with a significant difference for all doses tested, but not with the control treatment.

Table 4. Morphological parameters and Dickson quality index (DQI) of tomato seedlings (30 d after sowing) and plants (90 days after transplanting) treated with different doses (10^4, 10^5 and 10^6 spores per plant, (D1, D2 and D3, respectively) of *P. variotii*.

			Tomato seedling			
Treatment	**Length of stem (cm)**	**Diameter (mm)**	**N° leaves**	**Areal dry Weight (g)**	**Root dry Weight (g)**	**DQI**
p-value	*0.0000*	*0.0000*	*0.0002*	*0.0000*	*0.0028*	*0.0481*
Pae D1	16.12 ± 2.11a	3.72 ± 0.38b	3.58 ± 0.53b	0.27 ± 0.09a	0.039 ± 0.01a	0.045 ± 0.01b
Pae D2	14.46 ± 2.58b	3.50 ± 0.38c	3.71 ± 0.84ab	0.25 ± 0.08a	0.035 ± 0.01a	0.046 ± 0.01b
Pae D3	16.27 ± 3.05a	3.99 ± 0.28a	3.93 ± 0.78a	0.38 ± 0.09a	0.027 ± 0.01b	0.054 ± 0.01a
T0	12.5 ± 2.22b	3.45 ± 0.36c	3.29 ± 0.61c	0.25 ± 0.08b	0.036 ± 0.01a	0.045 ± 0.01b
			Tomato Plants			
Treatment	**Length of stem (cm)**	**Diameter (mm)**	**Internodes**	**Areal dry Weight (g)**	**Root dry Weight (g)**	**DQI**
p-value	*0.1876*	*0,1244*	*0.9759*	*0.0048*	*0.7796*	*0.0283*
Pae D1	82.75 ± 12.97ab	11.57 ± 1.08ab	14.20 ± 1.93a	50.67 ± 10.80a	4.22 ± 1.20a	24.84 ± 7.68a
Pae D2	87.20 ± 9.42a	11.59 ± 1.02ab	14.30 ± 1.25a	49.19 ± 10.21a	4.26 ± 1.64a	24.23 ± 8.16a
Pae D3	77.60 ± 7.75b	11.84 ± 1.08a	14.10 ± 1.28a	49.38 ± 10.53a	4.20 ± 1.37a	25.78 ± 7.66a
T0	84.00 ± 7.10ab	10.58 ± 0.91b	14.00 ± 1.63a	39.13 ± 7.99b	3.73 ± 0.91a	16.84 ± 3.79b

T0: control without *P. variotii*. Different letters indicate significant differences according to the one-way ANOVA test ($p = 0.05$). Treatment means were compared according to the Fisher's LSD statistical procedure (F-test at $p < 0.05$).

Once transplanted in soil, the findings showed that the three doses favored several morphological parameters studied in the development of tomato plants in relation to those of control plants, with higher values for dry weights and DQI but without significant differences for the three doses tested. The establishment of the endophytic fungus at the root (Figure 5), therefore, enabled its effects to persist after transplantation.

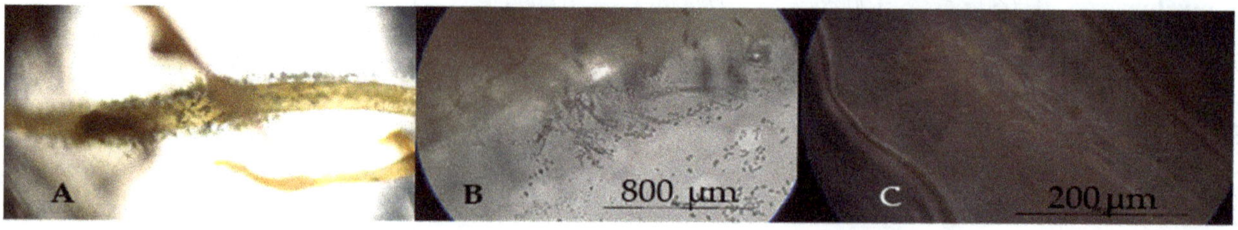

Figure 5. *P. variotii* mycelium and conidiophores colonizing tomato (**A**: 100×; **B**: 200×) and pepper roots (**C**: 400×).

4. Discussion and Conclusions

In this study, fungal isolates from surface-sterilized root segments of native plants and soil were collected from the CGNP. This park is located in an arid-to-semiarid Mediterranean region where the predominant fungi genera are *Penicillium* and *Aspergillus*, as well as isolates from the group termed Dark Septate Endophytic (DSE) fungi, which colonize plant roots. The DSE fungi isolated herein formed dark brownish microsclerotia-like structures. The role of DSE fungi in nature has been considered to be similar to that of mycorrhizal fungi [46,47]. These isolates will be analyzed in future studies. Among all isolates, the plant growth-promoting capacity of the only isolate from *P. variotii* was analyzed in this study.

Nevertheless, *P. variotii* was selected given the sparse literature on this species as a plant-growth promoter. The ability of *P. variotii* to stimulate plant growth is poorly studied, while for other species of the same genus, the growth-promoting effect is generally associated with an improvement in the plant status for its nematicide effect for the control of diseases caused by different nematode species [48].

A key requirement for selling microorganisms with a biostimulant capacity is that they have a high-spore production capacity in substrates with a low-production cost. Nevertheless, grains whose agitation leads to a rapid release of spores must be selected to obtain the formulation. For this reason, buckwheat husk and oats were selected for high performance in the spore production of *P. variotii*. This production occurred in the entire substrate, not only on the surface. Generally, commercial spore production methods often only use cereal grains, rice or other starch-based substrates [49]. The nutritional composition of the medium in which the growth promotor multiplies can affect its biocontrol or biostimulant capacity [50]. Based on our results, *P. variotii* can stimulate plant development under the production conditions described herein.

The *P. variotii* isolate has shown the ability to produce siderophores and IAA. Biostimulation is generally associated with increased nutrient availability, similar to biofertilization [51], but it is also caused by multiple other factors, such as mechanisms including enhancement of plant systemic resistance [52]. Thus, siderophore production plays a key role by enhancing the Fe uptake of plants and can be considered an ecofriendly alternative to the use of chemical fertilizers and pesticides in the agricultural sector [53]. Vala et al. [54] describe the production of both hydroxamate and carboxylate-type siderophores by *P. variotii* isolated from the surface of mangrove plants. Terrestrial *P. variotii* has previously been shown to secrete the trihydroxamate siderophore ferrirubin [55]. The *P. variotii* isolate has shown a relatively high IAA production in relation to its genus, and tryptophan incorporation slightly increased in vitro IAA biosynthesis by *P. variotii*. Other studies also reported that Trp considerably stimulated microbial IAA yield in vitro [56–58]. Ali [59] described IAA production by *P. variotii* lower than 2 ng mL^{-1} produced in Czapek medium supplemented with tryptophan. In our study, we described a much higher production for the same species. Waqas [60] described IAA production by *P. formosus* at 34.07 ± 3.92 μg mL^{-1}, a result very similar to the findings of this study, and consequently, a plant growth-promoting effect. Nevertheless, Waqas [61] analyzed the variability of IAA production as a function of the culture medium used and conversely found that *P. variotii* had virtually undetectable phosphatase activity. These results differed from the high phosphatase production by a *P. variotii* isolate from the medicinal plant *Caralluma acutangula* [59]. In soil, P-solubilizing fungi constitute approximately 0.1–0.5% of the total fungal population [62]. The use of microbial inoculants (biofertilizers) possessing P-solubilizing activities in crop productivity is considered as an alternative to further application of mineral P fertilizers [63]. Endophyte co-inoculation of plants exponentially improved the phosphatase activity of soil compared to that of the non-inoculated plants under stress conditions [64]. Etesami et al. [65] showed the relationship among ACC deaminase activity, IAA production, siderophore production and phosphate solubilization of bacterial strains and their effect on root elongation of rice seedlings. In this study, this correlation could not be determined since an isolate of *P. variotii* was analyzed.

The application of beneficial microorganisms (biopriming) may not only help to improve germination and vigor parameters but also relieve a wide range of physiological, abiotic and biotic stresses in both seeds and seedlings [66]. Nevertheless, biopriming can potentially lead to a more resistant plant after transplanting. This can depend on numerous factors, such as plant species, microorganisms, applied dose and substrates, among others. Thus, in this study, the *P. variotii* isolate enhanced germination and root vigor in tomato seeds. However, the results with pepper plants were different, with no significant differences from that of the controls and even with a 19.27% decrease in shoot length ($p < 0.05$). This may be because germination of pepper seeds is often more heterogeneous. Cochran [67] determined that the germination percentage and the accumulation of dry matter in the large seeds of bell pepper were higher in relation to the small seeds. *P. variotii* enhanced several tomato and pepper seedling parameters. The application doses for different crops must be studied for optimal outcomes. In this study, the application of a higher dose (10^6 spores mL^{-1} versus 10^5 spores mL^{-1}) improved the different morphological parameters value in tomato seedlings under commercial plant nursery conditions. After transplanting in the greenhouse, the biostimulant effects persisted 4 months after applying the three doses without any additional application. Similar benefits have been observed as a result of the application of other species of the genus *Paecilomyces*. Waqas [68] described improved soybean seedling germination and SVI when applying the endophyte *Paecilomyces* sp. The application of *P. variotii* extracts significantly increased cherry radish yield, dry matter accumulation, the root–shoot ratio and quality. This extract has a very high biological activity, with a low cost, which has a great application prospect [69]. Similar studies conducted by Anis [70] showed increased sunflower seedling vigor under biotic stress conditions when applying *P. variotii* and *Macrophomina phaseolina* spores, in vitro, with no favorable results when conducting the tests in pots. In turn, Maitlo [71] assessed chickpea plant biostimulation when inoculating them with *P. lilacinus* and *F. oxysporum f.* sp. *ciceris*, which also reduced chickpea wilt.

In intensive horticulture under plastic, the benefits of the application of biostimulants or biofungicides based on microorganisms, are in question, because farmers perceive low efficacy of these products as disease controllers when compared with the rapid response presented by chemical fertilizers or fungicide. The current changes in legislation regarding the reduction of active ingredients and the commercialization of biostimulants and biopesticides, together with the need to increase the sustainability of agriculture in terms of public health and the environment, require the use of PPMs as a key element in horticulture.

To the best of our knowledge, the present study demonstrated for the first time the biostimulant capacity of *P. variotii* in pepper and tomato plants under commercial plant nursery and greenhouse conditions.

5. Patents

This isolate was patented with a Spanish patent number ES2684858A1: New strain of *Paecilomyces variotii*, compositions and applications.

Author Contributions: F.D. and M.S. conceived and designed the experiments; A.M.-G. and B.S.-M. performed the experiments; F.D. analyzed the data; M.S. wrote the paper. All authors have read and agreed to the published version of the manuscript.

Acknowledgments: The present work benefited from the input of the project RTC-2017-6486-2 was supported by the Spanish Ministry of Science, Innovation and Universities.

References

1. Nguyen, T.T.T.; Paul, N.C.; Lee, H.B. Characterization of Paecilomyces variotii and Talaromyces amestolkiae in Korea Based on the Morphological Characteristics and Multigene Phylogenetic Analyses. *Mycobiology* **2016**, *44*, 248–259. [CrossRef] [PubMed]
2. Mioso, R.; Marante, F.J.T.; De Laguna, I.H.B. The Chemical Diversity of the Ascomycete Fungus Paecilomyces variotii. *Appl. Biochem. Biotechnol.* **2015**, *177*, 781–791. [CrossRef] [PubMed]
3. De Laguna, I.H.B.; Marante, F.J.T.; Mioso, R. Enzymes and bioproducts produced by the ascomycete fungus Paecilomyces variotii. *J. Appl. Microbiol.* **2015**, *119*, 1455–1466. [CrossRef] [PubMed]
4. Akinfala, T.O.; Houbraken, J.; Sulyok, M.; Adedeji, A.R.; Odebode, A.C.; Krska, R.; Ezekiel, C.N. Moulds and their secondary metabolites associated with the fermentation and storage of two cocoa bean hybrids in Nigeria. *Int. J. Food Microbiol.* **2020**, *316*, 108490. [CrossRef] [PubMed]
5. Lòpez-Fernàndez, S.; Campisano, A.; Schulz, B.J.; Steinert, M.; Stadler, M.; Surup, F. Viriditins from Byssochlamys spectabilis, their stereochemistry and biosynthesis. *Tetrahedron Lett.* **2020**, *61*, 151446. [CrossRef]
6. Al-Qasim, M.; Abu-Gharbieh, W.; Assas, K. Nematophagal ability of Jordanian isolates of Paecilomyces variotii on the root-knot nematode Meloidogyne javanica. *Nematol. Mediterr.* **2009**, *37*, 53–57.
7. Perveen, Z.; Shahzad, S. A comparative study of the efficacy of Paecilomyces species against root-knot nematode Meloidogyne incognita. *Pak. J. Nematol.* **2013**, *31*, 125–135.
8. Ahmad, R.Z.; Sidi, B.B.; Endrawati, D.; Ekawasti, F. Paecilomyces lilacinus and P. variotii as a predator of nematode and trematode eggs. *IOP Conf. Ser. Earth Environ. Sci.* **2019**, *299*, 012056. [CrossRef]
9. Rodrigo, S.; Santamaria, O.; Halecker, S.; Lledó, S.; Stadler, M. Antagonism betweenByssochlamys spectabilis(anamorphPaecilomyces variotii) and plant pathogens: Involvement of the bioactive compounds produced by the endophyte. *Ann. Appl. Boil.* **2017**, *171*, 464–476. [CrossRef]
10. Mohammadi, S.; Soltani, J.; Piri, K. Soilborne and invertebrate pathogenic Paecilomyces species show activity against pathogenic fungi and bacteria. *J. Crop. Prot.* **2016**, *5*, 377–387. [CrossRef]
11. Zhang, P.; Li, X.M.; Wang, J.N.; Wang, B.G. Oxepine-containing diketopiperazine alkaloids from the algal-derived endophytic fungus Paecilomyces variotii EN-291. *Helv. Chim. Acta* **2015**, *98*, 800–804. [CrossRef]
12. Hajano, J.U.D.; Lodhi, M.; Pathan, M.A.; Khanzada, A.; Shah, G.S. In-vitro evaluation of fungicides, plant extracts and bio-control agents against rice blast pathogen Magnaporthe oryzae couch. *Pak. J. Bot.* **2012**, *44*, 1775–1778.
13. Nakajima, M.; Itoi, K.; Takamatsu, Y.; Sato, S.; Furukawa, Y.; Furuya, K.; Honma, T.; Kadotani, J.; Kozasa, M.; Haneishi, T. Cornexistin: A new fungal metabolite with herbicidal activity. *J. Antibiot.* **1991**, *44*, 1065–1072. [CrossRef]
14. Song, X.; Zhang, L.H.; Peng, A.T.; Cheng, B.P.; Ling, J.F. First Report of Paecilomyces variotii Isolated From Citrus Psyllid (Diaphorina citri), the Vector of Huanglongbing of Citrus, in China. *Plant Dis.* **2016**, *100*, 2526. [CrossRef]
15. Hyung, J.J.; Kang, H.; Jong, J.J.; Soo, K.Y. Paecilomyces Variotii Extracts for Preventing and Treating Infections Caused by Fish Pathogenic Microorganisms. KR Patent 2013051523, 2013.

16. García-Peña, E.; Hernández, S.; Auria, R.; Revah, S. Correlation of Biological Activity and Reactor Performance in Biofiltration of Toluene with the Fungus Paecilomyces variotii CBS115145. *Appl. Environ. Microbiol.* **2005**, *71*, 4280–4285. [CrossRef]
17. García-Peña, E.; Ortiz, I.; Hernandez, S.; Revah, S. Biofiltration of BTEX by the fungus Paecilomyces variotii. *Int. Biodeterior. Biodegrad.* **2008**, *62*, 442–447. [CrossRef]
18. Liu, Z.; Liu, G.; Cai, H.; Shi, P.; Chang, W.; Zhang, S.; Zheng, A.; Xie, Q.; Ma, J. Paecilomyces variotii: A Fungus Capable of Removing Ammonia Nitrogen and Inhibiting Ammonia Emission from Manure. *PLoS ONE* **2016**, *11*, e0158089. [CrossRef]
19. Steiner, B.; Aquino, V.R.; Paz, A.A.; Silla, L.M.D.R.; Zavascki, A.; Goldani, L. Paecilomyces variotii as an Emergent Pathogenic Agent of Pneumonia. *Case Rep. Infect. Dis.* **2013**, *2013*, 1–3. [CrossRef]
20. Santamaria, O.; Lledó, S.; Rodrigo, S.; Poblaciones, M.J. Effect of Fungal Endophytes on Biomass Yield, Nutritive Value and Accumulation of Minerals in Ornithopus compressus. *Microb. Ecol.* **2017**, *74*, 841–852. [CrossRef]
21. Lu, C.; Liu, H.; Jiang, D.; Wang, L.; Jiang, Y.; Tang, S.; Hou, X.; Han, X.; Liu, Z.; Zhang, M.; et al. Paecilomyces variotii extracts (ZNC) enhance plant immunity and promote plant growth. *Plant Soil* **2019**, *441*, 383–397. [CrossRef]
22. Sánchez-Montesinos, B.; Diánez, F.; Moreno-Gavíra, A.; Gea, F.J.; Santos, M. Plant Growth Promotion and Biocontrol of Pythium ultimum by Saline Tolerant Trichoderma Isolates under Salinity Stress. *Int. J. Environ. Res. Public Health* **2019**, *16*, 2053. [CrossRef]
23. Raja, N. Biopesticides and Biofertilizers: Ecofriendly Sources for Sustainable Agriculture. *J. Fertil. Pestic.* **2013**, *4*, e112. [CrossRef]
24. Adesemoye, A.; Kloepper, J.W. Plant–microbes interactions in enhanced fertilizer-use efficiency. *Appl. Microbiol. Biotechnol.* **2009**, *85*, 1–12. [CrossRef] [PubMed]
25. Vandenberghe, L.; Garcia, L.M.B.; Rodrigues, C.; Camara, M.C.; Pereira, G.; De Oliveira, J.; Soccol, C.R. Potential applications of plant probiotic microorganisms in agriculture and forestry. *AIMS Microbiol.* **2017**, *3*, 629–648. [CrossRef]
26. Martínez, F.D.; Santos, M.; Carretero, F.; Marin, F. Trichoderma saturnisporum,a new biological control agent. *J. Sci. Food Agric.* **2015**, *96*, 1934–1944. [CrossRef]
27. Fernando, D.; Santos, M.; Francisco, C.; Francisco, M. Biostimulant Activity of Trichoderma saturnisporum in Melon (Cucumis melo). *HortScience* **2018**, *53*, 810–815. [CrossRef]
28. Smith, S.E.; Manjarrez, M.; Stonor, R.; McNeill, A.; Smith, F.A. Indigenous arbuscular mycorrhizal (AM) fungi contribute to wheat phosphate uptake in a semi-arid field environment, shown by tracking with radioactive phosphorus. *Appl. Soil Ecol.* **2015**, *96*, 68–74. [CrossRef]
29. Bhardwaj, D.; Ansari, M.; Sahoo, R.K.; Tuteja, N. Biofertilizers function as key player in sustainable agriculture by improving soil fertility, plant tolerance and crop productivity. *Microb. Cell Factories* **2014**, *13*, 66. [CrossRef]
30. Begum, N.; Qin, C.; Ahanger, M.A.; Raza, S.; Khan, M.I.; Ashraf, M.; Ahmed, N.; Zhang, L. Role of Arbuscular Mycorrhizal Fungi in Plant Growth Regulation: Implications in Abiotic Stress Tolerance. *Front. Plant Sci.* **2019**, *10*, 1068. [CrossRef]
31. Khan, M.R.; Mohiddin, F.A. Trichoderma: Its Multifarious Utility in Crop Improvement. In *Crop Improvement through Microbial Biotechnology*; Elsevier BV: Amsterdam, The Netherlands, 2018; pp. 263–291.
32. Murali, M.; Amruthesh, K.N. Plant Growth-promoting Fungus Penicillium oxalicum Enhances Plant Growth and Induces Resistance in Pearl Millet Against Downy Mildew Disease. *J. Phytopathol.* **2015**, *163*, 743–754. [CrossRef]
33. Shimizu, K.; Hossain, M.M.; Kato, K.; Kubota, M.; Hyakumachi, M. Induction of Defense Responses in Cucumber Plants by Using the Cell-free Filtrate of the Plant Growth-Promoting Fungus Penicillium simplicissimum GP17-2. *J. Oleo Sci.* **2013**, *62*, 613–621. [CrossRef] [PubMed]
34. Bitas, V.; McCartney, N.; Li, N.; Demers, J.; Kim, J.-E.; Kim, H.-S.; Brown, K.M.; Kang, S. Fusarium Oxysporum Volatiles Enhance Plant Growth Via Affecting Auxin Transport and Signaling. *Front. Microbiol.* **2015**, *6*, 351. [CrossRef] [PubMed]
35. Kojima, H.; Hossain, M.; Kubota, M.; Hyakumachi, M. Involvement of the salicylic acid signaling pathway in the systemic resistance induced in Arabidopsis by plant growth-promoting fungus Fusarium equiseti GF19-1. *J. Oleo Sci.* **2013**, *62*, 415–426. [CrossRef] [PubMed]

36. Zhou, L.S.; Tang, K.; Guo, S.-X. The Plant Growth-Promoting Fungus (PGPF) Alternaria sp. A13 Markedly Enhances Salvia miltiorrhiza Root Growth and Active Ingredient Accumulation under Greenhouse and Field Conditions. *Int. J. Mol. Sci.* **2018**, *19*, 270. [CrossRef]
37. Islam, S.; Akanda, A.M.; Sultana, F.; Hossain, M. Chilli rhizosphere fungus Aspergillus spp. PPA1 promotes vegetative growth of cucumber (Cucumis sativus) plants upon root colonisation. *Arch. Phytopathol. Plant Prot.* **2013**, *47*, 1231–1238. [CrossRef]
38. Naznin, H.A.; Kimura, M.; Miyazawa, M.; Hyakumachi, M. Analysis of Volatile Organic Compounds Emitted by Plant Growth-Promoting Fungus Phoma sp. GS8-3 for Growth Promotion Effects on Tobacco. *Microbes Environ.* **2012**, *28*, 42–49. [CrossRef]
39. Schwyn, B.; Neilands, J. Universal chemical assay for the detection and determination of siderophores. *Anal. Biochem.* **1987**, *160*, 47–56. [CrossRef]
40. Louden, B.C.; Haarmann, D.; Lynne, A.M. Use of Blue Agar CAS Assay for Siderophore Detection. *J. Microbiol. Boil. Educ.* **2011**, *12*, 51–53. [CrossRef]
41. Nautiyal, C.S. An efficient microbiological growth medium for screening phosphate solubilizing microorganisms. *FEMS Microbiol. Lett.* **1999**, *170*, 265–270. [CrossRef]
42. Lima-Rivera, D.L.; Lopez-Lima, D.; Desgarennes, D.; Velázquez-Rodríguez, A.S.; Carrión, G. Phosphate solubilization by fungi with nematicidal potential. *J. Soil Sci. Plant Nutr.* **2016**, *16*, 507–524. [CrossRef]
43. Fiske, C.H.; Subbarow, Y. The colorimetry determination of phosphorous. *J. Biol. Chem.* **1925**, *66*, 375–400.
44. Murali, M.; Amruthesh, K.N.; Sudisha, J.S.; Niranjana, R.; Shetty, H.S. Screening for plant growth promoting fungi and their ability for growth promotion and induction of resistance in pearl millet against downy mildew disease. *J. Phytol.* **2012**, *4*, 30–36.
45. Dickson, A.; Leaf, A.L.; Hosner, J.F. Quality Appraisal of White Spruce and White Pine Seedling Stock in Nurseries. *For. Chron.* **1960**, *36*, 10–13. [CrossRef]
46. Peterson, R.L.; Wagg, C.; Pautler, M. Associations between microfungal endophytes and roots: Do structural features indicate function? *Botany* **2008**, *86*, 445–456. [CrossRef]
47. Della Mónica, I.F.; Saparrat, M.C.; Godeas, A.M.; Scervino, J.M. The co-existence between DSE and AMF symbionts affects plant P pools through P mineralization and solubilization processes. *Fungal Ecol.* **2015**, *17*, 10–17. [CrossRef]
48. Huang, W.-K.; Cui, J.-K.; Liu, S.-M.; Kong, L.-A.; Wu, Q.-S.; Peng, H.; He, W.-T.; Sun, J.-H.; Peng, D.-L. Testing various biocontrol agents against the root-knot nematode (Meloidogyne incognita) in cucumber plants identifies a combination of Syncephalastrum racemosum and Paecilomyces lilacinus as being most effective. *Boil. Control.* **2016**, *92*, 31–37. [CrossRef]
49. Goettel, M.S.; Roberts, D.W. Mass production, formulation and field application of entomopathogenic fungi. In *Biological Control of Locusts and Grasshopper*; Lomer, C.J., Prior, C., Eds.; CAB International: Wallingford, UK, 1992.
50. Lane, B.S.; Trinci, A.P.; Gillespie, A.T. Influence of cultural conditions on the virulence of conidia and blastospores of Beauveria bassiana to the green leafhopper, *Nephotettix virescens*. *Mycol. Res.* **1991**, *95*, 829–833. [CrossRef]
51. Vessey, J.K. Plant growth promoting rhizobacteria as biofertilizers. *Plant Soil* **2003**, *255*, 571–586. [CrossRef]
52. Schuster, A.; Schmoll, M. Biology and biotechnology of Trichoderma. *Appl. Microbiol. Biotechnol.* **2010**, *87*, 787–799. [CrossRef]
53. Saha, M.; Sarkar, S.; Sarkar, B.; Sharma, B.K.; Bhattacharjee, S.; Tribedi, P. Microbial siderophores and their potential applications: A review. *Environ. Sci. Pollut. Res.* **2015**, *23*, 3984–3999. [CrossRef] [PubMed]
54. Vala, A.K.; Vaidya, S.Y.; Dube, H.C. Siderophore production by facultative marine fungi. *Indian J. Mar. Sci.* **2000**, *29*, 339–340.
55. Renshaw, J.C.; Robson, G.; Trinci, A.P.; Wiebe, M.; Livens, F.R.; Collison, D.; Taylor, R. Fungal siderophores: Structures, functions and applications. *Mycol. Res.* **2002**, *106*, 1123–1142. [CrossRef]
56. Ivanova, E.G.; Doronina, N.V.; Trotsenko, Y.A. Aerobic Methylobacteria Are Capable of Synthesizing Auxins. *Microbiology* **2001**, *70*, 392–397. [CrossRef]
57. Khalid, A.; Arshad, M.; Zahir, Z. Screening plant growth-promoting rhizobacteria for improving growth and yield of wheat. *J. Appl. Microbiol.* **2004**, *96*, 473–480. [CrossRef] [PubMed]
58. Egamberdieva, D.; Kucharova, Z. Selection for root colonizing bacteria stimulating wheat growth in saline soils. *Biol. Fertil. Soil* **2009**, *45*, 563–571. [CrossRef]

59. Ali, S. Endophytic fungi from Caralluma acutangula can secrete plant growth promoting enzymes. *Fresenius Environ. Bull.* **2019**, *28*, 2688–2696.
60. Waqas, M.; Khan, A.L.; Shahzad, R.; Ullah, I.; Khan, A.R.; Lee, I.-J. Mutualistic fungal endophytes produce phytohormones and organic acids that promote japonica rice plant growth under prolonged heat stress*. *J. Zhejiang Univ. Sci. B* **2015**, *16*, 1011–1018. [CrossRef]
61. Waqas, M.; Khan, A.L.; Lee, I.-J. Bioactive chemical constituents produced by endophytes and effects on rice plant growth. *J. Plant Interact.* **2013**, *9*, 478–487. [CrossRef]
62. Kucey, R.M.N. Effect of Penicillium bilaji on the Solubility and Uptake of P and Micronutrients from Soil by Wheat. *Can. J. Soil Sci.* **1988**, *68*, 261–270. [CrossRef]
63. Khan, M.S.; Zaidi, A.; Ahemad, M.; Oves, M.; Wani, P.A. Plant growth promotion by phosphate solubilizing fungi–current perspective. *Arch. Agron. Soil Sci.* **2010**, *56*, 73–98. [CrossRef]
64. Bilal, S.; Shahzad, R.; Khan, A.L.; Kang, S.-M.; Imran, Q.M.; Al-Harrasi, A.; Yun, B.-W.; Lee, I.-J. Endophytic Microbial Consortia of Phytohormones-Producing Fungus Paecilomyces formosus LHL10 and Bacteria Sphingomonas sp. LK11 to Glycine max L. Regulates Physio-hormonal Changes to Attenuate Aluminum and Zinc Stresses. *Front. Plant Sci.* **2018**, *9*. [CrossRef] [PubMed]
65. Etesami, H.; Alikhani, H.A.; Hosseini, H.M. Indole-3-acetic acid (IAA) production trait, a useful screening to select endophytic and rhizosphere competent bacteria for rice growth promoting agents. *MethodsX* **2015**, *2*, 72–78. [CrossRef] [PubMed]
66. Mastouri, F.; Björkman, T.; Harman, G.E. Seed Treatment withTrichoderma harzianumAlleviates Biotic, Abiotic, and Physiological Stresses in Germinating Seeds and Seedlings. *Phytopathology* **2010**, *100*, 1213–1221. [CrossRef]
67. Cochran, H.C. Effect of seed size on uniformity of pimento transplants (*Capsicum annuum* L.) at harvest time. *J. Am. Sot. Hort. Sci.* **1974**, *99*, 234–235.
68. Waqas, M.; Khan, A.L.; Hamayun, M.; Kamran, M.; Kang, S.M.; Kim, Y.H.; Lee, I.J. Assessment of endophytic fungi cultural filtrate on soybean seed germination. *Afr. J. Biotechnol.* **2012**, *11*, 15135–15143.
69. Jia, C.; Liu, Z.; Zhang, M.; Jia, J.; Zheng, L.; Wang, Q.; Wang, H. Effects of extracts from Paecilomyces variotii on the yield and quality of cherry radish (*Raphanus sativus* L. var. radculus pers). *J. Agric. Res. Environ.* **2019**, *36*, 176–183.
70. Anis, M.; Abbasi, W.; Zaki, M.J. Bioefficacy of microbial antagonists against Macrophomina phaseolina on sunflower. *Pak. J. Bot.* **2010**, *42*, 2935–2940.
71. Maitlo, S.A.; Rajput, N.A.; Syed, R.N.; Khanzada, M.A.; Rajput, A.Q.; Lodhi, A.M. Microbial control of Fusarium wilt of chickpea caused by *Fusarium oxysporum* f. sp. ciceris. *Pak. J. Bot.* **2019**, *51*, 2261–2268. [CrossRef]

5

Chlorophyll Fluorescence Parameters and Antioxidant Defense System can Display Salt Tolerance of Salt Acclimated Sweet Pepper Plants Treated with Chitosan and Plant Growth Promoting Rhizobacteria

Muneera D. F. ALKahtani [1], Kotb A. Attia [2], Yaser M. Hafez [3], Naeem Khan [4], Ahmed M. Eid [5], Mohamed A. M. Ali [6] and Khaled A. A. Abdelaal [7,*]

1 Biology Department, College of Science, Princess Nourah bint Abdulrahman University, Riyadh POX 102275-11675, Saudi Arabia; mdfkahtani@gmail.com

2 Center of Excellence in Biotechnology Research, King Saud University, Riyadh POX 2455-11451, Saudi Arabia; kattia1.c@ksu.edu.sa

3 Plant Pathology and Biotechnology Lab., Faculty of Agriculture, Kafrelsheikh University, Kafr El Sheikh 33516, Egypt; hafezyasser@gmail.com

4 Department of Agronomy, Institute of Food and Agricultural Sciences, University of Florida, Gainesville, FL 32611, USA; naeemkhan@ufl.edu

5 Botany and Microbiology Department, Faculty of Science, Al-Azhar University, Nasr City 11884, Egypt; aeidmicrobiology@azhar.edu.eg

6 Department of Horticulture, Faculty of Agriculture, New Valley University, El-Kharga, New Valley 72511, Egypt; Mohamedahmedali77@aun.edu.eg

7 Excellence Center (EPCRS), Plant Pathology and Biotechnology Lab., Faculty of Agriculture, Kafrelsheikh University, Kafr El Sheikh 33516, Egypt

* Correspondence: khaled.elhaies@gmail.com

Abstract: Salinity stress deleteriously affects the growth and yield of many plants. Plant growth promoting rhizobacteria (PGPR) and chitosan both play an important role in combating salinity stress and improving plant growth under adverse environmental conditions. The present study aimed to evaluate the impacts of PGPR and chitosan on the growth of sweet pepper plant grown under different salinity regimes. For this purpose, two pot experiments were conducted in 2019 and 2020 to evaluate the role of PGPR (*Bacillus thuringiensis* MH161336 10^{6-8} CFU/cm^3) applied as seed treatment and foliar application of chitosan (30 mg dm^{-3}) on sweet pepper plants (cv. Yolo Wonder) under two salinity concentrations (34 and 68 mM). Our findings revealed that, the chlorophyll fluorescence parameter (*Fv/Fm* ratio), chlorophyll *a* and *b* concentrations, relative water content (RWC), and fruit yield characters were negatively affected and significantly reduced under salinity conditions. The higher concentration was more harmful. Nevertheless, electrolyte leakage, lipid peroxidation, hydrogen peroxide (H_2O_2), and superoxide (O_2^-) significantly increased in stressed plants. However, the application of *B. thuringiensis* and chitosan led to improved plant growth and resulted in a significant increase in RWC, chlorophyll content, chlorophyll fluorescence parameter (*Fv/Fm* ratio), and fruit yield. Conversely, lipid peroxidation, electrolyte leakage, O_2^-, and H_2O_2 were significantly reduced in stressed plants. Also, *B. thuringiensis* and chitosan application regulated the proline accumulation and enzyme activity, as well as increased the number of fruit plant^{-1}, fruit fresh weight plant^{-1}, and total fruit yield of sweet pepper grown under saline conditions.

Keywords: sweet pepper; salinity; *Bacillus*; chitosan; chlorophyll fluorescence; fruit yield

1. Introduction

Sweet pepper belongs to Solanacease family. It is an annual plant in the cultivated lands in many countries, however it is grown as a perennial plant in tropical areas. It is one of the most widespread and popular vegetables, and has a greatest economic importance worldwide [1]. It is the richest source of different antioxidants and vitamins and has several health benefits [2]. However, salinity is a very significant factor that threatens the production of economic plants such as sweet pepper [1], strawberry plants [3], and cucumber plants [4]. Salinity damages plant growth and proliferation by creating water stress and cytotoxicity due to the excess in uptake of ions, such as sodium and chloride. Furthermore, salinity is usually accompanied by oxidative stress due to the generation of reactive oxygen species [5,6]. Salinity stress adversely affects morpho-physiological characters of sweet pepper such as plant height and leaf area which are significantly reduced [7]. Likewise, chlorophyll *a* and *b* as well as RWC were reduced under salinity in cucumber [4]. Photosynthesis is harmfully affected by salinity through the reduction in stomatal conductance. Also, salinity led to increased ion toxicity and negatively affected nutrients uptake, especially potassium uptake, so the salt stressed plants showed low membrane stability [8]. The chlorophyll fluorescence parameters were adversely affected with salinity and the content of chlorophyll pigments significantly decreased in cucumber [9]. Also, the study of Misra et al. [10] pointed out that salt stress causes photoinhibition in PSII and decreases its activity. Salt stress led to decreased chlorophyll concentrations, leaf area and mungbean yield [11] and led to an increase in the accumulation of Na^+, decreasing the uptake of mineral nutrients such as nitrogen and potassium [12]. The high level of Na^+ was associated with the ROS accumulation such as H_2O_2 and O^{2-}. The excessive formation of ROS causes protein oxidation and lipid peroxidation under several stresses mainly under salinity stress [1,13]. Previous studies have shown that the adverse effects of salinity stress on leaf number, plant length, fresh and dry weights of shoots, and plant yield also increases with the increase in NaCl concentration [14–16].

According to salinity concentrations, the plants are classified to euhalophytes or glycophytes. Euhalophytes have the salinity thresholds of 250 mM NaCl, i.e., euhalophytes are able to complete their life cycle upon salinities exceeding 250 mM NaCl. Glycophytes cannot grow under high salinity concentrations and their response to salinity differs in terms of osmotic regulation, photosynthetic electron transport, chlorophyll content, and reactive oxygen species (ROS) formation as well as antioxidant defense system [1,7]. The excessive accumulation of ROS under stress, such as salinity [1,17], drought [18,19], and biotic stress factors [20–23], results in the activation of the enzymatic and non-enzymatic antioxidant system to enhance stress tolerance in plants to cope with increased accumulation of ROS [24]. The antioxidative system also consists of some of the non-enzymatic systems, such as salicylic acid and carotenoids. Nonetheless, the enzymatic defense system contains ascorbate peroxidases (APX), glutathione reductases (GR), superoxide dismutases (SOD), catalases (CAT), and peroxidases (POD), which protect the plant tissues against stress factors [25]. Also, the plants have adaptive mechanisms to salinity stress through morphological, anatomical, and biochemical changes. Euhalophytes can cope with salinity stress through different mechanisms, such as salt exclusion, salt elimination, salt succulence and salt redistribution [7]. Furthermore, EL%, lipid peroxidation, and ROS were increased significantly under salinity, as these parameters are signals to various stresses, such as salinity, drought, and heat [26–29], that enable plants to respond to a particular stress. Some plants protect themselves from salinity stress by maintaining ion homeostasis and transportation of the excess salt to the vacuole or sequestering in the older tissues which ultimately are sacrificed, thereby defending itself from salinity stress [30]. Meanwhile, other plants keep the ion concentration in the cytoplasm at a low level. Membranes along with their linked components play an essential role in retaining ion concentration within the cytosol during the period of stress by regulating ion uptake and transport [31,32]. Chlorophyll fluorescence is a fast method for photosynthetic processes measurements [33] and provides a lot of information about the plant status under abiotic and biotic stresses to understand the mechanisms of photosynthesis and how plants respond to various stresses [34]. Chlorophyll fluorescence parameters are important indicators used to measure the quantum yield of photosystem II (PSII), display the plant response to stress and the harmful effects, particularly on photosynthesis and chlorophyll concentrations [35].

Chitosan or chitin is a natural polysaccharide consisting of two molecules of D-glucosamine and naturally present in the cell walls of many organisms such as crabs, shrimp, fungi, and the exoskeleton of insects [36]. In the agricultural field, it improves the morpho-physiological parameters and alleviates the injurious effect of abiotic stresses through stress transduction pathway [37]. Application of chitosan led to increased plant tolerance to many stresses in various plants [38,39], enhance growth characters and improve germination rate of many plants [38,40]. The fruit yield of tomato plants was improved with chitosan treatments [41]. Under drought, barley plants treated with chitosan showed a significant increase in chlorophyll, RWC, total soluble sugar, and grain yield [42]. Plant growth-promoting rhizobacteria (PGPR) can prompt plant tolerance to stress through some chemical and physical changes which are identified as induced systemic tolerance [43]. The application of PGPR led to improved growth and yield production [44]. Under stress conditions, PGPR can improve the injurious impacts and enhance the yield production under salt conditions [45], as a bio-fertilizer in sugar beet and sweet sorghum plants [20,46,47] and as a bio-control agent [48–50]. There are many PGPR strains, such as *Bacillus*, *Azotobacter*, *Azospirillum*, *Pseudomonas*, *Rhizobium*, and *Serratia*, which can be used in improving plant growth even under various stress factors [51,52] by the production of antioxidants, phytohormones and vitamins [53]. There is a lot of information about the effect of PGPR, nevertheless studies about chitosan and its effects on plants under salinity stress are still scarce and have not yet been fully understood. Hence, in this research, we focus on the effect of chitosan and *Bacillus thuringiensis* MH161336 in alleviating the harmful effect of salinity to improve chlorophyll fluorescence parameters, chlorophyll concentration, enzymes activity, and fruit yield of sweet pepper.

2. Materials and Methods

2.1. Experiments Preparation and Plant Materials

Two pot experiments were conducted at Kafrelsheikh University, Agricultural Botany Department during two summer seasons 2019 and 2020, to evaluate the effect of seed treatment with plant growth promoting rhizobacteria (*B. thuringiensis* MH161336 10^{6-8} CFU/cm^3) and foliar spray with chitosan 30 mg·dm^{-3} on sweet pepper plants under salinity (sodium chloride at 34 and 68 mM). The physio-biochemical characters were done at Plant Pathology & Biotechnology Lab., and EPECRS Excellence Center, Kafrelsheikh University. The seeds of sweet pepper (*Capsicum annuum* L.) cv. Yolo Wonder (obtained from a private agricultural company) were divided into three groups (the first group was treated with *B. thuringiensis* and the others without treatments). Seed treatment was done with *B. thuringiensis*. Thereby, the seeds underwent surface sterilization by sodium hypochlorite 2.5% for 5 min, 70% ethanol for 1 min, and were then washed 5 times by sterile distilled water. *B. thuringiensis* MH161336 which was isolated from the halophytic plant *Spergularia marina* (obtained from Dr. Ahmed Eid), *B. thuringiensis* pure cultures were grown in nutrient broth at 35 ± 2 °C on a shaker at 180× *g*. Bacterial cultures were diluted in sterilized distilled water to reach a final concentration of 10^{6-8} CFU/cm^3 [54]. Sterilized seeds were incubated with bacterial suspensions at room temperature for 6 h and sown in the nursery in foam trays on 7th and 3rd January in the two seasons, respectively. After forty-five days from the sowing, the transplantation was done in pots 50 cm^3 in diameter, each one containing two seedlings and the pots were divided into three groups (control, *B. thuringiensis* treatment and chitosan treatment 30 mg·dm^{-3}). The plants irrigated with two concentrations (34 and 68 mM) of saline water (was prepared from NaCl) and the group of chitosan treatment was treated with chitosan 30 mg·dm^{-3} twice after 20 and 40 days from transplanting. The compound fertilizer containing nitrogen, phosphorus, and potassium (NPK) (135:40:35 kg·ha^{-1}) was used as recommended in two doses, the first dose after 12 days from transplanting and the second at the flowering stage initiation. The experiments were in a completely randomized design with 4 replicates, the physiological and biochemical studies were done at 80 days from transplanting. The chemical and physical characters of experimental soil were determined [55] and are presented in Table 1.

Table 1. Chemical and physical characters of the experimental soil before conducting the experiments in 2019 and 2020 seasons.

Seasons	PH	* EC Ds/m	Mechanical Analysis			Soil Texture	Organic Matter (%)	Total N (%)	Total P (ppm)
			Sand%	Silt%	Clay%				
2019	8.11	0.464	21.96	23.98	47.4	Clay	1.79	0.158	8.8
2020	8.16	0.483	22.17	24.29	46.8	Clay	1.82	0.149	8.2

Seasons	Soluble Cations				Soluble Anions		
	Na^+	K^+	Ca^{++}	Mg^{++}	HCO_3^-	SO_4^{--}	Cl^-
2019	2.14	0.18	2.02	2.4	4.2	2.07	0.6
2020	2.19	0.17	2.04	2.3	4.3	1.93	0.5

* EC = Electrical conductivity.

2.2. *Biochemical and Physiological Characters*

The physiological and biochemical characters were recorded at 80 days from transplanting.

2.2.1. Relative Water Content (RWC%)

According to Sanchez et al. [56], twenty leaf discs were used to determine RWC, the fresh weight (FW) for the discs was determined, the same discs were saved in petri dishes containing distilled water for 1 h to determine the turgid weight (TW), after that the discs were dried for 24 h at 80 °C to determine the dry weight (DW). Relative water content (RWC%) was calculated as follows: RWC = (FW − DW)/(TW − DW) × 100.

2.2.2. Determination of Chlorophyll *a* and *b* Concentrations

The extraction was done using *N-N* Dimethyl formamide, whereby 5 cm^3 *N-N* Dimethyl formamide was added to 1 g fresh leaves and kept in the refrigerator overnight, and the absorbance was measured at 647 and 664 nm according to Moran [57].

2.2.3. Electrolyte Leakage Assay (EL%)

Ten discs (1 cm^2) of sweet pepper leaves were placed into flasks containing deionized water (25 cm^3). Flasks were shaken for 20 h, initial electrical conductivity was recorded for each vial and then flasks were immersed in a water bath at 80 °C for 1 h. The vials were shaken for 20 h at 21 °C. Final conductivity was measured for each flask. Electrolyte leakage % was calculated according to Szalai et al. [58] with the following formula: initial conductivity/final conductivity × 100.

2.2.4. Chlorophyll Fluorescence Parameter

Chlorophyll *a* fluorescence parameter (*Fv*/*Fm* ratio) was measured at 80 days from the sowing using a chlorophyll fluorometer (PEA, Hansatech Instrument Ltd., version 1.21, Norfolk, UK).

According to Schreiber [59], middle-aged sweet pepper leaves were placed in the dark for 30 min to stimulate the reaction of photosystem II. The minimum chlorophyll fluorescence (*Fo*) was measured using a measuring beam of <0.1 μmol $m^{-2}{\cdot}s^{-1}$. The maximum fluorescence (*Fm*) was determined after a 1 s saturating pulse (>3500 μmol·$m^{-2}{\cdot}s^{-1}$). Variable fluorescence (*Fv*) was determined by the difference between the maximum fluorescence and the minimum fluorescence (*Fm* − *Fo*). The maximum efficiency of PSII was determined as the ratio of (*Fv*) to (*Fm*) as follows: $Fv/Fm = (Fm - Fo)/Fm$.

2.2.5. Proline Determination

According to Bates et al. [60], proline was assayed in sweet pepper plants, 0.5 g fresh leaf in 3% sulphosalicylic acid and centrifuged for 20 min at 3000× *g*. Then, 2 cm^3 of glacial acetic acid and 2 cm^3 ninhydrin reagent was boiled with 2 cm^3 supernatant for 1 h, the reaction was completed in an ice bath, and proline was separated using toluene. Proline was determined as $\mu g \cdot g^{-1}$ FW using a spectrophotometer at 520 nm.

2.2.6. Determination of Lipid Peroxidation

According to Davenport et al. [61], lipid peroxidation was measured as malondialdehyde (MDA) using 100 mg fresh leaves in 1% trichloro acetic acid and centrifuged at 10,000× *g* for 5 min. 0.5% thiobarbituric acid was then added, and mixture was boiled at 95 °C for half an hour. The samples were placed on an ice bath and centrifuged for 5 min at 5000× *g*, the measurements were done using spectrophotometer at 532 and 600 nm. MDA ($nmol \cdot g^{-1}$ FW) = [6.45 × (A532 − A600) − (0.56 × A450)] × $V^{-1}W$, where V = volume (cm^3); W = weight (g).

2.2.7. Determination of Superoxide (O_2^-) and Hydrogen Peroxide (H_2O_2)

Sweet pepper leaves were vacuum infiltrated with 10 mM potassium salicylate buffer containing 0.1 *w/v*% nitro blue tetrazolium (NBT) or 0.1 *w/v*% 3,3-diaminobenzidine (DAB). The leaves were incubated in the light for 140 min and two hours, respectively. The samples were cleared with trichloroacetic acid in ethanol: chloroform 4:1 *v/v* for 1 day, the samples were washed and placed in 50% glycerol. O_2^- and H_2O_2 were determined as $nmol \cdot g^{-1}$ FW according to Huckelhoven et al. [62] using a ChemiImager 4000 digital imaging system (Alpha Innotech Corp., San Leandro, CA, USA).

2.2.8. Assay of Enzymes Activity

For the determination of enzymes, 0.5 g fresh leaves were homogenized in 3 cm^3 of 50 mM Tris buffer at 0–4 °C, containing 1 mM EDTA-Na_2 and 7.5% polyvinyl pyrrolidone. The samples were centrifuged 12,000× *g* for 20 min at 4 °C and the total soluble enzyme activities were measured using spectrophotometer in the supernatant [63]. Catalase activity (CAT) was determined through the decomposition of H_2O_2 by catalase results in the decrease of the ultraviolet absorption of H_2O_2 at 240 nm, catalase activity can be calculated from this decrease. The reaction mixture contained 2.15 cm^3, 2 cm^3 0.1 M Na-phosphate buffer, 100 µL H_2O_2, and 50 µL leaves extract. The solution is mixed, and the absorptions were recorded at 240 nm according to Aebi [64]. Peroxidase activity (POX) was calculated according to Hammerschmidt et al. [65]. The reaction mixture consisted of 2.9 cm^3 of a 100 mM sodium phosphate buffer containing 0.25% (*v/v*) guaiacol and 100 mM H_2O_2. The reaction was done with adding 100 µL of crude enzyme extract, the changes in absorbance were recorded every 30 s intervals for 3 min at 470 nm, the activity was determined for $min^{-1} \cdot g^{-1}$ fresh weight. Activity of superoxide dismutase (SOD) was measured according to Mishra et al. [66]. Then, we add 290 µL of a mixture containing 100 mM potassium phosphate buffer, 0.1 mM EDTA, 11 mm^3 xanthine, cytochrome-c, and 0.002 units of xanthine oxidase to 20 µg of protein extracts was prepared. Xanthine oxidase regulation produced an increase in the absorbance due to the reduction of cytochrome-c (0.025 ± 0.005 min^{-1}). Activity of SOD was stated by McCord and Fridovich [67]. According to Goldberg and Spooner [68], GR activity was measured, approximately 0.05 cm^3 enzyme extract was mixed with 1 cm^3 phosphate buffer combined with EDTA, 0.1 cm^3 glutathione, and 0.1 cm^3 NADPH, the absorbance was determined at 340 nm.

2.2.9. Fruit Yield

The harvest date starts at 120 days from transplanting to determine number of fruit $plant^{-1}$, fruit fresh weight $plant^{-1}$ (g), and total fruit yield (ton $hectare^{-1}$).

2.3. *Statistical Analysis*

Statistical analysis was done using analysis of variance (ANOVA) procedures according to the method of Gomez and Gomez [69] using the MSTAT-C statistical software package. The means between treatments were compared by Duncan [70] when the difference was significant ($P \leq 0.05$). The correlation analysis was done using XLSTAT 2014.5.03 statistical software.

3. Results

3.1. *Effect on Relative Water Content (RWC%)*

The presented results in Figure 1A showed a significant decrease in RWC in sweet pepper under two salinity concentrations (57.6% at the low concentration (34 mM) (S1) and 52% at the high concentration (68 mM) (S2) comparing with control plants (74.6%) as the mean of the two seasons. Likewise, the results in Figure 1 revealed that seed treatment with *B. thuringiensis* showed a significant increase in RWC in stressed plants (65.7% compared with 57.6% at the low concentration and 60.8% compared with 52% at the high concentration). Furthermore, chitosan application at 30 mg dm^{-3} caused a significant increase in RWC (71.5% compared with 57.6% at the low concentration of salinity) and (67.1% compared with 52% at the high concentration) as a mean of both seasons in the stressed plants. The best treatment under salinity conditions was chitosan at 30 mg·dm^{-3} which achieved 71.5% when compared with control plants 74.6% without any significant difference.

3.2. *Effect on Chlorophyll a and b Concentrations*

It is obvious from the achieved results in Figure 1B–C that chlorophyll was significantly reduced in stressed plants; chlorophyll *a* significantly decreased at low concentration of salinity (2 mg·g^{-1} FW^{-1}) compared with control (2.85 mg·g^{-1} FW^{-1}) as the mean of both seasons. Furthermore, the high salinity concentration caused a significant reduction in chlorophyll *a* (1.25 mg·g^{-1} FW^{-1}) in stressed plants compared to control (2.85 mg·g^{-1} FW^{-1}). Similarly, salinity stress led to a significant decrease in chlorophyll *b* concentration, the two concentrations caused significant decreases (0.84 and 0.55 mg·g^{-1} FW^{-1} respectively) compared with control (2.85 mg·g^{-1} FW^{-1}). Nonetheless, seed treatment with *B. thuringiensis* and chitosan application led to significant increases in chlorophyll *a* and *b*. The greatest result was obtained with chitosan (S1 + Chitosan) treatment (2.85 mg·g^{-1} FW^{-1}) in the stressed plants with the low salinity concentration compared to the stressed plants (S1) without treatments (2 mg·g^{-1} FW^{-1}).

3.3. *Effect on Electrolyte Leakage (EL%)*

The presented data in Figure 1D exhibited that EL% significantly increased in the stressed plants, the low salinity concentration caused significant increase (42.3%) comparing with control (13.8%) as the mean of two seasons. Furthermore, the high salinity concentration was more harmfully effective and caused a significant increase in EL% (52.6%) compared with control (13.8%). Nevertheless, chitosan application 30 mg dm^{-3} and seed treatment with *B. thuringiensis* led to significant decrease in EL% in the stressed plants under the two concentrations. Seed treatment with *B. thuringiensis* caused a positive effect and significant decrease in EL% (30.2% and 37.6%) in the stressed plants at the two concentrations compared with untreated plants (42.3% and 52.6%), respectively. Furthermore, EL% was reduced significantly in the stressed treated plants with chitosan 30 mg dm^{-3} (21.7% and 27.2%) that compared with the stressed untreated plants (42.3% and 52.6%).

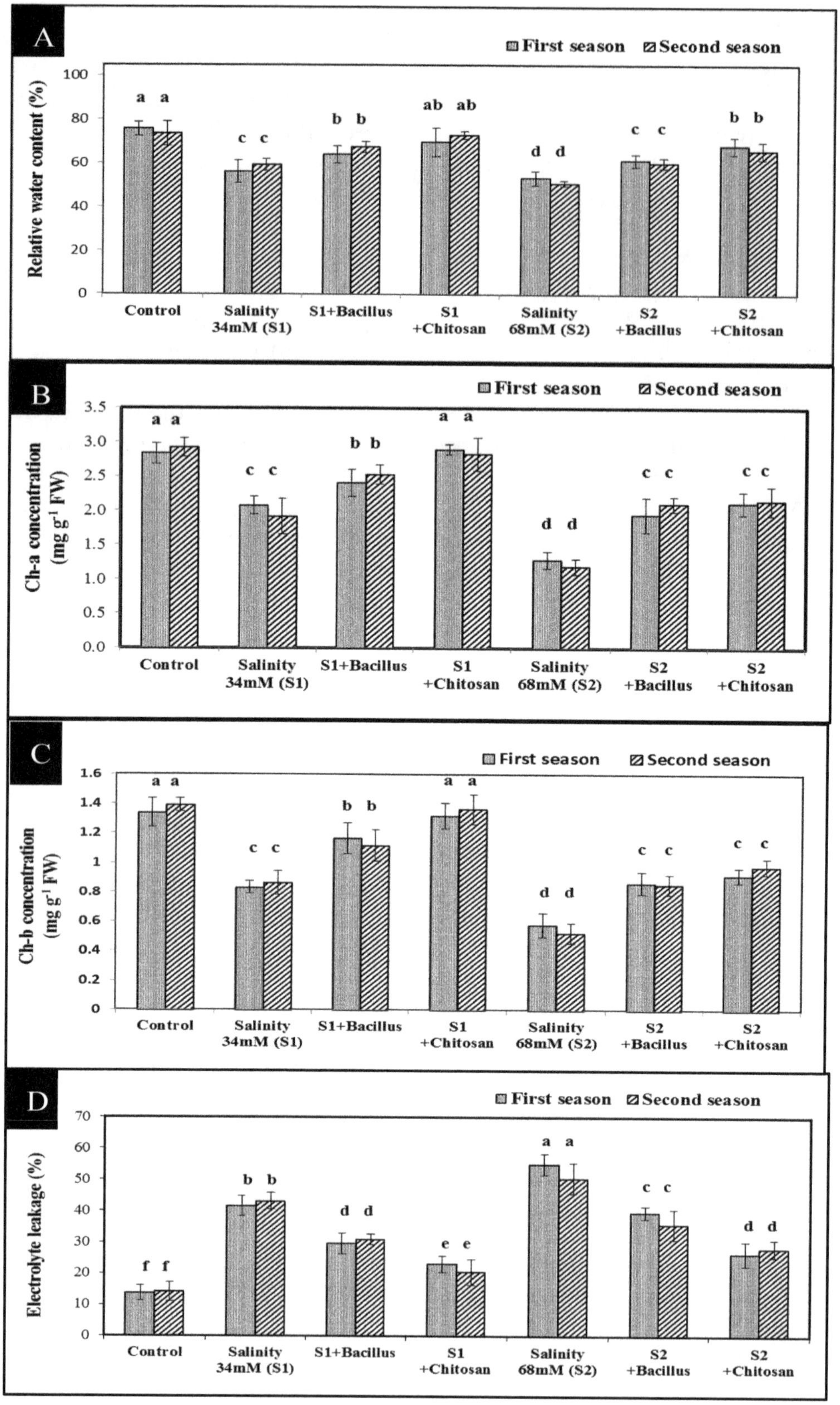

Figure 1. Effect of *B. thuringiensis* and chitosan on relative water content (**A**) chlorophyll *a*, (**B**) chlorophyll *b*, (**C**) and electrolyte leakage (**D**) under two salinity concentrations in sweet pepper plants during two seasons [first season (2019) and second season (2020)]. Data is the mean (±SE) of four replicates. Different letters above the data columns indicate significant differences between the samples determined by ANOVA, Duncan's multiple range test at 0.05 level.

3.4. Effect on Proline Concentration

It could be noted from Figure 2A that the exposed plants to salinity at (S1) and (S2) caused a significant increase in proline concentration, the high concentration of salinity (S2) achieved the high concentration of proline (24 $\mu g{\cdot}g^{-1}$ FW) comparing to control (9.1 $\mu g{\cdot}g^{-1}$ FW) as the mean of both seasons in sweet pepper. Application of seed treatment with *B. thuringiensis* and chitosan application in stressed plants led to regulate proline accumulation when compared with the control and the stressed untreated plants. *B. thuringiensis* seed treatment led to the regulation of proline accumulation in the stressed plants (12.7 $\mu g{\cdot}g^{-1}$ FW at the low concentration of salinity and 13.6 $\mu g{\cdot}g^{-1}$ FW at the high concentration comparing to the stressed untreated plants 17.4 and 24 $\mu g{\cdot}g^{-1}$ FW) at the two concentrations, respectively. Chitosan application had a significant effect on proline content (9.8 and 12.2 $\mu g{\cdot}g^{-1}$ FW) compared with stressed untreated plants (17.4 and 24 $\mu g{\cdot}g^{-1}$ FW) at the two concentrations, respectively. The difference was not significant between the both seasons.

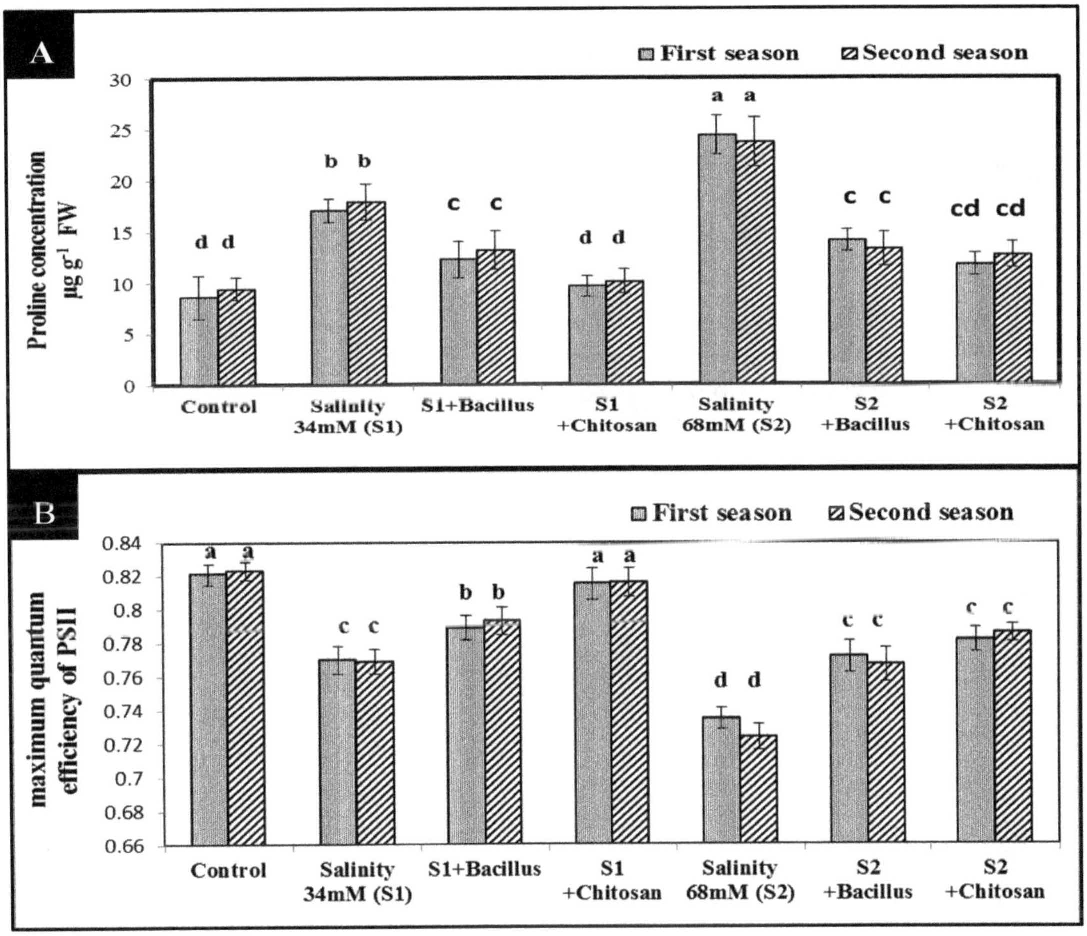

Figure 2. Effect of *B. thuringiensis* and chitosan on proline content (**A**) and maximum efficiency of PSII (*Fv/Fm*) (**B**) under two salinity concentrations in sweet pepper during two seasons. Data is the mean (±SE) of four replicates. Different letters above the data columns indicate significant differences between the samples determined by ANOVA, Duncan´s multiple range test at 0.05 level.

3.5. Effect on Chlorophyll Fluorescence Parameter (Fv/Fm)

Our results in Figure 2B indicated that chlorophyll fluorescence parameters were adversely affected under salinity conditions. The maximum efficiency of PSII (*Fv/Fm*) significantly reduced in sweet pepper (0.790) at the low salinity concentration and (0.729) at the high salinity concentration, respectively comparing to the control (0.822). However, seed treatment with *B. thuringiensis* caused significant increase in *Fv/Fm* ratio in the stressed plants (0.791) at the low concentration of salinity and (0.769) at the high salinity concentration when compared with the stressed untreated plants (0.790) at the low concentration and (0.729) at the high salinity concentration. Likewise, under the two concentrations, chitosan caused a significant increase in *Fv*/*Fm* ratio. The best treatment was chitosan at the low salinity concentration (0.815) compared with control (0.822).

3.6. *Effect on Lipid Peroxidation as Malondialdehyde*

According to the findings in Figure 3, lipid peroxidation (MDA) significantly increased in sweet pepper (11.35 and 13.8 μmol·g^{-1} FW) at the two salinity concentrations, respectively as the mean of both seasons when compared with control plants (6.75 μmol·g^{-1} FW). Nevertheless, MDA significantly decreased in the stressed plants according to seed treatment with *B. thuringiensis* and chitosan treatment. *B. thuringiensis* treatment had a positive effect on MDA and led to significant reduction in the MDA content at the two salinity concentrations (8.8 and 10.5 μmol·g^{-1} FW) when compared with the stressed untreated plants (11.35 and 13.8 μmol·g^{-1} FW). The application of chitosan significantly reduced MDA content in sweet pepper under the two salinity concentrations (7 and 7.85 μmol·g^{-1} FW) when compared with stressed untreated plants (11.35 and 13.8).

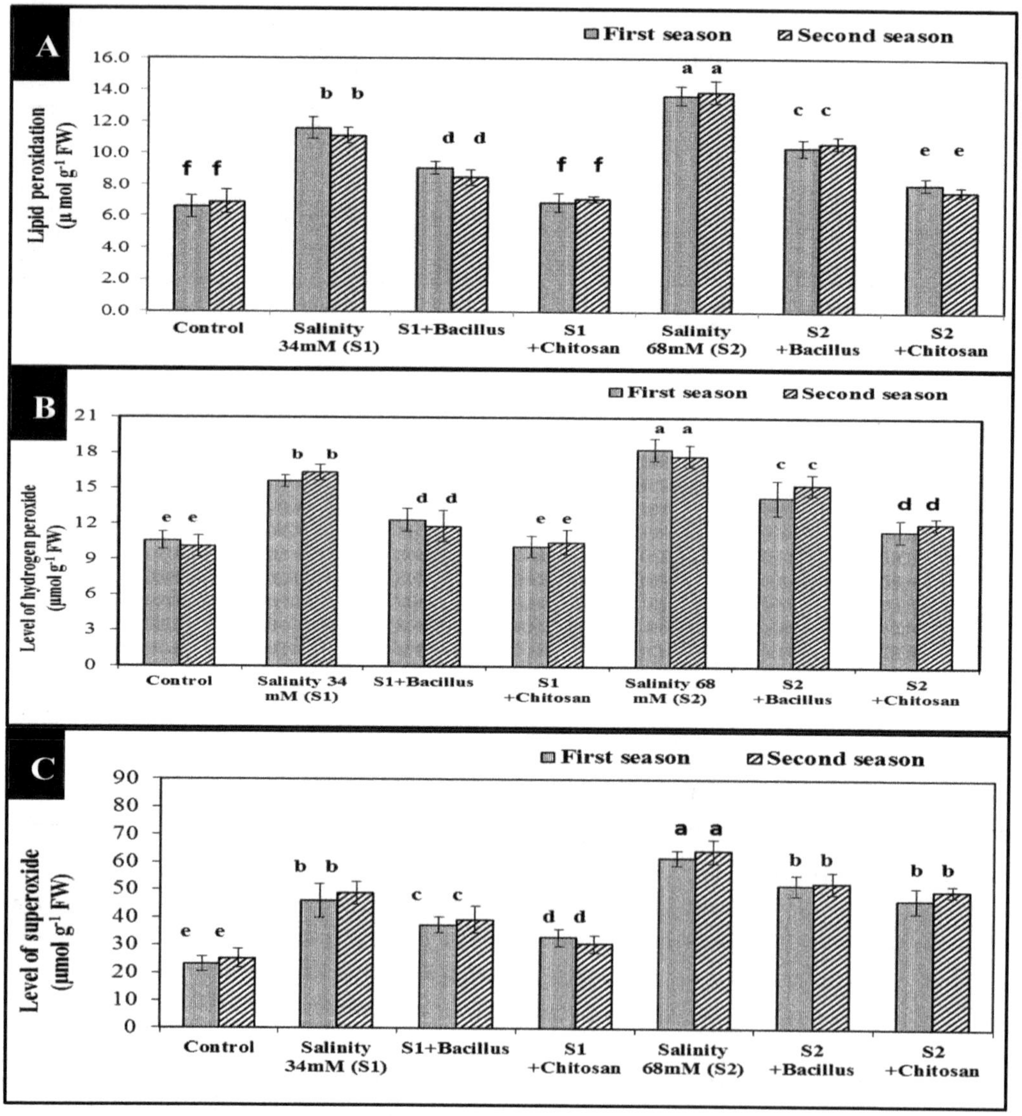

Figure 3. Effect of *B. thuringiensis* and chitosan on lipid peroxidation (**A**), H_2O_2 (**B**) and O_2^- (**C**) under two salinity concentrations in sweet pepper during two seasons. Data is the mean (±SE) of four replicates. Different letters above the data columns indicate significant differences between the samples determined by ANOVA, Duncan´s multiple range test at 0.05 level.

3.7. *Effect on O_2^- and H_2O_2*

ROS, mainly O_2^- and H_2O_2 significantly increased under the both salinity concentrations (Figure 3). O_2^- significantly increased (47.4 and 63 units) at the two salinity concentrations compared with control (24.16 units). Conversely, *B. thuringiensis* treatment caused a significant decrease in O_2^- in the salt stressed plants (38.3 and 52.3 units) in comparison with stressed untreated plants (47.4 and 63 units). Also, chitosan treatment caused a significant decrease in O_2^- (31.7 and 48.3 units) when compared with the stressed untreated plants (47.4 and 63 units).

Salinity stress caused a significant increase in H_2O_2 in sweet pepper (16 and 18.1 units) at the two concentrations, respectively as compared to control (10.3 units). However, the levels of H_2O_2 were decreased significantly according to *B. thuringiensis* seed treatment and chitosan application in the stressed plants at the two salinity concentrations. Chitosan application gave the best and most significant results (10.3 and 11.8 units) compared to stressed untreated plants (16 and 18.1 units) at the two salinity concentrations, respectively.

3.8. *Effect on the Activity of Catalase (CAT), Peroxidase Activity (POX), Superoxide Dismutase (SOD) and Glutathione Reductase (GR) Enzymes*

Salinity stress at both concentrations caused significant increases in CAT, POX, SOD and GR enzyme (Figure 4). CAT activity significantly increased in the stressed plants (124.8 and 149.3 mM H_2O_2 g^{-1} FW min^{-1}) at the two salinity concentrations, respectively, when compared with control (78.6 mM H_2O_2 g^{-1} FW min^{-1}).

However, chitosan treatment and *B. thuringiensis* seed treatment caused significant reduction in CAT activity at both salinity concentrations. Chitosan with the low salinity concentration (S1 + Chitosan) gave the best result (85.8 mM H_2O_2 g^{-1} FW min^{-1}) as compared to stressed untreated plants (124.8 mM H_2O_2 g^{-1} FW min^{-1}) and control plants (78.6 mM H_2O_2 g^{-1} FW min^{-1}). Moreover, POX, SOD and GR activities significantly increased in the stressed plants at the low salinity concentration (0.6 μmol tetra-gualacol g^{-1} FW min^{-1}, 74.5 and 0.59 unit/cm^3) as compared to control plants (0.24, 38.3 and 0.36), also, the enzymes activity significantly increased in the stressed plants at the high concentration (0.76, 98.7 unit mg^{-1} FW min^{-1} and 0.59 unit/cm^3) respectively. Nevertheless, chitosan application and seed treatment with *B. thuringiensis* caused a significant reduction in POX, SOD, and GR activity in the stressed plants at the two salinity concentrations compared to the stressed untreated plants.

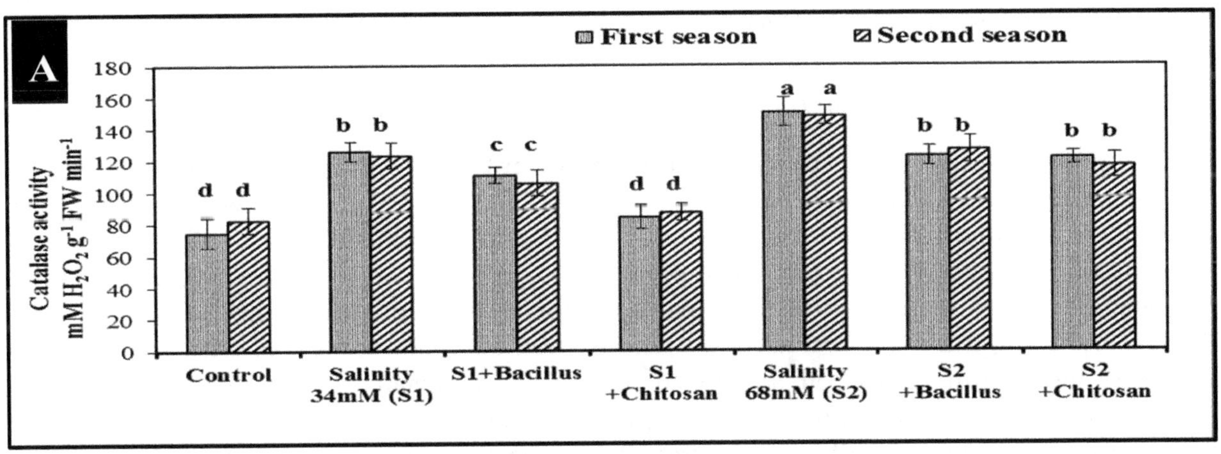

Figure 4. *Cont.*

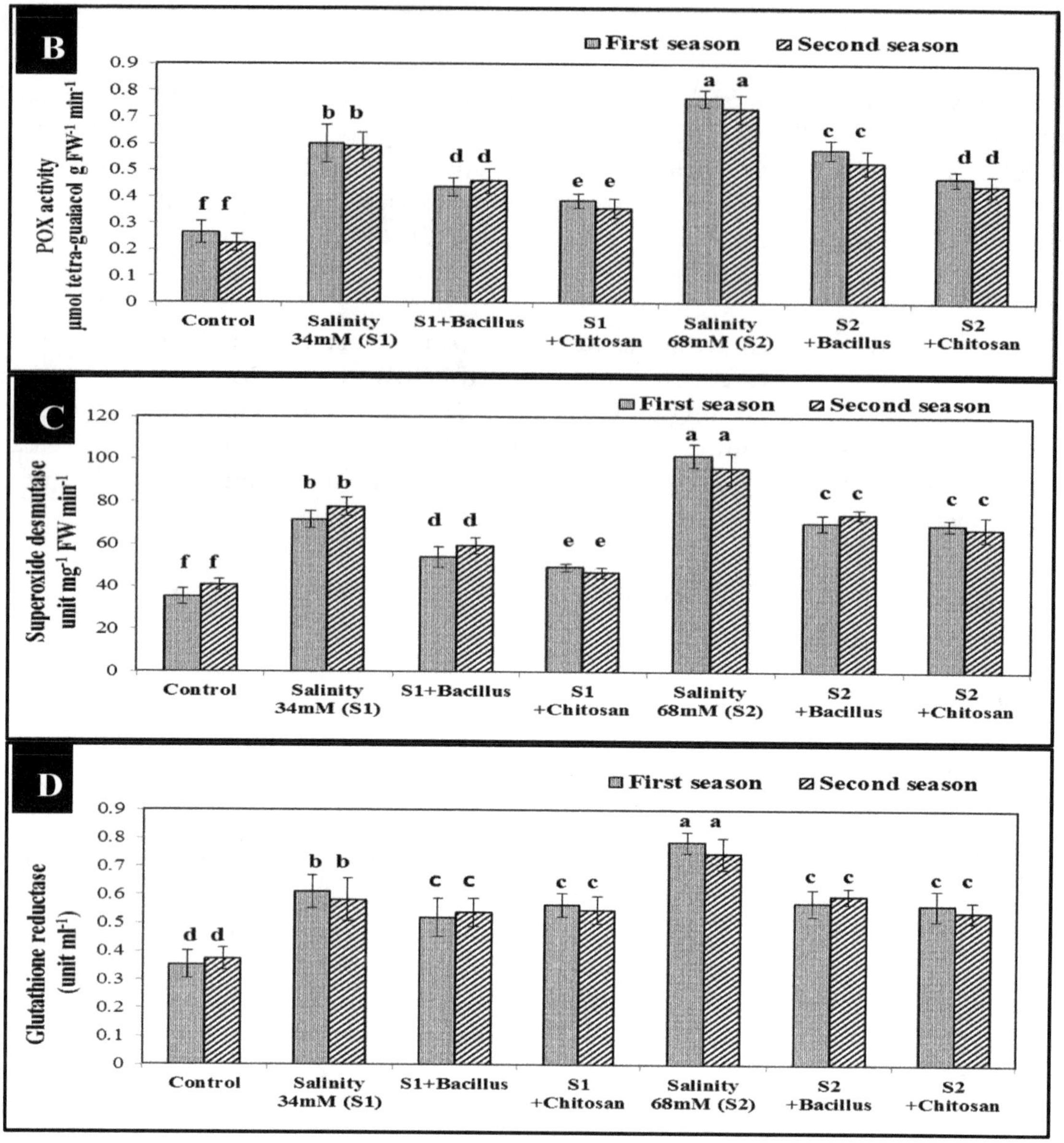

Figure 4. Effect of *B. thuringiensis* and chitosan on the activity of CAT (**A**), POX (**B**), SOD (**C**) and GR (**D**) under two salinity concentrations in sweet pepper during two seasons. Data is the mean (±SE) of four replicates. Different letters above the data columns indicate significant differences between the samples determined by ANOVA, Duncan´s multiple range test at 0.05 level.

3.9. Effect on Number of Fruit Plant^{-1}, Fruit Fresh Weight Plant^{-1} and Total Fruit Yield (Ton Hectare^{-1}).

In the present study, the results in Figure 5 point out that salinity at the both concentrations caused a significant decrease in number of fruit plant^{-1} (7.7 and 4.8 fruit), fresh weight of fruit plant^{-1} (524.5 and 356.4 g) and total fruit yield hectare^{-1} (7.05 and 5 ton) as the mean of the both seasons when compared to control plants (15.7 fruit plant^{-1}, 974 g plant^{-1} and 14.9 ton hectare^{-1}). However, *B. thuringiensis* and chitosan significantly increased the number of fruit plant^{-1}, fruit fresh weight (g plant^{-1}) and total fruit yield (ton hectare^{-1}) in the stressed plants compared with untreated plants. Interestingly enough, under the both salinity concentrations, chitosan application gave the best results and significantly increased the number of fruit plant^{-1} (14.9 and 12.7), fruit fresh weight plant^{-1} (911 and 527 g plant^{-1}), and total fruit yield (14 and 10.8 ton hectare^{-1}) as the mean of the both seasons.

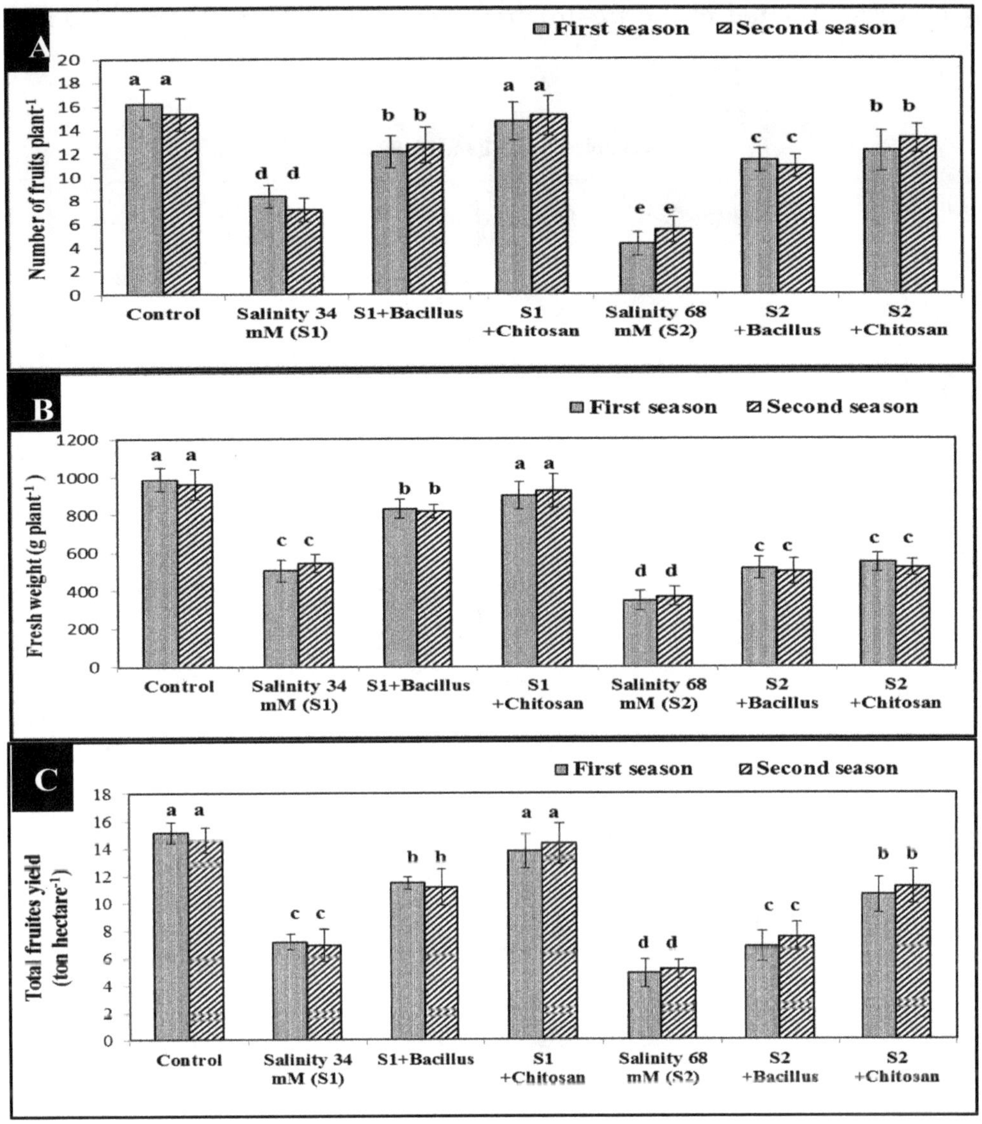

Figure 5. Effect of *B. thuringiensis* and chitosan on number of fruit plant^{-1}(**A**), fruit fresh weight plant^{-1} (**B**) and total fruit yield (ton hectare^{-1}) (**C**) under two salinity concentrations in sweet pepper during two seasons. Data is the mean (±SE) of four replicates. Different letters above the data columns indicate significant differences between the samples determined by ANOVA, Duncan's multiple range test at 0.05 level.

3.10. Correlation Studies

In the present study chlorophyll *a* was positively and significantly correlated with chlorophyll *b* (r = 0.99), number of fruits (r = 0.98), RWC (r = 0.97), GR (r = 0.80) and MDA (r = 0.75). Among the treatment it has a negative correlation with salinity stress @ 34 mM (r = −0.05), salinity stress @ 68 mM (r = −0.02), however, a positive correlation was noted among the chlorophyll *a* and treatments of Bacillus sp. and chitosan (Figure 6 and Supplementary Table S1). A similar trend of relationship was shown by chlorophyll *b*. Proline showed highly positive correlation with MDA (r = 0.98), H_2O_2 (r = 0.97), SOD (r = 0.96) and GR (r = 0.96) but was negatively correlated with the treatments Bacillus sp. (r = −0.04) and chitosan (r = −0.05). A very similar correlation was observed among all the studies. Antioxidant enzymes and H_2O_2 concentration that were highly correlated with each other also showed a negative correlation with the treatments of *Bacillus* sp. and chitosan. The number of fruits showed a highly significant correlation with chlorophyll *a* and *b* (r = 0.98) and with RWC (r = 0.95). However, this trait was inversely related to the treatments of salinity @ 34 mM (r = – 0.15) and @ 68 mM (r = – 0.25).

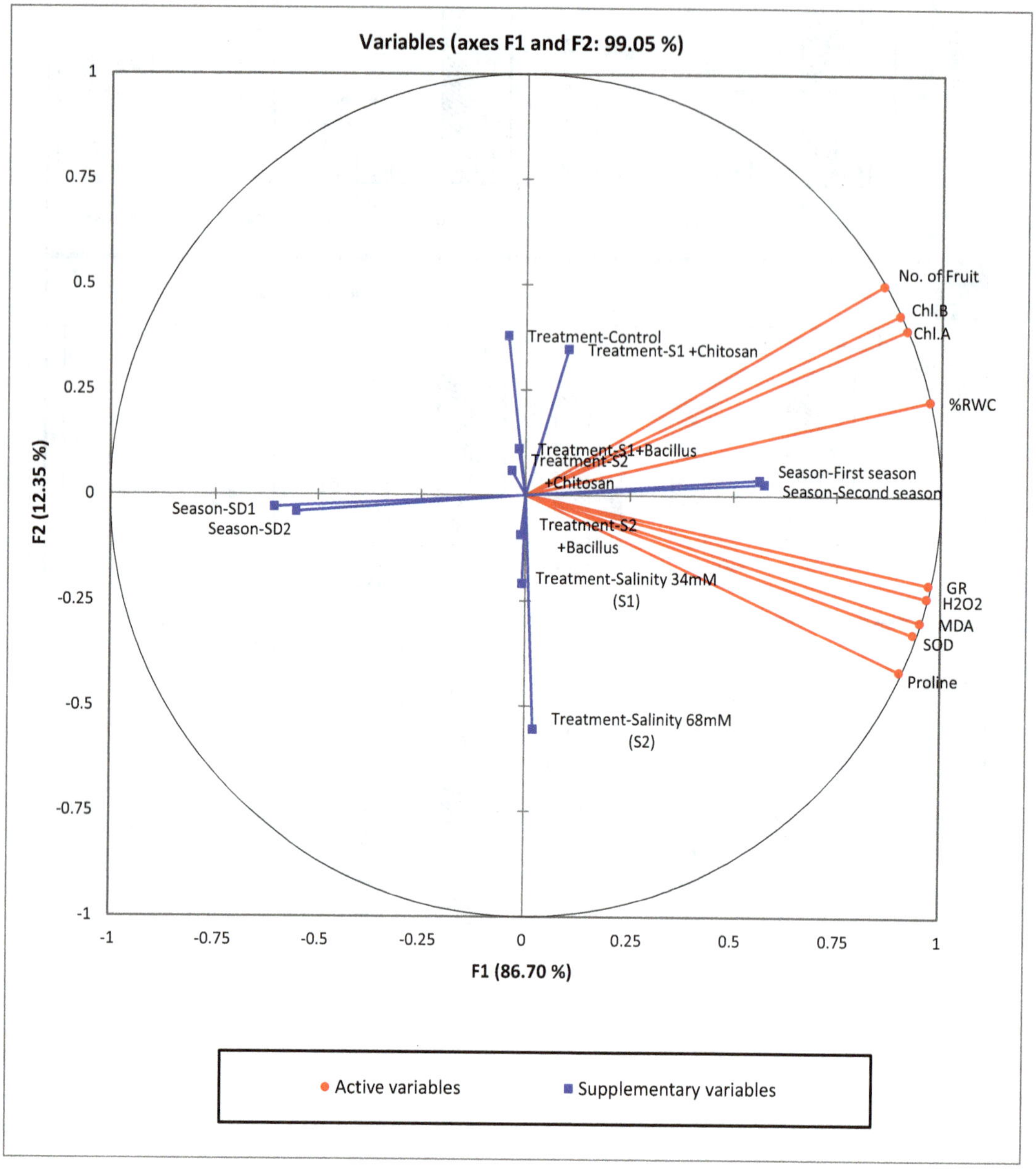

Figure 6. Circle of correlation between variables and factors for sweet pepper.

4. Discussion

Salinity stress adversely affects plant growth, inhibiting plant development and reducing fruit yield of sweet pepper. The present data revealed the deleterious effects of salinity at the two different concentrations (34 and 68 mM) on RWC. This might be due to the injurious influence of salinity on the cell wall structure [71], thereby increasing ethylene concentration, which reduces the growth of roots [44]. This effect causes changes in cell wall properties, the reduction in osmotic potential, and the decrease in water balance [72], consequently reducing RWC in sweet pepper [1]. These deleterious impacts of salinity were overcome by seed treatment with *B. thuringiensis* and treating stressed sweet pepper with chitosan. The pivotal role of *B. thuringiensis* under salinity stress could be due to the formation of Indole-acetic acid which causes enhancement of root growth and increased water uptake [73]. Likewise, PGPR can produce exopolysaccharides (EPSs) which aggregate with soil particles and improve soil structure as well as water uptake [74]. Further, the application of PGPR causes a decay in the soil bulk density and enhances the availability of soil water. Chitosan application positively affects RWC in stressed plants, this progressive effect of chitosan could be due to the positive role of chitosan on water availability in stressed plants. These valuable effects were documented in barley under drought [19].

Chlorophyll *a* and *b* are very important pigments in the process of photosynthesis, in this process, two reactions take place. One such reaction is the light reaction, in which NADPH and ATP are produced, and the second is the dark reaction, in which carbon dioxide is fixed [75]. Demonstrated data revealed a significant decrease in chlorophyll content under the two salinity concentrations, this decrease in chlorophyll was more considerable at the high concentration (68 mM) than at lower concentration (34.mM) and this might be due to the damaging effect of salinity on the chloroplast structure [3,76], that decrease energy transport from PSII to PSI [77] and, consequently, reduce the chlorophyll formation in stressed sweet pepper plants. The harmful effect of salinity on the content of chlorophyll was also due to reduction in stomatal conductance and destruction of biochemical processes [78]. These findings are in accordance with those reported by Abdelaal et al. [1] in sweet pepper under salinity stress. Also, Asrar et al. [79] indicated that a high salinity concentration caused harmful effects on PSII and decreased chloroplast proteins as well as chlorophyll concentrations. This decrease in chlorophyll concentrations is related to the reduction in RWC under high salt concentration.

Conversely, inoculation of seeds with *B. thuringiensis* mitigates the adverse effects of salinity on the content of chlorophyll that improve the overall growth and proliferation of plants under stressful environments [80]. Beside this, the application of chitosan had also synergistic effects on the contents of chlorophyll *a* and *b*. This increase in the content of chlorophyll with the application of chitosan may be attributed to the fact that chitosan is a rich source for amino acids which increase the chloroplast number and chlorophyll formation. These results are in harmony with the findings of Possingham [81]. During the present study, a significant increase was found in EL% under two different salt concentrations mainly. The higher salt concentration was more effective and significantly increased the EL%. This negative influence of salinity on EL% may be due to its damaging impacts on the cytoplasmic membrane and permeability process. Previously, a similar result was reported by Abdelaal et al. [1] in sweet pepper. Contrariwise, EL% significantly reduced in stressed plants as a result of seed treatment with *B. thuringiensis* and chitosan, these valuable effects of *B. thuringiensis* treatment and chitosan application is attributed to the positive roles of *B. thuringiensis* and chitosan on membrane stability and an improvement in the selective permeability of cell plasma membrane.

In the present study, the chlorophyll fluorescence parameter was adversely affected under two salinity concentrations. Salinity stress causes a significant decrease to maximum efficiency of PSII (*Fv/Fm*). This adverse effect of salinity on (*Fv/Fm*) might be due to its role in the inhibition of electron transport and the reaction centers at the PSII sites as well as destroys the oxygen-evolving complex [82–84]. Also, salinity stress has a negative effect on enzymes activity and decreases the activity of water splitting enzyme complexes and electron transport chains resulting in decrease *Fv/Fm* [85]. However, seed treatment with *B. thuringiensis* and the application of chitosan caused a significant increase *Fv/Fm* ratio in the stressed plants. These results are credited to the helpful role of *B. thuringiensis* and chitosan in increasing the production of protective metabolites, increasing N and K content as well as the number of chloroplasts under stress [81,86], and consequently, improving the chlorophyll fluorescence parameter. The obtained results indicated that proline significantly increased in the stressed plants under both the salinity concentrations (34 and 68 mM). This impact of salinity may be due to its role in reducing the proline oxidation to glutamate, consequently increasing the proline content [87]. Proline is one of the most important osmoprotectants, plays a key role in osmotic regulation, and protects the plants under stress [1,8]. Chitosan application and seed treatment with *B. thuringiensis* regulated proline content under salinity conditions. Seed inoculation with *B. thuringiensis* positively regulated proline content under stress because this species regulates the osmotic balance under saline conditions. Similar results for proline production under saline conditions were also reported by Egamberdieva et al. [88].

Salinity could hamper plant growth and increase lipid peroxidation, O_2^-, and H_2O_2. A significant increase was noted in the mentioned parameters during the present study. These reactive compounds can damage lipids and proteins, essential for the process of photosynthesis and electron transport chain. Islam et al. [18] noted similar results in two wheat cultivars grown under saline conditions. However, in the present study, a significant decrease was noted in the lipid peroxidation upon treatment with chitosan. This may be due to the involvement of chitosan in cell protection from oxidative stress under salinity conditions. Similarly, O_2^- and H_2O_2 were significantly reduced with chitosan due to the presence of hydroxyl and amino groups which react with ROS, thus chitosan can scavenge superoxide radicals [89]. Chitosan derived from the pathogen is recognized by a specific cellular receptor resulting in enhancing the defense response to abiotic and biotic stresses [90]. The positive effect of chitosan in the plant cell protection was also noted in plants under drought stress [20]. Interestingly, seed treatment with *B. thuringiensis* led to improved cell membrane stability and decreased the formation of MDA in the stressed sweet pepper, this effect of *B. thuringiensis* is due to its improved phenol content and defense enzyme system [91]. Also, *B. thuringiensis* causes decreases in O_2^- and H_2O_2 by increasing reactive oxygen scavenging enzyme activity [92].

Enzymes up-regulation (CAT, POX, SOD, and GR) is involved in the mitigation of salinity stress in sweet peppers compared with control plants. The significant increase in these enzymes is a natural defense system, which helps to cope with salinity stress and reduces the osmotic and toxic effects by scavenging ROS. Our results are in agreement with those reported by Abdelaal et al. [17] and Foyer et al. [93]. Nevertheless, it was clear from our results that the application of seed treatment with *B. thuringiensis* led to improved and regulated up-regulation of CAT, POX, SOD, and GR in the stressed sweet pepper. The induction of these enzymes is involved in the mitigation of salt stress in sweet pepper treated with *Bacillus*. A similar trend of enzyme activity was recorded in the findings of Kohler et al. [94]. Likewise, chitosan application causes an increase in enzymes activity to protect the plant from oxidative damage and reduce lipid peroxidation as well as scavenge O_2^- due to its structure and protective role in sweet pepper plants subjected to salinity stress. These results are in agreement with those reported by Hafez et al. [19]. The presented study showed that two salinity concentrations caused a significant reduction in the number of fruit plant^{-1}, fruit fresh weight plant^{-1}, and total fruit yield. This harmful impact of salinity may be due to the decrease in reproductive organs, such as pollen grains in stressed plants [95], and also due to the decrease in water absorption, nutrients uptake, and chlorophyll content [1,4], resulting in a significant decrease in fruit yield [96]. The vital role of *B. thuringiensis* might be due to the formation of growth regulators such as gibberellins, auxin, and cytokinins, as well as an increase in proline content [87], up-regulation of essential enzymes and solubilization of nutrients [89], and an increase in the number of fruits and fruit yield hectar^{-1} in sweet pepper. These findings are in agreement with the previous results reported by Hafez et al. [19], Katiyar et al. [36], and Hidangmayum et al. [37].

5. Conclusions

The present research concluded that seeds treated with *B. thuringiensis* and foliar application of chitosan 30 mg dm^{-3} on sweet pepper plants under two salinity concentrations (34 and 68 mM) led to an improvement of the adverse effects of salinity and enhanced the growth and yield of sweet pepper. RWC, chlorophyll *a* and *b* concentrations, chlorophyll fluorescence parameters, and fruit yield characters significantly increased with *B. thuringiensis* and chitosan treatments in sweet pepper under two salinity concentrations. Conversely, lipid peroxidation, electrolyte leakage, and reactive oxygen species (O_2^- and H_2O_2) were decreased significantly as a result of *B. thuringiensis* and chitosan treatments. Overall, seed treatment with *B. thuringiensis* and chitosan foliar application was an effective and cheaper approach to cope with the deleterious effects of salinity on sweet pepper by improving the chlorophyll fluorescence parameters, proline accumulation, and up-regulation of enzymes activity as well as the enhancement of fruit yield characters.

Author Contributions: Conceptualization, K.A.A.A., Y.M.H., K.A.A., and M.D.F.A.; methodology, K.A.A.A., Y.M.H., K.A.A., M.D.F.A., A.M.E., and M.A.M.A.; software, K.A.A.A., Y.M.H., K.A.A., M.D.F.A., and M.A.M.A.; formal analysis, K.A.A.A., Y.M.H., K.A.A., A.M.E., M.A., and N.K.; investigation, K.A.A.A., Y.M.H., K.A.A., A.M.E., M.A.M.A., and N.K.; resources, K.A.A.A., Y.M.H., K.A.A., A.M.E., M.A.M.A., and N.K.; data curation, K.A.A.A., Y.M.H., K.A.A., A.M.E., M.A.M.A., and N.K.; writing-original draft preparation, K.A.A.A., Y.M.H., K.A.A., A.M.E., M.A.M.A., M.D.F.A., and N.K.; writing-review and editing, K.A.A.A., Y.M.H., A.M.E., M.A., N.K., and M.D.F.A.; visualization, K.A.A.A., Y.M.H., and K.A.A.; funding acquisition, M.D.F.A., K.A.A.A., M.A., A.M.E., and K.A.A. All authors have read and agreed to the published version of the manuscript.

Acknowledgments: The authors extend their appreciation to all members of PPBL and EPCRS excellence center, Faculty of Agriculture, Kafrelsheikh University, and the Deanship of Scientific Research at Princess Nourah bint Abdulrahman University Fast-track Research Funding Program.

References

1. Abdelaal, K.A.A.; EL-Maghraby, L.M.; Elansary, H.; Hafez, Y.M.; Ibrahim, E.I.; El-Banna, M.; El-Esawi, M.; Elkelish, A. Treatment of Sweet Pepper with Stress Tolerance-Inducing Compounds Alleviates Salinity Stress Oxidative Damage by Mediating the Physio-Biochemical Activities and Antioxidant Systems. *Agronomy* **2020**, *10*, 26. [CrossRef]
2. Chavez-Mendoza, C.; Sanchez, E.; Muñoz-Marquez, E.; Sida-Arreola, J.P.; Flores-Cordova, M.A. Bioactive compounds and antioxidant activity in different grafted varieties of bell pepper. *Antioxidants* **2015**, *4*, 427–446. [CrossRef] [PubMed]
3. El-Banna, M.F.; Abdelaal, K.A.A. Response of Strawberry Plants Grown in the Hydroponic System to Pretreatment with H_2O_2 before Exposure to Salinity Stress. *J. Plant Prot. Mansoura Univ.* **2018**, *9*, 989–1001. [CrossRef]
4. Helaly, M.N.; Mohammed, Z.; El-Shaeery, N.I.; Abdelaal, K.A.A.; Nofal, I.E. Cucumber grafting onto pumpkin can represent an interesting tool to minimize salinity stress. Physiological and anatomical studies. *Middle East J. Agric. Res.* **2017**, *6*, 953–975.
5. Isayenkov, S.V.; Maathuis, F.J. Plant salinity stress: Many unanswered questions remain. *Front. Plant Sci.* **2019**, *10*, 80. [CrossRef] [PubMed]
6. Hernández, J.A.; Almansa, M.S. Short-term effects of salt stress on antioxidant systems and leaf water relations of pea leaves. *Physiol. Plant.* **2002**, *115*, 251–257. [CrossRef]
7. Koyro, H.W. Effect of salinity of growth, photosynthesis, water relations and solute composition of the potential cash crop halophyte *Plantago coronopus* (L.). *Environ. Exp. Bot.* **2006**, *56*, 136–146. [CrossRef]
8. Gorhman, J.; Hardy, C.; Wyn, R.G.; Jones, L.R.; Law, C.N. Chromosomal location of a K/Na discrimination character in the D genome of wheat. *Theor. Appl. Genet.* **1987**, *74*, 584–588.
9. Harizanova, A.; Koleva-valkova, L. Effect of silicon on photosynthetic rate and the chlorophyll fluorescence parameters at hydroponically grown cucumber plants under salinity stress. *J. Cent. Eur. Agric.* **2019**, *20*, 953–960. [CrossRef]
10. Misra, N.; Ansari, M.; Gupta, A. Differential response of scavenging of reactive oxygen species in green gram genotype grown under salinity stress. *Am. J. Plant Physiol.* **2006**, *1*, 41–53. [CrossRef]
11. Hasan, M.K.; El Sabagh, A.; Sikdar, M.S.; Alam, M.J.; Ratnasekera, D.; Barutcular, C.; Abdelaal, K.A.A.; Islam, M.S. Comparative adaptable agronomic traits of blackgram and mungbean for saline lands. *Plant Arch.* **2017**, *17*, 589–593.
12. Carmen, B.; Roberto, D. Soil bacteria support and protect plants against abiotic stresses. In *Abiotic Stress in Plants Mechanisms and Adaptations*; Shanker, A., Ed.; Pub. In Tech: London, UK, 2011; pp. 143–170.
13. Islam, F.; Yasmeen, T.; Ali, S.; Ali, B.; Farooq, M.A.; Gill, R.A. Priming induced antioxidative responses in two wheat cultivars under saline stress. *Acta Physiol. Plant.* **2015**, *37*, 153. [CrossRef]
14. Memon, S.A.; Hou, X.; Wang, L.J. Morphological analysis of salt stress response of pak Choi. *Electron. J. Environ. Agric. Food Chem.* **2010**, *9*, 248–254.
15. Khan, N.; Bano, A. Role of plant growth promoting rhizobacteria and Ag-nano particle in the bioremediation of heavy metals and maize growth under municipal wastewater irrigation. *Int. J. Phytorem.* **2016**, *18*, 211–221. [CrossRef]
16. Gama, P.B.S.; Inanaga, S.; Tanaka, K.; Nakazawa, R. Physiological response of common bean (*Phaseolus vulgaris* L.) seedlings to salinity stress. *Afr. J. Biotechnol.* **2007**, *6*, 79–88.

17. Abdelaal, K.A.; Mazrou, Y.S.; Hafez, Y.M. Silicon Foliar Application Mitigates Salt Stress in Sweet Pepper Plants by Enhancing Water Status, Photosynthesis, Antioxidant Enzyme Activity and Fruit Yield. *Plants* **2020**, *9*, 733. [CrossRef]
18. Abdelaal, K.A.; Hafez, Y.M.; Sabagh, A.E.L.; Saneoka, H. Ameliorative effects of Abscisic acid and yeast on morpho-physiological and yield characteristics of maize plant (*Zea mays* L.) under drought conditions. *Fresenius Environ. Bull.* **2017**, *26*, 7372–7383.
19. Hafez, Y.; Attia, K.; Alamery, S.; Ghazy, A.; Al-Doss, A.; Ibrahim, E.; Rashwan, E.; El-Maghraby, L.; Awad, A.; Abdelaal, K.A. Beneficial Effects of Biochar and Chitosan on Antioxidative Capacity, Osmolytes Accumulation, and Anatomical Characters of Water-Stressed Barley Plants. *Agronomy* **2020**, *10*, 630. [CrossRef]
20. Abdelaal, K.A.; Hafez, Y.M.; Badr, M.M.; Youseef, W.A.; Esmaeil, S.M. Biochemical, histological and molecular changes in susceptible and resistant wheat cultivars inoculated with stripe rust fungus *Puccinia striiformis* f.sp. *tritici*. *Egypt. J. Biol. Pest Control* **2014**, *24*, 421–429.
21. Abdelaal, K.A.; Omara, I.R.; Hafez, Y.M.; Esmail, S.M.; EL Sabagh, A. Anatomical, biochemical and physiological changes in some Egyptian wheat cultivars inoculated with *Puccinia graminis* f.sp. *tritici*. *Fresenius Environ. Bull.* **2018**, *27*, 296–305.
22. Esmail, S.M.; Omara, R.I.; Abdelaal, K.A.; Hafez, M. Histological and biochemical aspects of compatible and incompatible wheat-Puccinia striiformis interactions. *Physiol. Mol. Plant Pathol.* **2019**, *106*, 120–128. [CrossRef]
23. Omara, R.I.; El-Kot, G.A.; Fadel, F.M.; Abdelaal, K.A.; Saleh, E.M. Efficacy of Certain Bioagents on Patho-Physiological Characters of Wheat Plants under Wheat Leaf Rust Stress. *Physiol. Mol. Plant Pathol.* **2019**, *106*, 102–108. [CrossRef]
24. Hafez, Y.M.; Abdelaal, K.A. Investigation of susceptibility and resistance mechanisms of some Egyptian wheat cultivars (*Triticum aestivum* L.) inoculated with *Blumeria graminis* f.sp. *tritici* using certain biochemical, molecular characterization and SEM. *J. Plant Prot. Path. Mansoura Univ.* **2015**, *6*, 431–454.
25. Khan, N.; Bano, A.; Zandi, P. Effects of exogenously applied plant growth regulators in combination with PGPR on the physiology and root growth of chickpea (*Cicer arietinum*) and their role in drought tolerance. *J. Plant Interact.* **2018**, *13*, 239–247. [CrossRef]
26. Abdelaal, K.A.; Attia, K.A.; Alamery, S.F.; El-Afry, M.M.; Ghazy, A.I.; Tantawy, D.S.; Al-Doss, A.A.; El-Shawy, E.E.; Abu-Elsaoud, A.M.; Hafez Y., M. Exogenous Application of Proline and Salicylic Acid can Mitigate the Injurious Impacts of Drought Stress on Barley Plants Associated with Physiological and Histological Characters. *Sustainability* **2020**, *12*, 1736. [CrossRef]
27. Abdelaal, K.A. Effect of Salicylic acid and Abscisic acid on morpho-physiological and anatomical characters of faba bean plants (*Vicia faba* L.) under drought stress. *J. Plant Prod. Mansoura Univ.* **2015**, *6*, 1771–1788. [CrossRef]
28. Abdelaal, K.A.; Hafez, Y.M.; El-Afry, M.; Tantawy, D.S.; Alshaal, T. Effect of some osmoregulators on photosynthesis, lipid peroxidation, antioxidative capacity, and productivity of barley (*Hordeum vulgare* L.) under water deficit stress. *Environ. Sci. Pollut. Res.* **2018**, *25*, 30199–30211. [CrossRef]
29. Elkelish, A.; Qari, S.H.; Mazrou, Y.M.; Abdelaal, K.A.; Hafez, Y.M.; Abu-Elsaoud, A.M.; Batiha, G.; El-Esawi, M.; El Nahhas, N. Exogenous Ascorbic Acid Induced Chilling Tolerance in Tomato Plants Through Modulating Metabolism, Osmolytes, Antioxidants, and Transcriptional Regulation of Catalase and Heat Shock Proteins. *Plants* **2020**, *10*, 431. [CrossRef]
30. Zhu, J.-K. Regulation of ion homeostasis under salt stress. *Curr. Opin. Plant Biol.* **2003**, *6*, 441–445. [CrossRef]
31. Sairam, R.K.; Tyagi, A. Physiology and molecular biology of salinity stress tolerance in plants. *Curr. Sci.* **2004**, *86*, 407–421.
32. Gupta, B.; Huang, B. Mechanism of salinity tolerance in plants: Physiological, biochemical, and molecular characterization. *Int. J. Genom.* **2014**, 18. [CrossRef] [PubMed]
33. Maxwell, K.; Johnson, G.N. Chlorophyll fluorescence—A practical guide. *J. Exp. Bot.* **2000**, *51*, 659–668. [CrossRef] [PubMed]
34. Murchie, E.H.; Lawson, T. Chlorophyll fluorescence analysis: A guide to good practice and understanding some new applications. *J. Exp. Bot.* **2013**, *64*, 3983–3998. [CrossRef] [PubMed]
35. Kalaji, M.H.; Goltsev, V.; Zuk-Golaszewska, B.; Zivcak, M.; Brestic, M. *Chlorophyll Fluorescence: Understanding Crop Performance-Basics and Applications*; Taylor and Francis: New York City, NY, USA, USA 2017; p. 222. ISBN 9781498764490.

36. Katiyar, D.; Hemantaranjan, A.; Singh, B. Chitosan as a promising natural compound to enhance potential physiological responses in plant: A review. *Indian J. Plant Physiol.* **2015**, *20*, 1–9. [CrossRef]
37. Hidangmayum, A.; Dwivedi, P.; Katiyar, D.; Hemantaranjan, A. Application of chitosan on plant responses with special reference to abiotic stress. *Physiol. Mol. Biol. Plants* **2019**, *25*, 313–326. [CrossRef] [PubMed]
38. Al-Tawaha, A.M.; Seguin, P.; Smith, D.L.; Beaulieu, C. Foliar application of elicitors alters isofl avone concentrations and other seed characteristics of field-grown soybean. *Can. J. Plant Sci.* **2006**, *86*, 677–684. [CrossRef]
39. Kim, H.J.; Chen, F.; Wang, X.; Rajapakse, N.C. Effect of chitosan on the biological properties of sweet basil (*Ocimum basilicum* L.). *J. Agric. Food Chem.* **2005**, *53*, 3696–3701. [CrossRef]
40. Balal, R.M.; Shahid, M.A.; Javaid, M.M.; Iqbal, Z.; Liu, G.D.; Zotarelli, L.; Khan, N. Chitosan alleviates phytotoxicity caused by boron through augmented polyamine metabolism and antioxidant activities and reduced boron concentration in *Cucumis sativus* L. *Acta Physiol. Plant.* **2017**, *39*, 31. [CrossRef]
41. Monirul, I.M.; Humayun, K.M.; Mamun, A.N.K.; Monirul, I.; Pronabananda, D. Studies on yield and yield attributes in tomato and chilli using foliar application of oligo-chitosan. *GSC Biol. Pharm. Sci.* **2018**, *3*, 020–028.
42. Yang, J.; Kloepper, J.W.; Ryu, C.M. Rhizosphere bacteria help plants tolerate abiotic stress. *Trends Plant Sci.* **2009**, *14*, 1–4. [CrossRef]
43. Noel, T.C.; Sheng, C.; Yost, C.K.; Pharis, R.P.; Hynes, M.F. *Rhizobium leguminosarum* as a plant growth-promoting rhizobacterium: Direct growth promotion of canola and lettuce. *Can. J. Microbiol.* **1996**, *42*, 279–283. [CrossRef] [PubMed]
44. Mayak, S.; Tirosh, T.; Glick, B.R. Plant growth-promoting bacteria that confer resistance in tomato plants to salt stress. *Plant Physiol. Biochem.* **2004**, *42*, 565–572. [CrossRef]
45. Abdelaal, K.A.; Tawfik, S.F. Response of Sugar Beet Plant (*Beta vulgaris* L.) to Mineral Nitrogen Fertilization and Bio-Fertilizers. *Int. J. Curr. Microbiol. App. Sci.* **2015**, *4*, 677–688.
46. Abdelaal, K.A.; Badawy, S.A.; Abdel Aziz, R.M.; Neana, S.M.M. Effect of mineral nitrogen levels and biofertilizer on morphophysiological characters of three sweet sorghum varieties (*Sorghum bicolor* L. Moench). *J. Plant Prod. Mansoura Univ.* **2015**, *6*, 189–203. [CrossRef]
47. Abdelaal, K.A. Pivotal Role of Bio and Mineral Fertilizer Combinations on Morphological, Anatomical and Yield Characters of Sugar Beet Plant (*Beta vulgaris* L.). *Middle East J. Agric. Res.* **2015**, *4*, 717–734.
48. Hafez, Y.M.; Attia, K.; Kamel, S.; Alamery, S.; El-Gendyd, S.; Al-Doss, A.; Mehiar, F.; Ghazy, A.; Abdelaal, K.A. *Bacillus subtilis* as a bio-agent combined with nano molecules can control powdery mildew disease through histochemical and physiobiochemical changes in cucumber plants. *Physiol. Mol. Plant Pathol.* **2020**, *111*, 101489. [CrossRef]
49. Khan, N.; Bano, A.; Ali, S.; Babar, M.A. Crosstalk amongst phytohormones from planta and PGPR under biotic and abiotic stresses. *Plant Growth Regul.* **2020**, *90*, 189–203. [CrossRef]
50. Kousar, B.; Bano, A.; Khan, N. PGPR Modulation of Secondary Metabolites in Tomato Infested with *Spodoptera litura*. *Agronomy* **2020**, *10*, 778. [CrossRef]
51. Abou-Attia, F.A.M.; Abdelaal, K.A. Effect of Bio and Mineral fertilization on the main insect pests and some characters of sugar beet plants. *J. Agric. Sci. Mansoura Univ.* **2007**, *32*, 1471–1485.
52. Egamberdieva, D. Pseudomonas chlororaphis: A salt-tolerant bacterial inoculant for plant growth stimulation under saline soil conditions. *Acta Physiol. Plant.* **2012**, *34*, 751–756. [CrossRef]
53. Khan, N.; Bano, A.M.; Babar, A. Impacts of plant growth promoters and plant growth regulators on rainfed agriculture. *PLoS ONE* **2020**, *15*, e0231426. [CrossRef] [PubMed]
54. Zhao, S.; Zhou, N.; Zhao, Z.-Y.; Zhang, K.; Wu, G.-H.; Tian, C. Y. Isolation of endophytic plant growth-promoting bacteria associated with the halophyte *Salicornia europaea* and evaluation of their promoting activity under salt stress. *Curr. Microbiol.* **2016**, *73*, 574–581. [CrossRef] [PubMed]
55. Page, A.L. *Methods of Soil Analysis. Part 2, Chemical and Microbiological Properties*, 2nd ed.; Soil Science Society of America Inc.: Madison, WI, USA, 1982.
56. Sanchez, F.J.; de Andrés, E.F.; Tenorio, J.L.; Ayerbe, L. Growth of epicotyls, turgor maintenance and osmotic adjustment in pea plants (*Pisum sativum* L.) subjected to water stress. *Field Crop. Res.* **2004**, *86*, 81–90. [CrossRef]
57. Moran, R. Formulae for Determination of Chlorophyllous Pigments Extracted with N,N-Dimethylformamide 1. *Plant Physiol.* **1982**, *69*, 1376–1381. [CrossRef] [PubMed]

58. Szalai, G.; Janda, T.; Padi, E.; Szigeti, Z. Role of light in post-chilling symptoms in maize. *J. Plant Physiol.* **1996**, *148*, 378–383. [CrossRef]
59. Schreiber, U. *Pulse-Amplitude (RAM) Fluorometry and Saturation Pulse Method in Chlorophyll Fluorescence: A Signature of Photosynthesis*; Papageorgiou, G., Govindjee, Eds.; Springer: Dordrecht, The Netherlands, 2004; pp. 279–319.
60. Bates, L.S.; Waldren, R.P.; Teare, I.D. Rapid determination of free proline for water-stress studies. *Plant Soil* **1973**, *39*, 205–207. [CrossRef]
61. Davenport, S.B.; Gallego, S.M.; Benavides, M.P.; Tomaro, M.L. Behaviour of antioxidant defense system in the adaptive response to salt stress in *Helianthus annuus* L. cells. *Plant Growth Regul.* **2003**, *40*, 81–88. [CrossRef]
62. Huckelhoven, R.; Fodor, J.; Preis, C.; Kogel, K.H. Hypersensitive cell death and papilla formation in barley attacked by the powdery mildew fungus are associated with hydrogen peroxide but not with salicylic acid accumulation. *Plant Physiol.* **1999**, *1119*, 1251–1260. [CrossRef]
63. Hafez, Y.M.; Abdelaal, K.A.; Eid, M.E.; Mehiar, F.F. Morpho-physiological and biochemical responses of barley plants (*Hordeum vulgare* L.) against barley net blotch disease with application of non-traditional compounds and fungicides. *Egypt. J. Biol. Pest Cont.* **2016**, *26*, 261–268.
64. Aebi, H. Catalase in vitro. In *Methods in Enzymology; Oxygen Radicals in Biological Systems*; Academic Press: Cambridge, MA, USA, 1984; Volume 105, pp. 121–126.
65. Hammerschmidt, R.; Nuckles, E.M.; Kuć, J. Association of enhanced peroxidase activity with induced systemic resistance of cucumber to *Colletotrichum lagenarium*. *Physiol. Plant Pathol.* **1982**, *20*, 73–82. [CrossRef]
66. Mishra, N.P.; Mishra, R.K.; Singhal, G.S. Changes in the activities of anti-oxidant enzymes during exposure of intact what leaves to strong visible light at different temperatures in the presence of protein synthesis inhibitors. *Plant Physiol.* **1993**, *102*, 903–910. [CrossRef] [PubMed]
67. McCord, J.M.; Fridovich, I. Superoxide Dismutase: An enzymatic reaction for erythrocuprein (hemocuprein). *J. Biol. Chem.* **1969**, *244*, 6049–6055. [PubMed]
68. Goldberg, D.M.; Spooner, R.J. *Methods of Enzymatic Analysis*, 3rd ed.; Verlag Chemie: Weinheim, Germany, 1983; pp. 258–265.
69. Gomez, K.A.; Gomez, A.A. *Statistical Procedures for Agricultural Research*, 2nd ed.; Wiley Inter Science: New York City, NY, USA, 1984; pp. 1–690.
70. Duncan, B.D. Multiple ranges and multiple F-test. *Biometrics* **1955**, *11*, 1–42. [CrossRef]
71. Wang, Y.; Nil, N. Changes in chlorophyll, ribulose biphosphate carboxylase-oxygenase, glycine betaine content, photosynthesis and transpiration in *Amaranthus tricolor* leaves during salt stress. *J. Hortic. Sci. Biotechnol.* **2000**, *75*, 623–627. [CrossRef]
72. Parvin, K.; Hasanuzzaman, M.; Bhuyan, M.H.M.B.; Nahar, K.; Mohsin, S.M.; Fujita, M. Comparative Physiological and Biochemical Changes in Tomato (*Solanum lycopersicum* L.) under Salt Stress and Recovery: Role of Antioxidant Defense and Glyoxalase Systems. *Antioxidants* **2019**, *8*, 350. [CrossRef]
73. Wang, Y.; Li, K.; Li, X. Auxin redistribution modulates plastic development of root system architecture under salt stress in Arabidopsis thaliana. *J. Plant Physiol.* **2009**, *166*, 1637–1645. [CrossRef]
74. Naseem, H.; Ahsan, M.; Shahid, M.A.; Khan, N. Exopolysaccharides producing rhizobacteria and their role in plant growth and drought tolerance. *J. Basic. Microbiol.* **2018**, *58*, 1009–1022. [CrossRef]
75. Allakhverdiev, S.I.; Nishiyama, Y.; Miyairi, S.; Yamamoto, H.; Inagaki, N.; Kanesaki, Y.; Murata, N. Salt stress inhibits the repair of photodamaged photosystem II by suppressing the transcription and translation of psbA genes in Synechocystis. *Plant Physiol.* **2002**, *130*, 1443–1453. [CrossRef]
76. Khan, N.; Bano, A.; Curá, J.A. Role of Beneficial Microorganisms and Salicylic Acid in Improving Rainfed Agriculture and Future Food Safety. *Microorganisms* **2020**, *8*, 1018. [CrossRef]
77. Wang, W.; Wang, C.; Pan, D.; Zhang, Y.; Luo, B.; Ji, J. Effects of drought stress on photosynthesis and chlorophyll fluorescence images of soybean (*Glycine max*) seedlings. *Int. J. Agric. Biol. Eng.* **2018**, *11*, 196–201. [CrossRef]
78. Tavakkoli, E.; Rengasamy, P.; McDonald, G. High concentrations of Na^+ and Cl^- ions in soil solution have simultaneous detrimental effects on growth of faba bean under salinity stress. *J. Exp. Bot.* **2010**, *61*, 4449–4459. [CrossRef] [PubMed]
79. Asrar, H.; Hussain, T.; Midhat, S.; Hadi, S.; Gul, B.; Nielsen, B.L.; Khan, M.A. Salinity induced changes in light harvesting and carbon assimilating complexes of *Desmostachya bipinnata* (L.) staph. *Environ. Exp. Bot.* **2017**, *135*, 86–95. [CrossRef]

80. Shah, G.; Jan, M.; Afreen, M.; Anees, M.; Rehman, S.; Daud, M.K.; Malook, I.; Jamil, M. Halophilic bacteria mediated phytoremediation of salt-affected soils cultivated with rice. *J. Geochem. Explor.* **2017**, *174*, 59–65. [CrossRef]
81. Possingham, J.V. Plastid replication and development in the life cycle of higher plants. *Annu. Rev. Plant Physiol.* **1980**, *31*, 113–129. [CrossRef]
82. Plazek, A.; Rapacz, M.; Hura, K. Relationship between quantum efficiency of PSII and cold-induced plant resistance to fungal pathogens. *Acta Physiol. Plant.* **2004**, *26*, 141–148.
83. Mehta, P.; Jajoo, A.; Mathur, M.; Bharti, S. Chlorophyll a fluorescence study revealing effects of high salt stress on photosystem II in wheat leaves. *Plant Physiol. Biochem.* **2010**, *48*, 16–20. [CrossRef]
84. Kalaji, H.; Rastogi, A.; Živčgk, M.; Brestic, M. Prompt chlorophyll fluorescence as a tool for crop phenotyping: An example of barley landraces exposed to various abiotic stress factors. *Photosynthetica* **2018**. [CrossRef]
85. Yang, D.L.; Lian, J.R.; Ping, C.X.; Wei, L. Quantitative Trait Loci Mapping for Chlorophyll Fluorescence and Associated Traits in Wheat (*Triticum aestivum*). *J. Integr. Plant Biol.* **2007**, *49*, 646–654. [CrossRef]
86. Li, Z.; Zhang, Y.; Zhang, X.; Merewitz, E.; Peng, Y.; Ma, X.; Yan, Y. Metabolic pathways regulated by chitosan contributing to drought resistance in white clover. *J. Proteome Res.* **2017**, *16*, 3039–3052. [CrossRef]
87. Khan, N.; Bano, A.; Rahman, M.A.; Rathinasabapathi, B.; Babar, M.A. UPLC-HRMS-based untargeted metabolic profiling reveals changes in chickpea (*Cicer arietinum*) metabolome following long-term drought stress. *Plant Cell Environ.* **2019**, *42*, 115–132. [CrossRef]
88. Egamberdieva, D.; Davranov, K.; Wirth, S.; Hashem, A.; Abd_Allah, E.F. Impact of soil salinity on the plant-growth–promoting and biological control abilities of root associated bacteria. *Saudi J. Biol. Sci.* **2017**, *24*, 1601–1608. [CrossRef] [PubMed]
89. Prashanth, H.K.V.; Dharmesh, S.M.; Rao, K.S.; Tharanathan, R.N. Free radical-induced chitosan depolymerized products protect calf thymus DNA from oxidative damage. *Carbohydr. Res.* **2007**, *342*, 190–195. [CrossRef] [PubMed]
90. Malerba, M.; Cerana, R. Chitosan Effects on Plant Systems. *Int. J. Mol. Sci.* **2016**, *23*, 996. [CrossRef] [PubMed]
91. Sharma, I.P.; Sharma, A.K. Physiological and biochemical changes in tomato cultivar PT-3 with dual inoculation of mycorrhiza and PGPR against root-knot nematode. *Symbiosis* **2017**, *71*, 175–183. [CrossRef]
92. Bianco, C.; Defez, R. Medicago truncatula improves salt tolerance when nodulated by an indole-3-acetic acid-overproducing *Sinorhizobium meliloti* strain. *J. Exp. Bot.* **2009**, *60*, 3097–3107. [CrossRef]
93. Foyer, C.H.; Descourvieres, P.; Kunert, K.J. Protection against oxygen radicals: An important defense mechanism studied in transgenic plants. *Plant Cell Environ.* **1994**, *17*, 507–523. [CrossRef]
94. Kohler, J.; Caravaca, F.; Roldàn, A. An AM fungus and a PGPR intensify the adverse effects of salinity on the stability of rhizosphere soil aggregates of *Lactuca sativa*. *Soil Biol. Biochem.* **2010**, *42*, 429–434. [CrossRef]
95. Hu, Y.; Yu, W.; Liu, T.; Shafi, M.; Song, L.; Du, X.; Huang, X.; Yue, Y.; Wu, J. Effects of paclobutrazol on cultivars of Chinese bayberry (*Myrica rubra*) under salinity stress. *Photosynthetica* **2017**, *55*, 443–453. [CrossRef]
96. Hayat, R.; Ali, S.; Amara, U.; Khalid, R.; Ahmed, I. Soil beneficial bacteria and their role in plant growth promotion: A review. *Ann. Microbiol.* **2010**, *60*, 579–598. [CrossRef]

6

An Assessment of Seaweed Extracts: Innovation for Sustainable Agriculture

El Chami Daniel and Galli Fabio *

Timac Agro Italia S.p.A., S.P.13—Località Ca' Nova, I-26010 Ripalta Arpina (CR), Italy; daniel.elchami@roullier.com

* Correspondence: fabio.galli@roullier.com.

Abstract: Plant growth regulators (PGRs) are described in the literature as having a significant role in securing crop management of modern agriculture in conditions of abiotic and biotic stressors. A joint field experiment was carried out to assess the role of seaweed-based extracts in pear trees and to test the "less for more" theory, which consists of getting more and better agricultural produce using fewer innovative inputs. The trials took place on two production seasons (from March till September 2018–2019) and the selected case study was on a pear orchard (*Pyrus communis* L. cv. Abate Fètel) in Emilia Romagna (Italy) by Fondazione Navarra and Timac Agro Italia S.p.A. Results demonstrate that, depending on the yearly climate conditions, it was possible to substantially reduce the primary nutrients by 35–46% and total fertilisation units applied by 13% and significantly improve quantitative and qualitative production indicators (average weight of fruits (5%) and total yield (19–55%)). Results also confirm a positive correlation between plant growth regulators and agronomic efficiency of pears which increased between five and nine times compared to the conventional nutrition programme. These outcomes constitute scientific evidence for decision making in farm management.

Keywords: pear trees; PGR; sustainable development; crop nutrition; fertiliser; Timac Agro Italia

---

1. Introduction

For decades, plant nutrition has been under scrutiny for concerns about negative externalities generated by the use of fertilisers in agriculture, which emerged in the late 1960s [1]. Since then, a clear correlation has been found between plant nutrition, the eutrophication of surface water, the accumulation of nitrate in water bodies and energy consumption. Even more recently, global studies have warned about unprecedented nitrate contamination of water [2], which is creating direct irreversible damage to natural ecosystems and human health [3]. Further, the most universal form of deteriorated water quality in the world in recent decades is freshwater eutrophication from phosphorus loss [4,5].

Looking at the glass half-full, the importance of fertilisers in agriculture has been extensively documented in the literature for over 150 years of research and experiments. The relevance of plant nutrition is fundamental for (i) normal growth and reproduction of crops [6], (ii) average crop yield increase [7] and (iii) improving soil fertility [8]. However, the fertilising rates have reached the optimum in the developed world, and the new direction is to reduce them. This has been one of the European Green Deal recommendations, for example, as expressed by the "farm to fork" strategy (The Farm to Fork (F2F) Strategy is at the heart of the European Green Deal set out in 2019 to make Europe the first climate-neutral continent by 2050. The strategy comprehensively addresses the challenges of sustainable food systems and recognises the inextricable links between healthy people, healthy societies and a healthy planet.) with a target of diminishing nutrient losses by at least 50% and reducing fertiliser use by at least 20% by 2030 [9].

The focus of scientific innovation is currently on crop biostimulants to activate natural plant processes, which, according to the documented literature, improve nutrient uptake and efficiency, crop quality and yield and build plant tolerance to abiotic and biotic stressors [10–12]. A statutory definition of biostimulants was provided in 2018 by the primary agricultural and food policy tool of the United States federal government (Farm Bill: https://www.congress.gov/115/bills/hr2/BILLS-115hr2enr.pdf). This definition is consistent with the one currently proposed by the European Bio-stimulant Industry Council (EBIC) (http://www.bio-stimulants.eu/) and in line with the definition under review by the European Union in the context of revising the existing EU regulation (EC) No. 2019/1009 relating to fertilisers (https://eur-lex.europa.eu/legal-content/EN/TXT/HTML/?uri=OJ:L:2019:170:FULL&from=EN).

The definition sums up the scientific aspects raised in the literature and describes a plant biostimulant as "a substance or micro-organism that, when applied to seeds, plants, or the rhizosphere, stimulates natural processes to enhance or benefit nutrient uptake, nutrient efficiency, tolerance to abiotic stress, or crop quality and yield".

Nevertheless, du Jardin [13] identified in a review study seven categories of biostimulants: (i) humic and fulvic acids, (ii) protein hydrolysates and other N-containing compounds, (iii) seaweed extracts and botanicals, (iv) chitosan and other biopolymers, (v) inorganic compounds, (vi) beneficial fungi and (vii) beneficial bacteria. This emerging field of research is very promising and represents one of the fundamental management aspects of agro-systems to reach sustainable agriculture that is more resilient to climate change and able to feed the increasing population [14].

Therefore, the literature still needs to explore different research aspects related to the biostimulant categories and their use in agriculture to answer evolving enquiries that arise with the technological advances in this field. Recently, algae have proved to contain natural active compounds with biostimulation and/or bioregulation effects [15], e.g., phytohormones, hormone-like substances, vitamins, antibiotics, amino acids, and primary, secondary and micro-nutrients.

Even though several forms of applying algal constituents have been reported in the literature, algal extracts from seaweed have proved to be the most efficient in terms of growth enhancement and stress tolerance [16–18]. Indeed, Mutale-joan et al. [16] have tested 15 different Crude Bio-Extracts (CBEs) obtained from acid hydrolysis of microalgae on tomato plant growth, chlorophyll content, nutrient uptake and metabolite profile. The authors have recorded positive effects on plant development, particularly, significant root and shoot length improvement and increased nutrients uptake. Further, Shukla et al. [17] have reviewed the ability of Ascophyllum Nodosum Extracts (ANE) to improve plant growth and agricultural productivity and have confirmed the plant growth promotion, the improvement of root/microbe interactions and nutrient use efficiency in plants and enhancement of plant tolerance to abiotic and/or biotic stresses. Finally, Michalak et al. [18] have successfully tested different seaweed extracts to enhance carotenoid and chlorophyll content in plant shoots and develop root thickness and above-ground biomass.

In this context, this paper proposes to explore the category of seaweed extracts produced by Timac Agro Italia, the Italian holding of the French multinational Groupe Roullier, a world leader in the field of plant nutrition, with the largest private research centre in Europe dedicated to plant physiology and nutrition and investing in these technologies. The selection of the trial crop also has significance, because pears are one of the major fruits of temperate climates, grown in almost all four corners of the world, reaching a total harvested area of 1.5 million hectares in 2018 and over 23.5 million tons of production according to FAOSTAT [19] (Figure 1). The tree belongs to two species: the common pear cultivated mainly in Europe, the Near East, America and Australia and known as the European pear (*Pyrus communis L.*) given its European descendants and the Nashi or Oriental pear (*Pyrus pyrifolia*) widely grown in Asia.

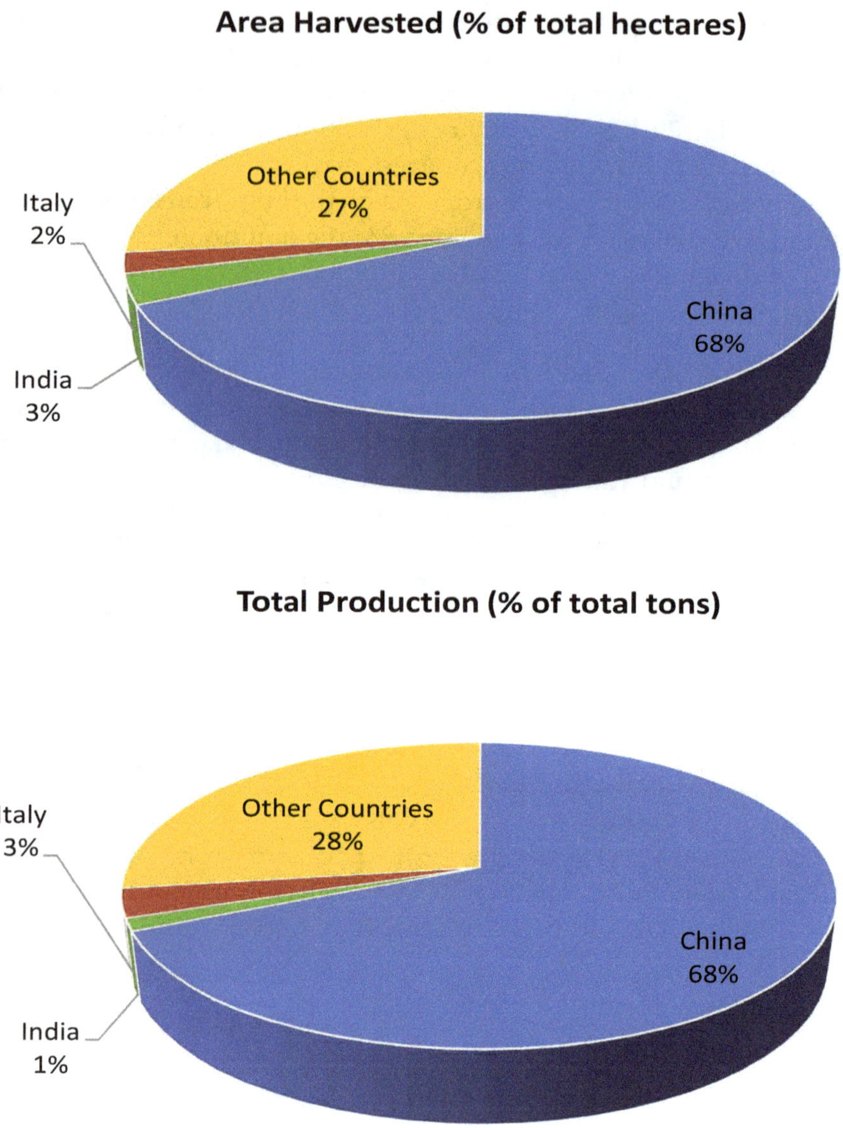

Figure 1. Harvested area of pear (**top**) and total production (**bottom**) in the top 3 countries and the world.

2. Materials and Methods

The experiment was carried out in Emilia Romagna (Italy) at an Abate Fetel orchard for the relevance of this cultivar in Italy, which happens to be the main producer of pears in Europe [20], the third producer in the world in terms of area harvest and the second after China in terms of total production.

There are over 3000 identified pear cultivars worldwide [21], and in Italy, Abate Fetel (also known as Abbé Fetel) and 3 others (Conference, Beurrè Bosc, Doyenne du Comice) are the major cultivars commercially grown, providing more than 70% of the total annual production [22]. Further, the selection of Emilia Romagna has local importance, given that this region, in terms of fruit trees, is the first ranked in harvest area, production and the average size of farms (Table 1).

Table 1. Top 5 regions in fruit production and farm size in Italy.

Region	Area Harvested		Number of Farms	
	(ha)	(%)	(N)	(%)
Emilia Romagna (EMR)	67,454.3	15.9	18,355	7.8
Campania (CAM)	58,836.7	13.9	32,133	13.6
Sicily (SIC)	54,295.5	12.8	36,055	15.3
Piedmont (PIE)	43,673.3	10.3	20,168	8.5
Lazio (LAZ)	36,318.8	8.6	15,323	6.5
Sum of top 5 regions	**260,578.5**	**61.4**	**122,034**	**51.7**
Other regions	**163,725**	**38.6**	**114,206**	**48.3**
Total Italian fruit farms	**424,303.5**	**100**	**236,240**	**100**
Average Italian fruit farm size	**1.8 ha**			
Average fruit farm size in EMR	**3.7 ha**			
Farm size	**Area Harvested**		**Number of Farms**	
	(ha)	(%)	(N)	(%)
Small (<10 ha)	222,270.4	52.4	201,324	85.2
Medium (10–50 ha)	150,171.9	35.4	30,674	13.0
Large (>50 ha)	51,861.5	12.2	4242	1.8
Total Italian fruit farms	**424,303.8**	**100**	**236,240**	**100**

Source: Istat [23]

2.1. Case Study

The experiment took place at the experimental field of the Navarra Foundation, a reference of agricultural knowledge for the Navarra agricultural technical institute and farmers of the north-east of Italy, given its contribution to the development of the region's agri-food sector through research, experiments, innovation and knowledge transfer.

The experimental field has a total area of ≈2.5 ha, similar to the average size of fruit tree farms in the area, and is located in Ferrara (Table 2), characterised by a warm and temperate climate classified as Cfa by the Köppen-Geiger system. The historic precipitation and temperature measured at a weather station in Ferrara (Table 2) between 1961 and 1990 revealed a yearly average temperature of 13.1 °C and rainfall of around 689.5 mm [24], with considerable rain at high temperatures in the driest months (Figure 2).

The trials took place over two consecutive seasons in 2018–2019 on a V-shaped orchard system planted in 2005 using 3.8 m spacing between rows and variable in-row spacing of 0.5 m, with a tree density of 5263 trees per hectare. The orchard was evenly irrigated with a drip system without any variation between the rows and was covered with black anti-hail netting.

The soil structure is silty clay loam according to the classification system of the United States Department of Agriculture (USDA) and silt clay according to the International Soil Sciences Society (ISSS). The general composition of the soil is about 60% silt, 30% clay and 10% sand; it presents low compaction risk, high fertility indicators (organic matter 2.21%, C/N = 8.87) and a high content of available nutrients, given that the field is experimental with continuous trials carried out yearly. Soil tests were carried out before and during the experiment to guide the definitions of annual fertilisation programmes.

Table 2. Geospatial coordinates of experimental field and weather station of reference.

Location	Latitude	Longitude	Altitude
Experimental field	44.857 N	11.653 E	5 m
Weather station	44.861 N	11.656 E	4 m

Source: [24,25].

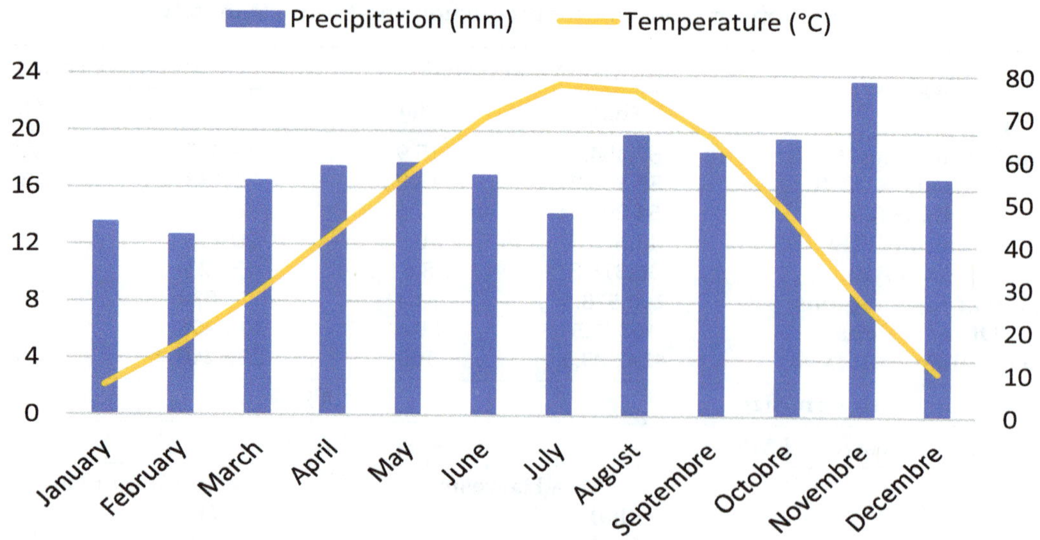

Figure 2. Historic average of precipitation and temperature in Ferrara (1961–1990) (after [24]).

The fertilisation programme was divided into 3 treatments: control, grown without any fertilisation; conventional treatment (CF), representing empirical nutritional treatment (primary, secondary and micro-nutrients and organic matter of animal or vegetable origin) conceived from the available products in the region (to simulate a conventional nutritional programme); and Timac Agro treatment (TIMAC), corresponding to a programme based on integrating conventional nutrients (primary, secondary and micro-nutrients and organic matter of animal or vegetable origin) with cutting-edge technologies created to reduce the environmental burden of fertilisers, increase the profit of farms and improve their well-being. The total fertilisation units per hectare for each treatment are reported in Table 3, which shows a great difference in the quantity of fertiliser applied in the 2 years. This difference is mainly due to the general climate conditions during the year, which considerably determines the quality and quantity of agricultural yield.

Table 3. Fertilisation units (FU) applied per treatment and hectare in 2018 and 2019 seasons.

Type	Element	2018 FU ha^{-1}			2019 FU ha^{-1}		
		CF	TIMAC	*TIMAC/CF*	CF	TIMAC	*TIMAC/CF*
Primary Nutrients	Nitrogen (N)	205.1	141.3	**68.9%**	174.9	120.7	**69.0%**
	Phosphorus (P_2O_5)	184.4	79.0	**42.8%**	103.1	77.6	**75.3%**
	Potassium (K_2O)	292.7	145.1	**49.6%**	246.0	140.8	**57.2%**
	Total Primary Nutrients	**682.2**	**365.4**	**53.6%**	**524**	**339.1**	**64.7%**
Secondary Nutrients	Calcium (CaO)	42.5	46.4	**109.2%**	4.8	43.2	**900.0%**
	Magnesium (MgO)	21.6	24.1	**111.6%**	3.0	35.8	**1193.3%**
	Sulphur (SO_3)	109	200.2	**183.7%**	49.5	197.6	**399.2%**
	Total Secondary Nutrients	**173.1**	**270.7**	**156.4%**	**57.3**	**276.6**	**482.7%**
Micro-Nutrients	Boron (B)	0.45	0.88	**197.5%**	0	1.45	–
	Copper (Cu)	0.43	0.06	**13.9%**	0	0.05	–
	Iron (Fe)	5.83	3.90	**66.9%**	2.25	1.50	**66.7%**
	Manganese (Mn)	0.24	0	–	0.03	0.07	**233.3%**
	Molybdenum (Mo)	0.01	0.04	**655.7%**	0	0.30	–
	Zinc (Zn)	0.31	0.10	**32.3%**	0.02	0.11	**550.0%**
	Total Micro-Nutrients	**7.26**	**4.98**	**68.6%**	**2.30**	**3.48**	**151.3%**
OM	**Total Organic Matter**	**43.8**	**44.4**	**101.3%**	**48.7**	**41.2**	**106.1%**

A supplement of complexed seaweed-based extracts was added to the TIMAC treatment (58 L ha^{-1}) in different growth stages and concentrations (Table 4) of three technologies: Fertiactyl®, NMX®, Seactiv®. These technologies are registered in the European Patent Office (EPO) under the numbers EP0609168, EP1147706 and EP0855375, respectively. The species involved in the extraction are *Lithothamnion corallioides, Lithothamnion glaciale, Lithothamnion tophiforme* and *Phymatolithon calcareum* and the general composition of the technologies are as follows:

Table 4. Quantity of seaweed-based technologies applied in Timac Agro treatment (TIMAC) treatment and corresponding growth stage.

Technology	Quantity (L ha^{-1})	Growth Stage
Fertiactyl®	8	Vegetative growth
NMX®	3	
Seactiv®	3	
NMX®	6	Fruit set
Seactiv®	6	
Seactiv®	3	Post-harvest

Fertiactyl®: based on organic substrates acids whose fulvic and humic acids have been mobilized through the formation of soluble inorganic salts, phenolic and polyphenolic acids and zeatin;

NMX®: based on precursor, inhibitor and simulating enzymes (e.g., precursor compound of cyclic AMP; inhibitor compound of the enzymes of the Phosphodiesterases, stimulating compound of the enzymes of the Adenyl-Cyclase, etc.), mixed with mineral fertilisers and plant growth regulators (PGRs) (e.g., auxins, cytokinins, gibberellins, n-ethanolamines, polyamines, sugars, etc.);

Seactiv®: based on adenine derivatives to help, as foliar treatment agent, the migration and distribution within plants of nitrogenous nutrients and/or micro-nutrients.

The concentration in the different stages did not vary between 2018 and 2019.

2.2. Statistical Analysis

The experimental design was a randomised block design to minimise the effects of systematic errors. This design consisted of dividing the experimental block into 3 fertilisation treatments randomly selected within the block, with 2 replicates of 5 trees for each treatment. In total, 60 trees were used for data collection (all performed manually) and statistical analysis to determine whether mean scores differed significantly across the treatments. The measurements performed were divided into 3 pillars as follows:

Total harvest (t ha^{-1});

Average fruit weight (g);

$$\text{Flower density} = \text{FBT} = \text{Number of Floral Buds per Tree} \tag{1}$$

$$\text{Fruit density} = \text{NFT} = \text{Number of Fruits per Tree} \tag{2}$$

$$\text{Fruit set} = \text{FS} = (\text{NFT}/\text{FBT}) \times 100 \tag{3}$$

$$\begin{aligned}&\text{Agronomic efficiency AE}\left(\text{kg kg}^{-1}\right)\\&= \text{Yield}_{\text{fertilised}}\left(\text{kg ha}^{-1}\right) - \text{Yield}_{\text{not fertilised}}\left(\text{kg ha}^{-1}\right)/\text{N Applied}\left(\text{kg ha}^{-1}\right)\end{aligned} \tag{4}$$

The collected field data were statistically examined, separately for each year, using analysis of variance (one-way ANOVA) with statistical probability ($p - value \leq 0.05$) and Tukey's honestly significant difference (HSD) test, which is a single-step multiple comparison procedure to find significantly different means. An excel spreadsheet has been used for this purpose.

The assumptions of both tests are essentially the same and they are three: normality (experimental errors of the data are normally distributed), homogeneity (equal variances between treatments) and independence (each sample is randomly selected and independent).

3. Results

Plants within a population often vary in the numbers of open flowers and fruits. The correlation between these two indicators is calculated by the fruit set, a ratio defined as the transition from flower to young fruit. These quantitative indicators in the development process of any plant are correlated to the rate of pollination [26] and determine the final yield quantities (or total harvest).

Field data for two consecutive years demonstrated an increase in all quantitative indicators under the TIMAC treatment compared to the conventional treatment and the control, which generated the highest harvest for TIMAC treatment. The literature has mentioned different abiotic and biotic stressors which could lead to flower drop in the early season [27]. In this experiment, the difference between the two years of trials is substantial and can be explained by the yearly climate variability in Italy. Indeed, 2019 was a dry winter with many clear-sky mornings generating diamond dust and leading to frost flowers, this generated a lower flower density trend in 2019 compared to 2018. Yet, TIMAC treatment has participated in avoiding further damage with higher flower density compared to the other treatment.

Even though the numerical difference is considerable, statistical significance is present only between the control and TIMAC treatments. Complete statistical results are listed in a final table (Table 5).

Table 5. Statistical results of selected indicators.

Indicator	Treatment	First Year, 2018			Second Year, 2019		
		Mean	Std Dev.	Variance	Mean	Std Dev.	Variance
Flower Density	Control	223.5 [a]	25.4	681.1	61.0 [a]	33.9	1211.0
	CF	219.4 [a]	28.7	865.8	53.1 [a]	28.3	844.6
	TIMAC	223.6 [a]	27.7	807.6	74.0 [a]	36.3	1384.2
Fruit Density	Control	38.2 [a]	10.5	115.4	11.1 [a]	6.1	39.8
	CF	42.4 [a,b]	18.1	343.2	13.9 [a]	9.4	93.7
	TIMAC	50.4 [b]	17.1	308.4	25.1 [b]	11.4	136.2
Fruit Set	Control	17.4 [a]	5.3	29.4	25.2 [a]	22.9	551.6
	CF	20.1 [a]	10.2	109.5	30.1 [a]	17.7	328.1
	TIMAC	22.3 [a]	6.2	40.9	37.9 [a]	18.7	368.0
Fruit Weight	Control	238.9 [a]	29.0	882.9	225.2 [a]	25.5	684.6
	CF	237.1 [a]	31.8	1064.4	228.6 [a]	24.7	643.9
	TIMAC	242.7 [a]	24.5	633.8	230.6 [a]	30.1	952.4
Total Harvest	Control	47.5 [a]	13.1	180.7	13.5 [a]	8.1	69.7
	CF	51.6 [a,b]	19.8	413.5	16.1 [a]	9.9	104.0
	TIMAC	63.3 [b]	19.1	384.6	29.8 [b]	12.7	169.3

[a,b] Letters indicate statistically significant means of different indicators.

However, high plant fertility would negatively affect the quality of the harvest, which, in conventional agriculture, would require flower removal intervention to boost the quality [28–30]. The role of plant regulators has been limited mainly to this growth stage of plants to control fruit set [31] and simultaneously boost the quantity and quality of yield [32].

While the average fruit weight of different pear cultivars can vary according to the genetic characteristics [33], within the same variety, fruit fresh weight is considered to be one of the most important quality indicators [34] determining the market value of the harvest. The effect of climate variability between 2018 and 2019, which initially generated a reduction in fruit density, led to a drop in the total harvest; however, the yield gap between the two treatments (CF vs. TIMAC) confirms the role of plant growth regulators (PGRs) in reducing crop response to abiotic stress. Further, fruit quality did not undergo any variation. Figure 3 shows the qualitative and quantitative improvement of yield in the TIMAC treatment compared to the control and CF.

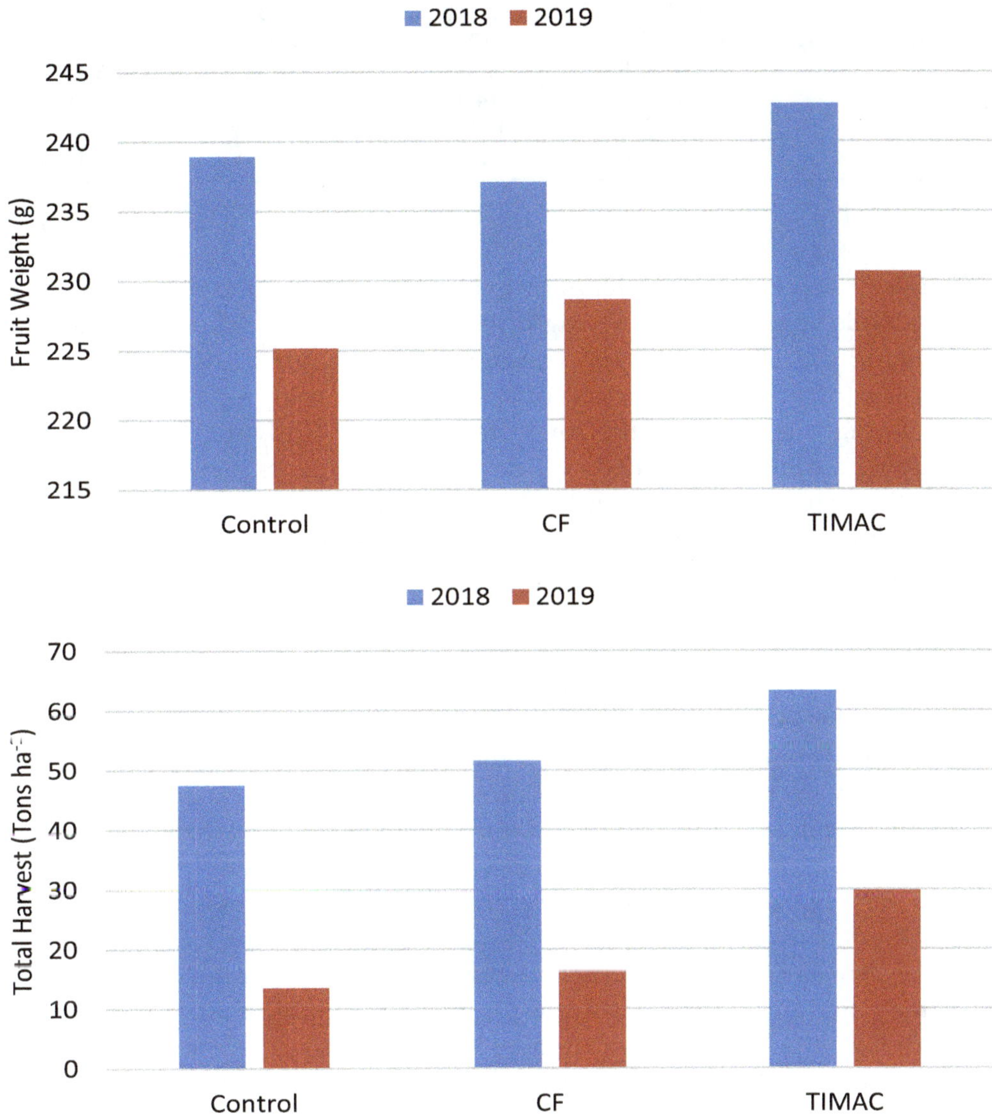

Figure 3. Improvement of fruit weight (**top**) and total harvest (**bottom**) under TIMAC treatment.

The outcomes of this experiment confirm the results of An et al. [31], showing a positive correlation between phytohormones application (auxin and ethylene) and fruit quantity and quality, and Bons and Kaur [32], who reviewed the effects of different PGRs on fruit set. The review included the impacts of 6-Benzyladenine (BA), 6-Benzylaminopurine (BAP), Gibberellic acid (GA3) and Naphthalene acetic acid (NAA).

Nutrient agronomic efficiency measures the technical performance of a crop and is calculated based on the yield difference between fertilised and unfertilised treatments, divided by the fertiliser units applied. Specifically, AE estimates productivity improvement gained by the use of nutrient input. Initially, it was used to evaluate nitrogen performance [35,36] and to improve the environmental and economic performance of agriculture, then it was extended to include the performance of phosphorus nutrition [37] to have a broader meaning and use agronomic efficiency correlated inputs for agro-system performance as an indicator of a transition to sustainable agriculture [38].

In this study, we calculated nitrogen, phosphorus and total nutrient efficiency for conventional fertilisation and TIMAC treatments (Table 6). The results show that the efficiency of the TIMAC treatment varied from 5.18 to 9.37 times higher than the conventional treatment (CF).

Table 6. Nutrient agronomic efficiency of different treatments during experimental years.

Year	Treatment	$AE_{(N)}$	$AE_{(P_2O_5)}$	$AE_{(Total)}$
	CF	20.1	22.3	4.8
2018	TIMAC	112.1	200.4	24.7
	TIMAC/CF	**5.58**	**8.98**	**5.18**
	CF	14.4	24.4	4.3
2019	TIMAC	134.9	209.8	26.3
	TIMAC/CF	**9.37**	**8.59**	**6.10**

4. Discussion

The results over two consecutive years of experiments show the role of PGRs in increasing crop tolerance to abiotic stress and improving physiological activities such as nutrient uptake and assimilation, reducing the total fertilisation units (FU) to around 13% (≈35.6 FU), which is an encouraging outcome towards the reduction of fertilisers according to the European Farm-to-Fork Strategy. Extrapolating this result to the total area of pears cultivated worldwide (≈1.4 million ha), this means saving a minimum of 49.8 million fertilisation units annually, without considering overfertilisation still practised in many countries around the world.

The results also reveal a substantial reduction in P_2O_5 use (over 45%), which is a significant result impacting the AE of phosphorus, and global efforts to reduce and/or substitute the use of phosphate rock, a mineral fundamental for food security, is expected to end in a short time [39,40]. The agronomic efficiency was higher in the second experimental, which presented higher abiotic stress due to climate conditions, affecting total yield. This confirms studies describing the important role of biostimulants as abiotic stress alleviators [41,42].

The field experiment confirmed the reviewed literature in Bons and Kaur [32], which assessed the positive correlation between plant growth regulators and the quality and quantity of harvests, as the TIMAC treatment improved both the quality and quantity of pears. Therefore, these results disproved the results of Dicenta et al. [43], which did not show a correlation between fruit set and total harvest.

There is not a consensus in the scientific community on one process to explain this because the mechanisms underlining seaweed extract-induced stimulation are still not completely revealed, though, several factors can be attributed to their activity. Literature has demonstrated that components within seaweed extracts may modulate innate pathways for the biosynthesis of phytohormones in plants [44]. It is also proved that marine bioactive substances (IPA extracts) enhance nutrient flux in [45] and increase endogenous antioxidant activity in plants [46]. Furthermore, the presence of phytohormones in seaweed extracts, as regulators of various cellular processes and responses, is presumably another factor that may have improved metabolic activity of crops [47] leading to better nutrition and yields.

Some questions that the research has raised and some future recommendations are mainly related to the importance of a balanced nutrition programme for sustainable management of crops. This can be defined by Liebig's law of the minimum, which is a fundamental principle in plant nutrition. This research partially demonstrates the importance of this law to the overall agronomic efficiency of crops (AE was not assessed in this study). Furthermore, it is recommended to follow the framework suggested by El Chami et al. [14], who proposed a methodology to reached sustainable agro-systems based on a life cycle study [48]; future studies will be intensified and address these questions, and will implement that methodology towards fulfilling the European Farm-to-Fork Strategy and the United Nations Sustainable Development Goals.

5. Conclusions

In conclusion, the experiment has confirmed the role of PGRs in reducing the total fertilisation units by 13% while improving the harvest between 19 and 55% and the quality of the fruit by about 5%. This improvement has positively affected the agronomic efficiency from five to nine times compared to the conventional nutrition programme. PGRs are, indeed, the precursors of a new agricultural revolution which will transform the sector to sustainably face the present and future challenges of humanity. Thus, trials will continue to explore all aspects raised in the discussion to add applicable high-impact scientific evidence about the agronomic performance of PGRs.

Author Contributions: Conceptualisation, methodology, analysis and writing, E.C.D. Experiment and data curation G.F. All authors have read and agreed to the published version of the manuscript.

Acknowledgments: The authors would like to thank the Navarra Foundation, with special recognition of Alessandro Zago and Michele Mariani for their support in the experimental field.

References

1. Viets, F.G.; Lunin, J. The environmental impact of fertilizers. *Crit. Rev. Env. Control* **1975**, *5*, 423–453. [CrossRef]
2. Zhou, Z. *A Global Assessment of Nitrate Contamination in Groundwater*; Internship Report; Wageningen University: Wageningen, The Netherlands, 2015.
3. Ward, M.H.; Jones, R.R.; Brender, J.D.; de Kok, T.M.; Weyer, P.J.; Nolan, B.T.; Villaneuva, C.M.; van Breda, S.G. Drinking Water Nitrate and Human Health: An Updated Review. *Int. J. Environ. Res. Public Health* **2018**, *15*, 1557. [CrossRef] [PubMed]
4. MacDonald, G.K.; Bennett, E.M.; Potter, P.A.; Ramankutty, N. Agronomic phosphorus imbalances across the world's croplands. *Proc. Natl. Acad. Sci. USA* **2011**, *108*, 3086–3091. [CrossRef] [PubMed]
5. Sharpley, A.N.; McDowell, R.W.; Kleinman, P.J.A. Phosphorus loss from land to water: Integrating agricultural and environmental management. *Plant Soil* **2001**, *237*, 287–307. [CrossRef]
6. Barker, A.V.; Pilbeam, D.J. Introduction. In *Handbook of Plant Nutrition*, 2nd ed.; Barker, A.V., Pilbeam, D.J., Eds.; CRC Press: Boca Raton, FL, USA, 2007; pp. 3–18.
7. Colla, G.; Cardarelli, M.; Bonini, P.; Rouphael, Y. Foliar applications of protein hydrolysate, plant and seaweed extracts increase yield but differentially modulate fruit quality of greenhouse tomato. *Hortscience* **2017**, *52*, 1214–1220. [CrossRef]
8. Dong, W.; Zhang, X.; Wang, H.; Dai, X.; Sun, X.; Qiu, W.; Yang, F. Effect of Different Fertilizer Application on the Soil Fertility of Paddy Soils in Red Soil Region of Southern China. *PLoS ONE* **2012**, *7*, e44504. [CrossRef] [PubMed]
9. EU. *Farm to Fork Strategy: For a Fair, Healthy and Environmentally-Friendly Food System*; EU: Brussels, Belgium, 2020; p. 22. Available online: https://ec.europa.eu/food/sites/food/files/safety/docs/f2f_action-plan_2020_strategy-info_en.pdf (accessed on 12 June 2020).
10. Carillo, P.; Ciarmiello, L.F.; Woodrow, P.; Corrado, G.; Chiaiese, P.; Rouphael, Y. Enhancing sustainability by improving plant salt tolerance through macro- and micro-algal biostimulants. *Biology* **2020**, *9*, 253. [CrossRef]
11. Drobek, M.; Frąc, M.; Cybulska, J. Plant Bio-stimulants: Importance of the Quality and Yield of Horticultural Crops and the Improvement of Plant Tolerance to Abiotic Stress—A Review. *Agronomy (Basel)* **2019**, *9*, 335.
12. Battacharyya, D.; Zamani, M.; Babgohari, M.Z.; Rathor, P.; Prithiviraj, B. Seaweed extracts as biostimulants in horticulture. *Sci. Hortic.* **2015**, *196*, 39–48. [CrossRef]
13. Du Jardin, P. Plant bio-stimulants: Definition, concept, main categories and regulation. *Sci. Hortic.* **2015**, *196*, 3–14. [CrossRef]
14. El Chami, D.; Daccache, A.; El Moujabber, M. How can sustainable agriculture increase climate resilience? A systematic review. *Sustainability (Basel)* **2020**, *12*, 3119. [CrossRef]
15. Michalak, I.; Chojnacka, K.; Saeid, A. Plant Growth Biostimulants, Dietary Feed Supplements and Cosmetics Formulated with Supercritical CO_2 Algal Extracts. *Molecules (Basel)* **2017**, *22*, 66. [CrossRef] [PubMed]
16. Mutale-joan, C.; Redouane, B.; Elmernissi, N.; Kasmi, Y.; Lyamlouli, K.; Sbabou, L.; Zeroual, Y.; El Arroussi, H. Screening of microalgae liquid extracts for their bio-stimulant properties on plant growth, nutrient uptake and metabolite profile of *Solanum lycopersicum* L. *Sci. Rep.* **2020**, *10*, 2820. [CrossRef] [PubMed]
17. Shukla, P.S.; Mantin, E.G.; Adil, M.; Bajpai, S.; Critchley, A.T.; Prithiviraj, B. Ascophyllum nodosum-Based Biostimulants: Sustainable Applications in Agriculture for the Stimulation of Plant Growth, Stress Tolerance, and Disease Management. *Front. Plant Sci.* **2019**, *29*, 655. [CrossRef]

18. Michalak, I.; Górka, B.; Wieczorek, P.P.; Rój, E.; Lipok, J.; Łęska, B.; Messyasz, B.; Wilk, R.; Schroeder, G.; Dobrzyńska-Inger, A.; et al. Supercritical fluid extraction of algae enhances levels of biologically active compounds promoting plant growth. *Eur. J. Phycol.* **2016**, *51*, 243–252. [CrossRef]
19. FAOSTAT. *Crop Production Data*; The United Nations Food and Agricultural Organisation (FAO): Rome, Italy, 2020; Available online: http://www.fao.org/faostat/en/#data (accessed on 12 June 2020).
20. Eurostat. *Database—Agricultural Production*; European Statistics (Eurostat): Brussels, Belgium, 2020; Available online: https://ec.europa.eu/eurostat/data/database (accessed on 12 June 2020).
21. Elzebroek, A.T.G.; Wind, K. Edible fruits and nuts. In *Guide to Cultivated Plants*; Elzebroek, A.T.G., Wind, K., Eds.; CAB International: Wallingford, UK; Boston, MA, USA, 2008; pp. 25–131.
22. Eccher, T.; Pontiroli, R. Old pear varieties in Northern Italy. *Acta Hortic.* **2005**, *671*, 243–246. [CrossRef]
23. Istat. *The 2010 Agricultural Census*; The National Institute of Statistics (Istat): Rome, Italy, 2020; Available online: https://www.istat.it/it/censimenti-permanenti/censimenti-precedenti/agricoltura/agricoltura-2010 (accessed on 12 June 2020).
24. Harris, I.C.; Jones, P.D. *CRU TS4.01: Climatic Research Unit (CRU) Time-Series (TS) Version 4.01 of High-Resolution Gridded Data of Month-by-Month Variation in Climate (Jan. 1901–Dec. 2016)*; University of East Anglia, Climatic Research Unit, Centre for Environmental Data Analysis: Norwich, UK, 2017.
25. Google Earth. Navarra Foundation and Weather station in Ferrara (44.857 N, 11.653 E and 44.861 N, 11.656 E). Google Earth Pro. v. 7.3, 3 D Map, Building Data Layer. 2020. Available online: http://www.google.com/earth/index.html (accessed on 12 June 2020).
26. Essenberg, C.J. Explaining Variation in the Effect of Floral Density on Pollinator Visitation. *Am. Nat.* **2012**, *180*, 153–166. [CrossRef]
27. Alburquerque, N.; Burgos, L.; Egea, J. Influence of flower bud density, flower bud drop and fruit set on apricot productivity. *Sci. Hortic.* **2004**, *102*, 397–406. [CrossRef]
28. Balkic, R.; Gunes, E.; Altinkaya, L.; Gubbuk, H. Effect of male bud flower removal on yield and quality of 'Dwarf Cavendish' banana. *Acta Hortic.* **2016**, *1139*, 587–590. [CrossRef]
29. Dong, H.; Zhang, D.; Tang, W.; Li, W.; Li, Z. Effects of planting system, plant density and flower removal on yield and quality of hybrid seed in cotton. *Field Crop. Res.* **2005**, *93*, 74–84. [CrossRef]
30. Daugaard, H. The effect of flower removal on the yield and vegetative growth of A+ frigo plants of strawberry (Fragaria×ananassa Duch). *Sci. Hortic.-Amsterdam* **1999**, *82*, 153–157. [CrossRef]
31. An, J.; Althiab Almasaud, R.; Bouzayen, M.; Zouine, M.; Chervin, C. Auxin and ethylene regulation of fruit set. *Plant Sci.* **2020**, *292*, 110381. [CrossRef]
32. Bons, H.K.; Kaur, M. Role of plant growth regulators in improving fruit set, quality and yield of fruit crops: A review. *J. Hortic. Sci. Biotechnol.* **2019**, *95*, 137–146. [CrossRef]
33. Lāce, B.; Lācis, G.; Blukmanis, M. Average fruit weight variability of Pear cultivars under growing conditions of Latvia. *Acta Hortic.* **2015**, *1094*, 189–195. [CrossRef]
34. Zhang, C.; Tanabe, K.; Wang, S.; Tamura, F.; Yoshida, A.; Matsumoto, K. The Impact of Cell Division and Cell Enlargement on the Evolution of Fruit Size in *Pyrus pyrifolia*. *Ann. Bot.-London* **2006**, *98*, 537–543. [CrossRef] [PubMed]
35. Craswell, E.T.; Godwin, D.C. The efficiency of nitrogen fertilisers applied to cereals in different climates. *Adv. Plant. Nutr.* **1984**, *1*, 1–55.
36. Novoa, R.; Loomis, R.S. Nitrogen and plant production. *Plant Soil* **1981**, *58*, 177–204. [CrossRef]
37. Casanova, E.F. Agronomic evaluation of fertilisers with special reference to natural and modified phosphate rock. *Fertil. Res.* **1995**, *41*, 211–218. [CrossRef]
38. De Koeijer, T.J. *Efficiency Improvement for a Sustainable Agriculture: The Integration of Agronomic and Farm Economics Approaches*; Thesis Wageningen University: Wageningen, The Netherlands, 2002; p. 143.
39. Shepherd, J.G.; Kleemann, R.; Bahri-Esfahani, J.; Hudek, L.; Suriyagoda, L.; Vandamme, E.; van Dijk, K.C. The future of phosphorus in our hands. *Nutr. Cycl. Agroecosyst.* **2016**, *104*, 281–287. [CrossRef]
40. Herring, J.R.; Fantel, R.J. Phosphate rock demand into the next century: Impact on world food supply. *Nat. Resour. Res.* **1993**, *2*, 226–246. [CrossRef]
41. Bulgari, R.; Franzoni, G.; Ferrante, A. Biostimulants Application in Horticultural Crops under Abiotic Stress Conditions. *Agronomy (Basel)* **2019**, *9*, 306. [CrossRef]
42. Van Oosten, M.J.; Pepe, O.; De Pascale, S.; Silletti, S.; Maggio, A. The role of biostimulants and bioeffectors as alleviators of abiotic stress in crop plants. *Chem. Biol. Technol. Agric.* **2017**, *4*, 5. [CrossRef]

43. Dicenta, F.; Ortega, E.; Egea, J. Influence of flower density on fruit set rate and production in almond. *Acta Hortic.* **2006**, *726*, 307–310. [CrossRef]
44. Wally, O.S.D.; Critchley, A.T.; Hiltz, D.; Craigie, J.S.; Han, X.; Zaharia, L.I.; Abrams, S.R.; Prithiviraj, B. Regulation of phytohormone biosynthesis and accumulation in *Arabidopsis* following treatment with commercial extract from the marine macroalga *Ascophyllum nodosum*. *J. Plant Growth Regul.* **2013**, *32*, 324–339. [CrossRef]
45. Mancuso, S.; Azzarello, E.; Mugnai, S.; Briand, X. Marine bioactive substances (IPA extract) improve foliar ion uptake and water stress tolerance in potted Vitis vinifera plants. *Adv. Hort. Sci.* **2006**, *20*, 156–161.
46. Fan, D.; Hodges, D.M.; Zhang, J.; Kirby, C.W.; Ji, X.; Locke, S.J.; Critchley, A.T.; Prithiviraj, B. Commercial extract of the brown seaweed Ascophyllum nodosum enhances phenolic antioxidant content of spinach (Spinacia oleracea L.) which protects Caenorhabditis elegans against oxidative and thermal stress. *Food Chem.* **2011**, *124*, 195–202. [CrossRef]
47. Fahad, S.; Hussain, S.; Matloob, A.; Khan, F.A.; Khaliq, A.; Saud, S.; Hassan, S.; Shan, D.; Khan, F.; Ullah, N.; et al. Phytohormones and plant responses to salinity stress: A review. *Plant Growth Regul.* **2015**, *75*, 391–404. [CrossRef]
48. El Chami, D.; Daccache, A. Assessing sustainability of winter wheat production under climate change scenarios in a humid climate—An integrated modelling framework. *Agric. Syst.* **2015**, *140*, 19–25. [CrossRef]

7

Insights into the Physiological and Biochemical Impacts of Salt Stress on Plant Growth and Development

Muhammad Adnan Shahid [1,*], Ali Sarkhosh [2,*], Naeem Khan [3], Rashad Mukhtar Balal [4], Shahid Ali [5], Lorenzo Rossi [6], Celina Gómez [7], Neil Mattson [8], Wajid Nasim [9] and Francisco Garcia-Sanchez [10]

1 Department of Agriculture, Nutrition and Food Systems, University of New Hampshire, Durham, NH 03824, USA
2 Horticultural Sciences Department, IFAS-University of Florida, Gainesville, FL 32611, USA
3 Department of Agronomy, IFAS-University of Florida, Gainesville, FL 32611, USA; naeemkhan@ufl.edu
4 Department of Horticulture, College of Agriculture, University of Sargodha, Sargodha 40100, Pakistan; uaf_rashad@yahoo.com
5 Plant Epigenetic and Development, Northeast Forestry University Harbin, Heilongjiang 150040, China; shahidsafi926@gmail.com
6 Horticultural Sciences Department, University of Florida, IFAS-Indian River Research and Education Center, Fort Pierce, FL 34945, USA; l.rossi@ufl.edu
7 Environmental Horticulture Department, IFAS-University of Florida, Gainesville, FL 32611, USA; cgomezv@ufl.edu
8 Horticulture Section, School of Integrative Plant Science, Cornell University, Ithaca, NY 14853, USA; nsm47@cornell.edu
9 Department of Agronomy, Faculty of Agriculture and Environmental Sciences, The Islamia University of Bahawalpur (IUB), Punjab 63100, Pakistan; wajid.nasim@iub.edu.pk
10 CEBAS-CSIC, Departamento De Nutrición Vegetal, Campus Universitario de Espinardo, Espinardo, 30100 Murcia, Spain; fgs@cebas.csic.es
* Correspondence: muhammad.shahid@unh.edu (M.A.S.); sarkhosha@ufl.edu (A.S.)

Abstract: Climate change is causing soil salinization, resulting in crop losses throughout the world. The ability of plants to tolerate salt stress is determined by multiple biochemical and molecular pathways. Here we discuss physiological, biochemical, and cellular modulations in plants in response to salt stress. Knowledge of these modulations can assist in assessing salt tolerance potential and the mechanisms underlying salinity tolerance in plants. Salinity-induced cellular damage is highly correlated with generation of reactive oxygen species, ionic imbalance, osmotic damage, and reduced relative water content. Accelerated antioxidant activities and osmotic adjustment by the formation of organic and inorganic osmolytes are significant and effective salinity tolerance mechanisms for crop plants. In addition, polyamines improve salt tolerance by regulating various physiological mechanisms, including rhizogenesis, somatic embryogenesis, maintenance of cell pH, and ionic homeostasis. This research project focuses on three strategies to augment salinity tolerance capacity in agricultural crops: salinity-induced alterations in signaling pathways; signaling of phytohormones, ion channels, and biosensors; and expression of ion transporter genes in crop plants (especially in comparison to halophytes).

Keywords: abiotic stresses; cell membrane stability; climate change; osmolytes; polyamines

1. Introduction

1.1. Overview of Salinity

Abiotic stresses like salinity, drought, and high temperature have undesirable effects on crop productivity and quality, and negative trends in sustainable agriculture [1]. Salinity in particular is an important limiting factor, causing low yield with inferior quality. Climate change is considered one of the major contributing factors to soil salinization, leading to land degradation and desertification [2]. According to Flowers et al. [3], high salt concentration is responsible for negative impacts on 7% of total land surface, and 5% of cultivated land. Poor irrigation water quality is another important factor contributing to soil salinization [4]. For these reasons, soil salinization is a reported major cause of reductions in the productivity of irrigated and rainfed lands of the world [5,6].

The adverse effects of salinization on plants are evident from negative growth trends from alteration or inhibition of biochemical and physiological processes. Plants can be classified as glycophytes or halophytes by their ability to survive under high salt concentrations [7]. Glycophytes are plants that are severely affected by saline conditions both at the cellular and whole-plant level. Under saline conditions, these plants exhibit greater accumulation of solutes, and ionic and osmotic stresses confer nutritional imbalances, which limit the productivity of these plants. The majority of terrestrial plants are glycophytes, including crop plants [1,8].

Conversely, halophytes regulate their biochemical and physiological processes through ionic compartmentalization, production of osmolytes and compatible solutes, enzymatic changes, and absorption of selective ions. These adaptations promote seed germination, succulence, and salt exclusion for these plants in a saline environment [9,10]. Halophyte succulence keeps the proportion of ions to water in balance under high salt conditions by maintaining high water contents. Succulence is expressed as large cell size, reduced growth and surface area per tissue volume, and increased water constituents. Interestingly, halophytes also have a greater number of mitochondria, indicating that more energy is required to survive under saline conditions [11,12]. Halophytes also have less sodium and chloride ion accumulation in their cytoplasms, allowing their chloroplasts to survive even while the plant experiences salinity shocks [13,14]. Halophytes also have a specialized system for salt excretion from the plant tissues via specific glands. These glands are characteristic of halophytic leaves. The leaves will remove the salts onto the leaf surface before the salts can reach the shoots of the plant. The presence of halophytes is limited to habitats with plentiful water (e.g., salt marshes, etc.). "Salt hairs," which regulate water loss, replace "secretary glands", if a plant is adapted to a relatively drier climate as compared to marshes [15,16]. Hydathodes are another adaptation by plants to remove excessive salts, with less stomatal conductance and transpiration water loss [17,18].

Various undesirable effects appear because of high salt concentration. Ion imbalance is one of the major consequences. A high concentration of Na and Cl ions, as an example, can lead to biochemical processes which can prove to be fatal for the plants [19–21]. Sodium and chloride toxicity not only induce nutritional disorders but also cause physiological drought by lowering the osmotic potential of the soil solutions [22]. Soil salinity prevents the plant from taking up water from the soil, resulting in a decline in cellular water, thus affecting cell turgor. Soil salinity also adversely affects photosynthetic activity in the plant and encourages the production of reactive oxygen species (ROS), thus reducing plant growth [23,24].

The identification of salt stress by plant species and their subsequent response is controlled by signals—signals which are generated by ions, osmotic differential, hormones, or ROS [25]. These signals bind to their respective receptors and initiate the physiological mechanisms which enable a plant to adapt to stress conditions (Figure 1). Under abiotic stress conditions, three types of signal transduction have been categorized, i.e., the ionic signaling pathway, the osmolyte regulation pathway, and the gene regulation pathway [26]. For signal transduction under salinity stress, the ionic stress signaling pathway has been elucidated. Calcium (Ca) occupies a central position in this regard. It induces signal transduction in plants to adapt to stress conditions [27]. High cytosolic Ca concentration initiates many processes involving enzymatic activity regulation, ion channel performance, and gene expression [28]. Exogenously applied calcium regulates K^+/Na^+ selectivity, and thus confers salt adaptation by improving signal transduction. Glycinebetaine is reported to maintain signal transduction and ion homeostasis under salt-stressed conditions [29,30].

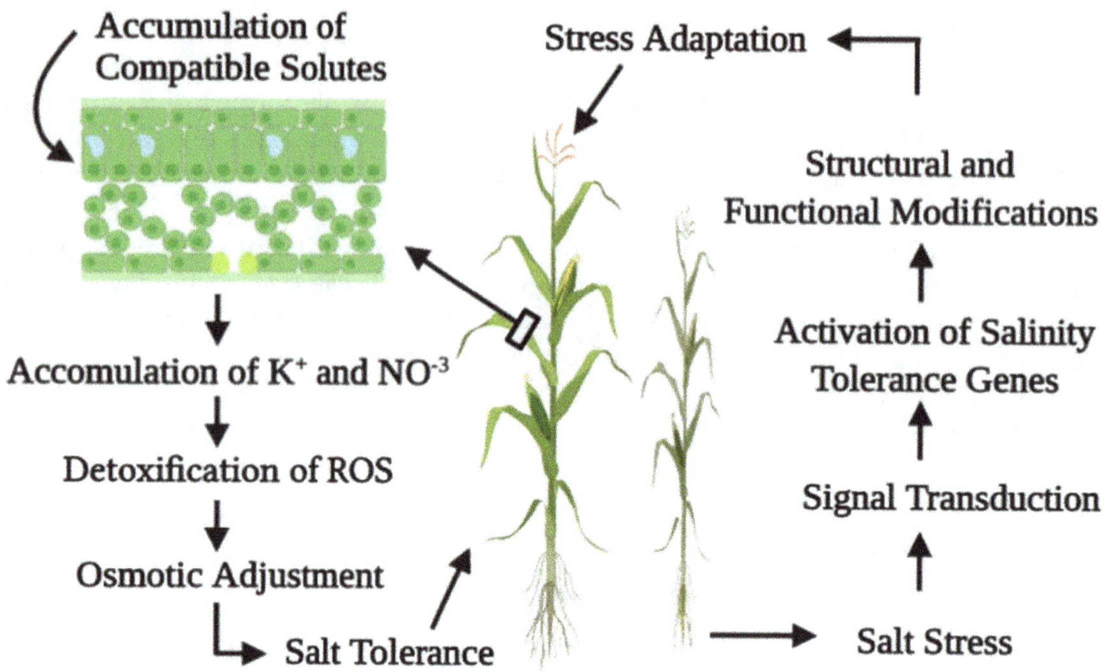

Figure 1. Salt stress signals that bind to their respective receptors and initiate the physiological and molecular mechanisms to enable a plant to survive under stressed conditions.

Glycophytes and halophytes mediate serious effects of salinity at the cellular level by inducing changes in the plasma membrane and cytoplasm. As a tolerance mechanism, the plant alters the structure and composition of their plasma membrane, especially lipid and protein contents. The cell membrane is usually the foremost target of any stress [31]. Salt stress also alters the cell cytoplasmic viscosity and composition [32,33].

It is essential to understand each tolerance mechanism at the cellular level in order to understand each tolerance mechanism at the plant level. The protoplasmic features studied at the cellular level include plasma membrane permeability, cytoplasmic viscosity, cytoplasmic streaming, and cell solute potential. Cytoplasmic viscosity describes the water contents of the cytoplasm in conjunction with its inter-macromolecular interactions. Cell solute potential represents the solute contents of the cell. Cell membrane permeability significantly increases with the increase in salinity [34,35]. The ability of the plasma membrane to repair, regenerate, and maintain its integrity stabilizes the cell structure and function under stress conditions. It is mainly dependent on the composition of the plasma membrane (i.e., mainly lipid contents). Saline conditions result in enhanced lipid peroxidation [36].

The cell membrane stability technique is widely utilized to judge the behavior of various plant genotypes in response to salt stress [37]. Thus, it can contribute to assessing the salt tolerance of plant genotypes. Cell membrane stability is also reported to correlate with potassium (K) ions, osmotic potential, osmotic adjustment, and relative water contents [38,39]. The differing patterns of cell membrane permeability help in characterizing genotypes as tolerant or sensitive. In a saline environment, salt-sensitive genotypes show marked alterations, whereas salt-tolerant genotypes show minor changes. Salinity also alters the degree of saturation of membrane fatty acids and membrane fluidity [40].

The salt-tolerant genotypes have high cytoplasmic viscosity due to augmentation in hydrophilic cytoplasmic proteins and other macromolecules [41]. In more sensitive genotypes, a saline environment results in a high concentration of salts in plant cells, lowering the solute potential [42]. Salinity inflicts serious irregularities during cell division, one of the various metabolic processes which face severe alterations in a saline environment. A saline environment especially alters the leaf anatomy by affecting mitochondria and vacuoles [43,44], plant leaf area, and stomatal thickness [45]. One way plants exhibit tolerance to saline environments is portioning or compartmentalization of toxic ions. This mechanism enables salt-tolerant plant species to retain toxic levels of harmful ions in vacuoles and inhibit their interference with cytoplasmic metabolic activities [46]. The Na^+ and Cl^- partitioning in the vacuole stimulates higher concentrations of K^+ and organic osmolytes in the cytoplasm in order to adjust osmotic pressure of the ions in the vacuole [47].

1.2. Salinity and Morphological Attributes

Halophytes have the unique feature of succulence, a feature which keeps the ionic uptake in proper proportion with water, by maintaining high water contents (Figure 2). Succulence results in large cell size, reduced growth [48,49], reduced surface area per tissue volume, and increased water constituents. The salt tolerance is evident as maintenance of vegetative growth and yield and lower necrotic percentage [50,51]. Moreover, halophytes also have a greater number of mitochondria, indicating that more energy is required to survive under saline conditions [11,21]. Sodium (Na) accumulation causes necrosis in old leaves, initiating from tips and then extending towards the leaf base. It also decreases leaf life span, net productivity, and crop yield [52]. Biomass reduction and foliar damage become more prominent with time and at higher salinity levels. High salt concentration caused a reduction in fresh fruit yield in various vegetables [31,53–55]. However, salinity treatments did not prove harmful for vegetative growth and the number of flowers [53,55]. In contrast to the above report, Chartzoulakis and Klapaki, [56], Kaya et al. [57], and Giuffrida et al. [58], noted a reduction in fruit numbers and fruit weight under salt stress. Saline conditions were reported as producing non-significant results on some growth attributes, water status, and tissue concentration of major nutrients [59]. Salt treatments caused a significant reduction in plant height, root length, and dry weight [60–62]. Salt stress caused reduction in fresh and dry weight of cotton seedlings [63,64] and seed germination percentage in wheat varieties [35,65]. Tomato (*Solanum lycopersicum*) and pepper plants grown under salt stress experienced a reduction in dry weight, plant height [57,66], fruit weight, and relative water contents [67]. Broad bean, which is a green vegetable, experienced a significant reduction in plant height, leaf area, pod weight, number of pods per plant, seed yield, number of seeds, and product quality due to salt stress. However, a significant positive trend was observed between dry leaf matter, specific leaf weight, and salinity [68]. Suppression in seedling growth and dry matter accumulation was observed in Indian mustard (*Brassica juncea* L.) because of salt stress, which was ameliorated by putrescine application [69]. Exogenously applied sugar beet extract and shikimic acid on salt stressed eggplant and tomato crops respectively displayed a marked influence on fresh fruit weight, number of fruits, and shoot and root fresh and dry weight [54,70].

According to Caines and Shennan [71], root growth is more susceptible to saline conditions than shoot growth, but both are affected, making them reasonable indicators of salinity damage. A similar suppression in the shoot and root growth was also observed by Evers et al. [72] and Gao et al. [73] in the case of *Solanum tuberosum* L. Under high salt concentrations, potato root and shoot development was hindered [74,75]. High salt concentrations reduced the leaf area and increased the root:shoot ratio in wild type (Ailsa Craig) and ABA-deficit mutant (*notabilis*) tomato genotypes [76]. Suppression of fresh and dry weight of tomato plants because of salt stress can be alleviated by promoting the growth of *Achromobacter piechaudii* in the tomato growth media. *A. piechaudii* also caused a reduction in ethylene production by tomato seedlings, the opposite effect of salt stress on tomato seedlings [77].

High ethylene levels proved to be harmful to the growth of the plant [78]. Interestingly, Hu et al. [79] reported that potato root growth could be improved through brassinosteroid application. Salt stress caused a reduction in marketable yield of pepper plants grown hydroponically. Saline conditions also imparted negative features to the fruit quality in terms of fruit pulp thickness and firmness. It also resulted in increased fructose, glucose, and *myo*-inositol fruit concentrations [4,42]. The salt-sensitive pepper genotypes showed maximum damage and experienced severe chlorosis and necrosis, whereas tolerant genotypes were slightly less affected. Sodium exclusion can be regarded as a criterion to allocate salt stress tolerance status to pepper genotypes [80]. Korkmaz et al. [81] have reported that the exogenous application of glycine betaine can reduce the effect of salinity in pepper plants.

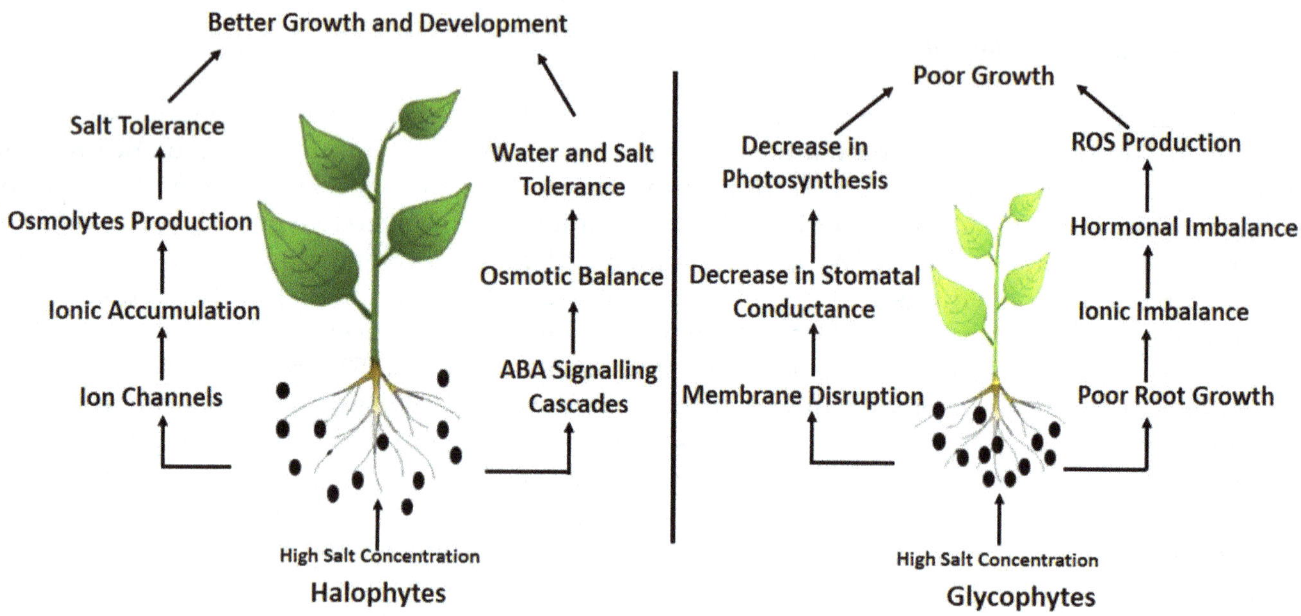

Figure 2. Regulation of physiological and biochemical process in halophytes through ionic compartmentalization, osmotic adjustment, enzymatic activities, polyamines, and stress signaling regulation.

High salt concentration has been observed to cause detrimental effects on leguminous crops. In fact, saline conditions induced smaller sized nodules, reduced nodule volume per plant, less nodulation, and inferior plant growth. At the cellular level, saline conditions caused drastic alteration in the mechanism of nodule formation. It reduced the turgor of the peripheral cells of the nodule, altered its zonation, enhanced the infection thread enlargement, reduced the release of bacteria from infection threads, and stimulated the electron-dense material (phenolics) and its accumulation in vacuoles. Salinity caused a reduction in nitrogen-fixing ability of the nodules, which has the outcome of reduced respiration rate and protein synthesis [82–84]. Salinity tolerance patterns vary considerably among leguminous crops. It is distinctly apparent in *Vicia faba*, *Glycine max*, *Pisum sativum*, and *Casuarina glauca*, which can be categorized as salt-sensitive [85,86]. In addition, germination of seeds faces serious limitations upon exposure to salt stress [87,88]. Halophytes and glycophytes differ significantly in their germination behavior. Halophytes have the ability to maintain their germination mechanism to an extent with the advent of salinity. However, a sharp decline in germination occurs for glycophytes under salt stress. Imbibition is affected by the lower solute potential of the soil solution. It results in enzymatic deregulations and imbalances in source-sink relationships and ratios of different plant growth regulators present in the seed and required for efficient seed germination [89,90].

1.3. Salinity and Physiological Attributes

Photosynthesis is of prime importance in the production of adenosine triphosphate (ATP), which provides the energy required for CO_2 fixation to sugars. Various abiotic stresses alter photosynthetic mechanisms [91] by disrupting thylakoid membranes, modifying the electron transport chain, altering enzymatic activity and protein synthesis, and changing Calvin cycle patterns. All these abnormalities cause deregulation in ATP synthesis [92], which can lead to the deficiency of certain ions due to ion degradation and synthesis inhibition [93–96].

Stepien and Klobus [97] reported a decline in the relative water content of cucumber (*Cucumis sativus*) leaves after exposure to saline solutions. They attributed their results to higher Na and reduced K content, a situation which reduces photosynthesis because of the antagonistic competition of Na for ion uptake. Gas exchange is inversely related to the concentration of Na and chloride ions [98,99], and parameters like photosynthesis, stomatal conductance, and transpiration tend to be negatively influenced by saline treatments [100,101]. Photosynthesis of pepper plants has been reported to be lower when the entire root system is affected by salt stress compared to partial root exposure, and factors like stomatal conductance and transpiration are similarly affected by complete or partial salt stress [99,102]. A low stomatal conductance was reported in wild type (Ailsa Craig) and ABA-deficit mutant (*notabilis*) tomato genotypes, which was negatively correlated with increasing xylem ABA for both genotypes [76]. Stomatal conductance is often correlated with photosynthetic efficiency, which is a prerequisite for higher biomass production and yield [10]. In addition, salinity is reported to reduce the maximum quantum efficiency of photosystem II (PSII) [48,94,103–106]. A distinct correlation has been found between Na ion contents and chlorophyll fluorescence, which is often used to estimate salt tolerance of plants [48]. Higher salt levels are also known to alter photosynthesis via non-stomatal limitations, including variations in photosynthetic enzyme activity and changes in the concentration of chlorophyll and carotenoids [107,108]. Pepper leaves have exhibited a significant reduction in chlorophyll pigment under saline conditions [109]. A similar reduction in chlorophyll *a* and *b* contents has been reported in melon by Kaya et al. [57]. This chlorophyll degradation under salt stress can be attributed to an enzyme called chlorophyllase [110–113] and to the absolute concentration of chloride and Na in the leaves [98,114–118]. Although carotenoid content has been reported to decline in response to salinity, anthocyanin pigments typically increase as a result of salinity.

Plant antioxidants include secondary metabolites (phenolic compounds), which are generated in response to stress conditions. These secondary metabolites may include tocopherol, which serves to stabilize membrane integrity [119,120], ascorbic acid, carotenoids, flavonoids, and glutathione [90,121]. Tocopherol plays a key role as a signaling molecule between cells [93]. Ascorbic acid is a significant antioxidant involved in plant adaptation [122] and occurs abundantly in cell organelles and apoplasts [99]. It has the ability to scavenge superoxide, hydroxyl, and singlet oxygen. Carotenoids are found in chloroplasts and reported to aid in light reception for photosynthesis. Moreover, they are also protective compounds which scavenge ROS [95,123]. Putrescine is reported to increase the level of carotenoids and glutathione in Indian mustard (*B. juncea* L.) against salt stress. Putrescine supplementation inhibits ROS generation by accelerating antioxygenic enzymes and therefore aiding in the maintenance of chloroplastic membranes and the $NADP^+/NADPH$ ratio [124–126].

C_4 plants are reported to be more resistant against salinity stress than C_3 plants by having a better capacity to preserve the photosynthetic apparatus against oxidative stress [127,128]. Other physiological responses that are indirectly related to salinity stress include changes in water use efficiency and evapotranspiration, which can benefit from the use of beneficial bacteria [77,129–131]. In addition, higher salt concentration can stimulate the accumulation of spermine and spermidine, which contribute to the induction of salt tolerance in plants and lead to maintenance of fruit quality (Figure 3) [132–135].

Figure 3. Bunches, grapevine cultivar Shiraz. Control bunch (**A**); Salt-stressed bunch (**B**); showing symptoms of coulure and millerandage. Salt treatment was applied from budburst until veraison via fertigation with 35 mM NaCl added to control nutrient solution [136].

2. Salinity and Water Relations

Water is an essential constituent of plant cells, supporting almost all physiological and biochemical processes contributing to plant growth and development [8]. High salt concentration hinders the movement of water from the soil to the plants by reducing the water conductivity of roots [9] and affecting relative water content at the cellular level [31,99]. Alterations in rootzone water status have profound effects in the way plants respond to higher salt concentration [1,93], but adverse effects of salinity stress can be mitigated to an extent with proper irrigation and nutrient management and by controlling the evaporation rate of plants. A recent review by van Zelm et al. [137] listed root hydraulic conductivity, osmotic potential, relative water content, leaf water potential, water-use efficiency, stomatal conductance, and transpiration, among others, as typical parameters that are commonly used to indicate water-relation responses to salinity stress, which are determinant for plant growth.

Others have shown a relationship between water potential and photosynthetic parameters. For example, a decline in CO_2 assimilation has been reported with the depression of water potential [126,138]. To achieve active growth, plants maintain positive turgor pressure and regulate osmotic potential. Under saline conditions plants face osmotic stress due to alterations in water potential [48,97,139]. Osmotic stress triggers cellular and whole plant responses [44,140–143], and plants cope with that osmotic stress through regulating ionic homeostasis, which induces tolerance against toxic ions and growth under unfavorable conditions [8,144]. In order to maintain osmotic adjustment, plants lower the cellular osmotic potential to help improve water uptake and adjust ionic concentration in cells [42,145]. For that purpose, osmolytes appear to be of major consideration. These may be sugars, polyols, or amino acids. Osmotic adjustment appears to be a significant and effective salinity resistant mechanism in crop plants, which can be exploited efficiently through selection and breeding efforts that target salinity tolerance in different plant species.

3. Salinity and Biochemical Attributes

Plants have the ability to sustain their life under a saline environment through synthesis and accumulation of compatible solutes in the cytosol. These are soluble compounds with low molecular mass. These chemical compounds can maintain physiological and biochemical processes, without having interference in these processes. The chief components of compatible solutes are sugar alcohols (mannitol, sorbitol, ononitol), quaternary ammonia compounds (glycine betaine, proline betaine), proline, and tertiary sulfonium compounds. They act to scavenge the reactive oxygen species (ROS) and inhibit lipid peroxidation, hence preventing damage at the cellular level. These compatible solutes act in favor of osmotic adjustment and prevent ROS damage at the cellular level [146,147]. These maintain macromolecular conformation in the cytosol, which may be changed due to the accumulation of charged ions, under saline conditions [8,97]. These organic compounds are termed as compatible due to their consistency with the cell's metabolism [148], and their lowering of the water potential without altering cell water contents. These organic compounds are hydrophilic in nature, having the ability to replace water present on protein surfaces [149,150], without interfering with their structure and function. These solutes play a key role in preventing the drastic effects of high ion concentration on enzymatic activities [24,151–153]. Such important roles of compatible solutes lead to osmoregulation of plant cells under osmotic stress. In addition to osmoregulation, these organic compounds have a distinct role in protein stabilization, maintenance of membrane integrity, protection of OEC of PSII from dissociation [154], and scavenging of reactive oxygen species (ROS). Mannitol, sorbitol, glycerol, proline, ononitol, and pinitol have been reported to scavenge ROS species [155].

Mannitol (sugar alcohol) metabolism in higher plants is a superior attribute, contributing to salt and osmotic stress tolerance while playing a significant role as a compatible solute. It also improves plant responses under biotic stress as well, like under pathogen infestation [156]. Mannitol is reported to be synthesized at the same time with either sucrose or raffinose saccharide. In salt-tolerant species, mannitol accumulation increases, indicating that high mannitol levels contribute to salt tolerance. Mannitol acts to scavenge the reactive oxygen species, thus protecting protein molecules [157,158]. Pinitol and ononitol have been reported to accumulate under various stresses, predominantly drought and salt stress [159,160]. Interestingly, polyols can be used as a potential biochemical marker for genetically engineered stress resistance plant genotypes [161,162].

3.1. *Salinity and Proline*

Proline is an osmolyte, an amino acid, which is thought to play a significant role in inducing tolerance in plants against stressed conditions [163]. Salt stress can result in elevation in proline levels [74]. Ethephon, when used with sodium chloride in spinach, also increased proline levels [164]. The importance of proline is highlighted by its existence in bacteria with a relationship to plants experiencing water or salinity stress. High proline levels can serve as a nitrogen source for plants during recovery [165]. The precursor of proline synthesis is glutamate, involving pyrroline carboxylic acid synthetase and pyrroline carboxylic reductase [166]. An increase was noted in the activity of pyrroline-5-carboxylate synthetase (P5CS) and a decline was recorded in proline dehydrogenase activity in potato seedlings under salt stress. These changes of enzymatic activity were more pronounced in salt-sensitive cultivars [74,167]. However, an increase in proline contents of potato clones was recorded upon salt exposure [50,168]. It serves to stabilize ultra-structural changes in cells, scavenge ROS (reactive oxygen species), and maintain cellular redox potential. Under stress conditions, a higher accumulation of proline is reported in cell cytosol, strengthening the ability of the cell to make ionic adjustments. Its accumulation is linearly related to stress tolerance in plants [169]. Proline biosynthesis is reported to be mediated by Ca [170–172] and abscisic acid [8]. Previously, contrasting views about proline accumulation were also reported in plants under stress [19,158], where it appeared as a salt stress injury symptom, e.g., rice [173] and sorghum [174].

Some plant genotypes do not respond to proline accumulation, but their salt tolerance potential can be enhanced through the exogenous application of proline [31,175]. It may be helpful in counteracting the harmful effects through osmo-protection, resulting in a higher growth rate. Proline also increases the activities of antioxidant enzymes like SOD (superoxide dismutase) and POD (peroxidase) [176]. Proline is not reported to scavenge ROS directly, but through enhanced antioxidant enzyme activity. It is reported to be more effective in mitigating the drastic effects of salinity than glycine betaine [177]. Proline used at higher concentrations may prove to be lethal for the plant, causing ultra-structural damages leading to ROS generation [178]. The effective dose of proline varies with genotype and plant developmental stage [179–182]. Proline accumulation has been reported for drought sensitive and tolerant barley genotypes grown under saline conditions. Under salt stress, a considerable amount of proline was present, with relatively lower quantities in root tissues. Proline accumulation is reported to be more prominent in tolerant genotypes [183].

3.2. *Salinity and Polyamines*

Polyamines are multivalent compounds consisting of two or more amino groups. In higher plants, these are identified as putrescine, spermidine, and spermine [147,184]. These are involved in various physiological mechanisms including rhizogenesis, somatic embryogenesis, maintenance of cell pH and ionic balance [29], pollen and flower formation, abscission, senescence, and dormancy. Endogenous polyamine synthesis can be stimulated by cytokinin [185,186]. These compounds act to stabilize macromolecules like DNA and RNA. Moreover, polyamines have a significant role in numerous abiotic and biotic stresses [151,187]. At the cellular level, polyamines contribute to regulating the plasma membrane potential, ionic homeostasis, and tolerance against salinity [188,189]. Exogenously applied polyamine or ornithine caused a reduction in proline accumulation in plant tissues under salt stress. However, an alternate trend was observed in the case of non-stressed beans [99,190]. Putrescine is characterized as a de-stressor agent and a nitrogen source under stressed conditions [191]. Putrescine has been reported to reverse the biomass reduction in Indian mustard [68,192]. Its production in plant cells follows two alternative pathways: conversion from ornithine or arginine. Putrescine is then converted to spermidine and subsequently to spermine by addition of an aminopropyl group. Spermine deficiency caused Ca ion imbalance in *Arabidopsis thaliana*, thus indicating spermine as a maintainer of plant cell ionic homeostasis under salt stress [193].

3.3. *Salinity and Glycine-Betaine*

Glycine-betaine (GB) is present in a wide range of organisms, from bacteria to higher plants and animals. In addition to being involved in osmoregulation, it maintains and regulates the performance of PSII protein complexes by protecting extrinsic regulatory protein against denaturation. It also stabilizes macromolecules, due to its ability to form strong bonds with water [136]. It protects these macromolecules during drought and thermal stress, which is why it is sometimes called an "osmoprotectant" [129,194]. Glycine-betaine accumulates in some crops under stress, like members of family *Poaceae* and *Chenopodiaceae* [195], and is absent entirely from other plants, like rice and tobacco. This directed the scientists to develop transgenic plants that have the ability to produce GB. In transgenic plants, the reproductive organs are capable of tolerating abiotic stresses if they can accumulate GB [195]. The precursor for GB is choline, and the conversion is managed by enzymes like choline monooxygenase and betaine-aldehyde dehydrogenase [151,196]. Choline supplementation to the growth media of the salt-stressed plant can act to restore the suppressed growth [197]. GB is water soluble, is not harmful at higher concentrations, and accumulates mainly in plastids and chloroplasts. Exogenous application of GB promotes salinity tolerance in plant species which do not naturally produce GB. A plant can utilize exogenously applied GB via leaves [198], as well as roots [199]. After absorption, GB is translocated in phloem [99,200]. GB is not directly involved in scavenging ROS species, but it alleviates the damaging effects of ROS by promoting enzymes responsible for the destruction or production suppression of ROS [201].

The reproductive stage of any plant during GB application is considered critical to ensure maximum yield. In various studies, it was reported that the plant reproductive organs acquire higher levels of

GB than the vegetative parts under stressed conditions. This indicates that high GB accumulation is more necessary for protecting the reproductive organs than it is for protecting the vegetative tissues from abiotic stresses, indicating that application timing is key [181,202]. The natural GB accumulating species include sugar beets, spinach, wheat, barley, and sorghum. High GB concentration is linearly linked with increased tolerance. Osmotic adjustment is the major mechanism involved in increased tolerance to abiotic stresses, especially salt stress. GB is responsible for turgor maintenance through osmotic adjustment [54,182,203]. However, this relationship is not satisfactory in some cases like *Triticum* spp. and *Agropyron* spp. [204]. Thus, this relationship varies with genotype [158]. The plant species which do not produce GB naturally can give a satisfactory yield and survival rate under salt stress conditions through the exogenous application of GB [42,205]. Exogenous GB, once applied, is transported rapidly throughout the plant. Exogenous application of GB has been reported in many plant species, including tobacco, rice, soybean, barley, and wheat. In barley, GB application improved stress tolerance by lowering water potential, which improved survivability. GB plays a role in osmotic adjustment and ionic homeostasis by maintaining high K^+ concentration compared to Na^+ ions. Exogenous application of GB also increased the K^+/Na^+ ratio [206–208]. GB also protects the photosynthetic apparatus. It enhances photosynthetic activity through increased stomatal conductance and reduced photorespiration [42,209,210].

In contrast to its positive influence, some researchers have also suggested neutral or somewhat negative responses to exogenously applied GB in some plant genotypes. For example, it appeared to have a neutral influence on growth in cotton [211], turnip, rapeseed, and tomato [212]. For the commercial application of GB, the rate, duration, timing, and frequency should be considered [158,213]. It can be used for seed treatment as well as foliar application. The application method is dependent on the plant material on which it will be applied, the timing of the application relative to plant developmental stage, and environmental conditions during the time of application [214].

Exogenously applied GB improved salt tolerance in rice by improving relative water contents in the leaves and increasing antioxidant levels, including superoxide dismutase, ascorbate peroxidase, catalase, and glutathione reductase (GR) [213]. Reduction in peroxidase activity was reported in a salt-tolerant rice genotype under salt stress. GB is also reported to reduce lipid peroxidation [215–217]. GB can prevent membrane adulterations due to osmotic stress more efficiently than proline [218]. Proline accumulation in leaves of salt-stressed plants is not reported to be correlated with exogenously applied glycine betaine [54,180]. Sugar beet is identified as the foremost source of GB [158,219]. It is appreciated as a valuable source of GB along with other beneficial compounds and is useful in inducing tolerance against salt stress in eggplant (*Solanum melongena* L.) as compared to pure GB. It has a marked influence on the morphological (growth and yield) as well as physiological and biochemical (gas exchange, photosynthetic rate, transpiration, GB accumulation) attributes [186,220].

4. Salinity and Enzymatic Attributes

Generation of reactive oxygen species (ROS) like singlet oxygen, superoxide radical, hydrogen peroxide, and hydroxyl radical as a consequence of exposure to various abiotic stresses causes injury to plants. Molecular oxygen is non-reactive and requires electron donors for the production of reactive oxygen species. These electron donors are typically metal ions [24,221]. Increased production of ROS is an indicator of plant stress [222] for proteins, lipids, pigments, DNA, and other molecules at the cellular level [223,224]. Plants combat ROS through enzymatic and non-enzymatic mechanisms. The enzymes used for scavenging ROS include superoxide dismutase (SOD), catalase (CAT), peroxidase (POD), ascorbate peroxidase (APX), glutathione reductase (GR), and glutathione-synthesizing enzymes [197,205,225]. However, increased generation of ROS scavenging enzymes does not always correspond to higher salinity tolerance. Many factors contribute to the effectiveness of antioxidant systems, including the site of antioxygenic enzyme production, enzyme action, and interaction of different antioxidant enzymes, according to Blokhina et al. [226]. Das and Roychoudhury [227] and Sairam et al. [228] reported increased SOD, APX, CAT, and GR activities in intolerant wheat plants under abiotic stresses. The site of superoxide radical synthesis is reported to be chloroplasts [229,230], mitochondria [231], and microbodies. Ascorbate peroxidase (APX) is recognized as a reducing agent for H_2O_2 to water by the utilization of ascorbate and release of monodehydroascorbate (MDHA). MDHA reductase contributes to the conversion of MDHA to AsA (ascorbate). Ascorbate peroxidase causes effective regeneration of AsA and interrupts the cascade of oxidation caused by H_2O_2 [24,232]. The antioxidant response of plants can be regulated by compounds like H_2O_2 under stress conditions. H_2O_2 contributes almost 50% of the destruction caused by oxygen radicals in photosynthetic reduction. In terms of negative impacts on plant physiology, H_2O_2 is the most harmful of all reactive oxygen radicals. APX exists in four different forms: chloroplast stromal soluble form (sAPX), chloroplast thylakoid bound form (tAPX), cytosolic form (cAPX), and glyoxysome membrane form (gmAPX) (Figure 4). H_2O_2 increased the level of other antioxidants and caused a decline in lipid peroxidation in maize under stressed conditions [210,211,233]. Saline conditions applied to potato seedlings and *Broussonetia papyrifera* showed an increase in APX activity, new POD and SOD isoenzyme activities, and alterations in isoenzyme composition [74,212]. Similarly, seed treatment of wheat by H_2O enhanced the salinity tolerance of young seedlings of wheat through prevention of oxidation damage and induction of stress proteins [234]. Exogenously applied GB has been reported to enhance antioxidant activity in terms of SOD, ascorbate peroxidase, CAT, and GR in salt-tolerant rice and wheat genotypes under salt stress conditions [197,214,235].

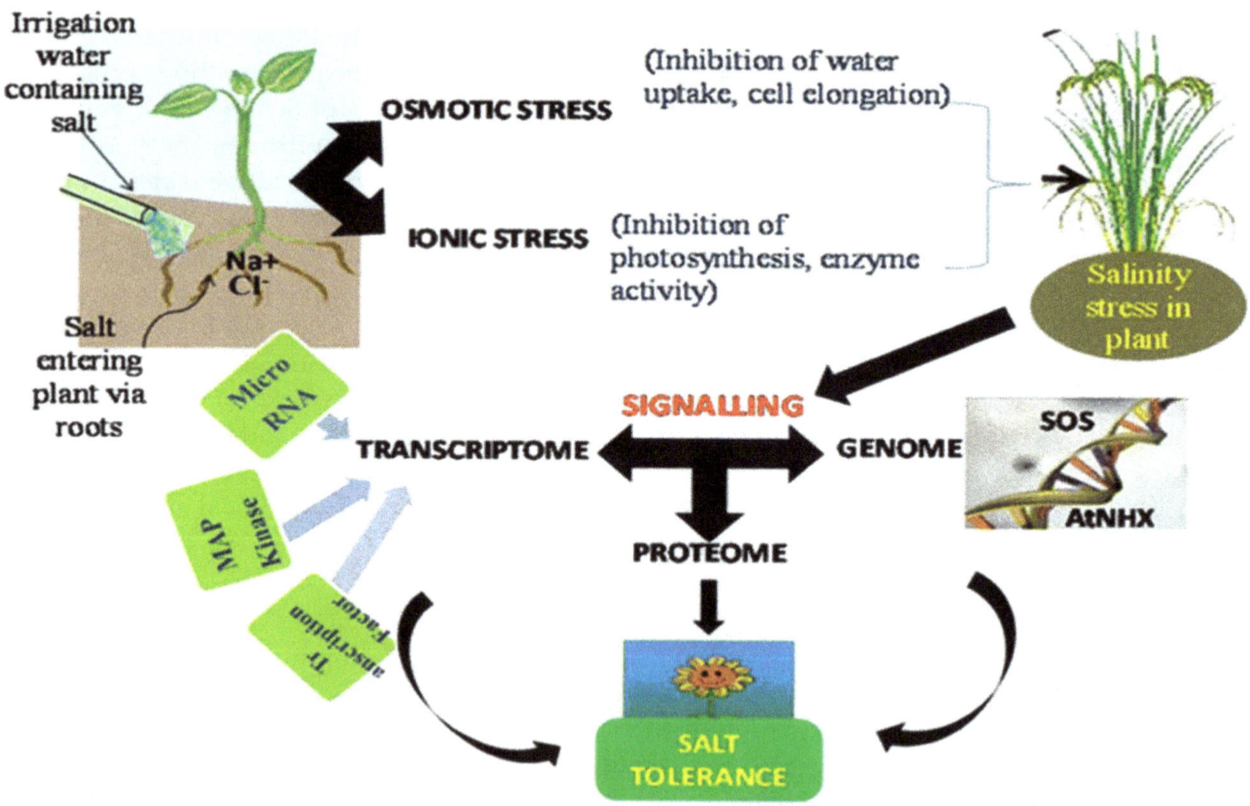

Figure 4. Schematic diagram showing routes of salt stress toxicity and various tolerance strategies in plants. Putative roles and action of genes, transcription factors, mitogen-activated protein kinases, microRNAs, and metabolites are shown [236].

Superoxide dismutase (SOD) activity is reported in almost all plant cell types. It causes disproportionation of singlet oxygen to molecular oxygen and hydrogen peroxide. The isoforms of SOD are categorized as copper- and zinc-containing superoxide dismutase (Cu/Zn-SOD), manganese-containing superoxide dismutase (Mn-SOD), and nickel- and iron-containing superoxide dismutase as Ni-SOD and Fe-SOD [79]. A negative correlation was observed between proline accumulation and SOD activity in wild halophytes, so SOD activity does not necessarily induce salt tolerance in plants [216,237]. Salt treatments caused a reduction in SOD activity in potato cultivars [74]. Catalases (CATs) contribute to the plant defense system, and are synthesized in peroxisomes and glyoxysomes. CAT is responsible for the conversion of hydrogen peroxide into water and oxygen. Under saline conditions, SOD activity was reported to be more pronounced in C_3 (wheat) as compared to C_4 (maize). Both types of plants showed the same level of elevation in APX. However, an increase in GR level was more prominent in maize plants [118,238].

Exogenous application of potassium nitrate alleviated salinity effects in winter wheat by promoting some antioxidant enzymes [60,239]. Similarly, coronatine, which has properties similar to methyl jasmonate, is reported to promote the activities of antioxidant enzymes including superoxide dismutase (SOD), catalase (CAT), peroxidase (POD), and glutathione reductase (GR), and DPPH (1,1-diphenyl-2-picrylhydrazyl) scavenging behavior in cotton leaves. DPPH-radical scavenging determines non-enzymatic antioxidant activity. COR causes a decline in generation of reactive oxygen species by increasing the activity of antioxidant enzymes [240–243]. COR is functionally similar to jasmonic acid, which mitigates salt stress [244–246]. Polyamines applied to salt-stressed plants can act to ameliorate the negative impact of salinity (i.e., growth suppression) in many crop plants [247].

5. Salinity and Phytohormones

Plant hormones are signaling molecules with the ability to alter the physiological mechanisms of the plant, even if present in very minute quantities [248]. Common plant hormones are auxins, gibberellins, cytokinins, abscisic acid, and ethylene [249]. Plants growing under salt stress experience imbalances in hormonal homeostasis. Stressed conditions drastically alter physiological mechanisms of the plant, creating massive changes in endogenous hormonal contents. Higher concentrations of toxic ions are negatively correlated with the levels of plant hormones like gibberellins, auxins, and cytokinin and are positively associated with the abscisic acid level [250,251]. Exogenous plant hormone application on salt stressed plants was found to alleviate the negative effects of salinity on the morphological (leaf area, dry mass), physiological (chlorophyll content, stomatal conductance, photosynthetic rate), and yield characteristics of crops [252,253].

Abscisic acid (ABA) and ethylene are involved in signaling under stress conditions. An increase in ABA concentration in plant cells has been reported under saline conditions [254]. Carotenoids are the precursor for ABA synthesis, with roots and leaves the sites of synthesis [255]. Water deficit in the root zone causes ABA generation in roots, and ABA transport to shoots is xylem mediated. The increase in pH of xylem sap increases the transport of ABA to the guard cells, where it regulates the stomatal opening and closing through the involvement of Ca ions [256]. Alterations in ABA levels indirectly affect photosynthesis through disruption in stomatal opening and closing. The photosynthetic efficiency of the plant cell declines along with deregulation of translocation and assimilate partitioning of photosynthates [8,257].

Exogenously applied plant growth regulators have been widely reported to enhance stress tolerance in numerous plant species [258]. ABA is reported to be useful in alleviating plant salt stress under low water potential. ABA production in the plant cell is also related to ethylene synthesis under salt stress. The interaction between these two stress hormones is apparent from vegetative growth and seed germination under salt stress. During root inhibition by salt stress, ethylene regulates the ABA concentration. However, the reverse has been reported in the case of seed germination [259]. Salicylic acid has displayed a defensive role in plants experiencing stress, signaling the plant to adapt to the stressful environment [260].

Salinity and Growth Regulation

Brassinosteroids (BRs) are growth regulators which mitigate adverse growth patterns caused by salinity. It improves the germination of seeds experiencing salt stress. The improved germination rate has been reported for rice [261] and tobacco. Application of BRs as a seed treatment enhanced the growth of rice seedlings under salt stress [262]. It helps the plant to retain its green pigments and enhances nitrate reductase activity [263–265] and nitrogen-fixing capability. Brassinosteroids (28-homoBL) increased dry matter accumulation and seed yield [266]. Foliar application of brassinosteroids (24-epibrassinolide) on pepper plants grown with saline water greatly affected shoot growth parameters and leaf water contents as compared to roots. However, its effect on chlorophyll fluorescence was non-significant [267,268]. Similar patterns of brassinosteroidal effects were observed in wheat grown under salt stress. 24-epibrassinolide application on salt-stressed wheat seedlings exhibited non-significant results in terms of plant biomass, chlorophyll content, photosynthesis rate, substomatal CO_2 concentration, and water use efficiency. The incremented water use efficiency can be related to higher transpiration rate shown by salt-stressed wheat seedlings, as a result of 24-epibrassinolide application [269,270]. The efficiency of exogenously applied brassinosteroids to mitigate salinity effects varies with plant species, appropriate growth stage, dose, frequency, and method of brassinosteroidal application [271–273].

The results of BR application also vary with climatic conditions—mainly temperature, light duration, and applied fertilizers [274]. Brassinolide application to salt stressed *Vigna radiata* caused enhancement in growth, photosynthetic rate, and maximum quantum yield of PSII. Generally, brassinolide has the potential to protect the photosynthetic apparatus under salt stress. It also contributed to improving the membrane stability index and leaf water potential. However, no significant results were recorded in the case of lipid peroxidation and electrolyte leakage. Brassinolide increases antioxidant enzyme and proline contents [275–278]. Increases in proline level create a protective shield when the plant is under stress by acting as a source of carbon and nitrogen, a stabilizer of the plasma membrane, and an oxygen radical scavenger [279]. Brassinosteroids also increase pigment levels in the plant [280]. BRs also improve the nitrate and nitrite reductase activity in *Vigna radiata* under salt stress. This effect can be attributed to the ability of BRs to modulate transcription and translation at the gene level, and to increase cell nitrate uptake. Increased stress tolerance caused by brassinolide application is observable as improved growth parameters such as shoot length, root length, and plant biomass [106,281].

6. Salinity and Carbohydrate (Sugars) Metabolism in Plants

Salinity stress progressively depletes carbohydrates in plant leaves and roots [282]. Young leaves have more hexose and starch accumulation compared to older leaves. Carbon metabolism can be used to assess the salt tolerance of a plant species. Salt-tolerant genotypes accumulated more sucrose than salt-sensitive genotypes. The role of carbohydrates in osmotic adjustment has also been authenticated [231,283]. Previous studies have reported a significant correlation between sugar accumulation levels and stress tolerance in various plant species [284]. This augmentation in sugar contents may be related to the high rate of sugar hydrolysis via hydrolytic enzymes [127,285]. Sugars stabilize the plasma membrane during plant stress by interacting with phospholipids. Under salt stress, a higher accumulation of carbohydrates in leaves may contribute to osmotic adjustment [122,286]. The higher chloride contents can cause increases in carbohydrate levels in plant tissues or starch degradation in sensitive species, whereas salt-tolerant species have less starch accumulation [287]. The carbohydrate accumulation rate varies among salt-tolerant species [147,288].

7. Salinity and Root Apoplastic Barriers

Plant roots are the main organ involved in the uptake of water and nutrients from a solution, whereas xylem vessels allow nutrient transport to the aerial tissues. In addition, plant roots function as the primary site for sensing salinity levels so that the plant can respond rapidly to maintain functionality. Roots can exclude and/or counteract potentially harmful substances by modifying their anatomy [289].

In particular, the endodermis that separates the cortex from the central cylinder is characterized by the development of specific wall modifications, called "apoplastic barriers" [290], formed by a combination of Casparian strips and suberin lamellae. A significant anatomical change in the root system due to salt stress is the deposition of hydrophobic polymers such as cutin and suberin on the cell wall, polymers that are often associated with hydrophobic compounds (e.g., waxes). Rossi et al. [291] reported that different apoplastic adjustments in roots modify Na^+ fluxes to the shoots of olive trees exposed to up to 120 mM NaCl. Similarly, Krishnamurthy et al. [292] showed that the Na^+ bypass flow in rice roots was reduced by the deposition of apoplastic barriers. These findings substantiated the role of root apoplastic barriers in plants' tolerance to salt stress. Afterwards, several studies have confirmed the formation of plant root apoplastic barriers as a response to different environmental stresses such as heavy metals [268], salt, and drought stress [291]. Overall, the literature indicates that plants react to environmental constraints by developing apoplastic barriers close to the root apex to mitigate the intrusion of toxic ions. This is a specific anatomical response by roots when exposed to hostile environments.

8. Salinity and Ionic Attributes

Abiotic stress alters the patterns of nutrient availability and transport, thus causing enormous changes in plant growth. Na^+ and Cl^- ions are the chief competitors which restrict nutrients like Ca^{2+}, K^+, and NO_3^-. The increase in uptake of Na^+ and Cl^- was more prominent in mature leaves compared to young actively growing leaves [59,293]. Plants experience a deficiency of both macro and micronutrients under salt stress. However, the extent to which salinity affects micronutrient availability is determined by plant type, growing conditions, and nutrient concentration [294,295]. Hence an effective fertilization regime in salt-affected areas can be utilized to overcome the negative effects of salinity [296,297]. The increase in nutrient availability to plants enhances the plant's ability to survive under stressed conditions [1].

8.1. *Salinity and Ionic Homeostasis*

Salinity impairs the ionic balance of the cells. Maintaining ionic balance in plants is important for increasing plant survivability under salt stress. Plants maintain ionic balance by various processes, including cellular uptake, sequestration, and ion inclusion and exclusion [8,298]. Early maturing clones of diploid potato accumulated Na ions in their leaves while late maturing varieties excluded Na ions from their leaves (Figure 5). Late maturing varieties of potato were more salt-tolerant compared to early cultivars. The lower leaves of the potato plant accumulated more Na ions than the higher leaves in both the tested potato cultivars. Late maturing genotypes also established higher K to Na ratio, conferring salt tolerance to them [50,298]. Saline conditions alter the ionic concentration of soil solutions, with an increase in Na^+ [74] and Cl^- ions, particularly [54,299]. With the increased absorption rate of Na and chloride ions, a significant decline in other ions (e.g., K, Ca, and Mg) occurs [54,297,300].

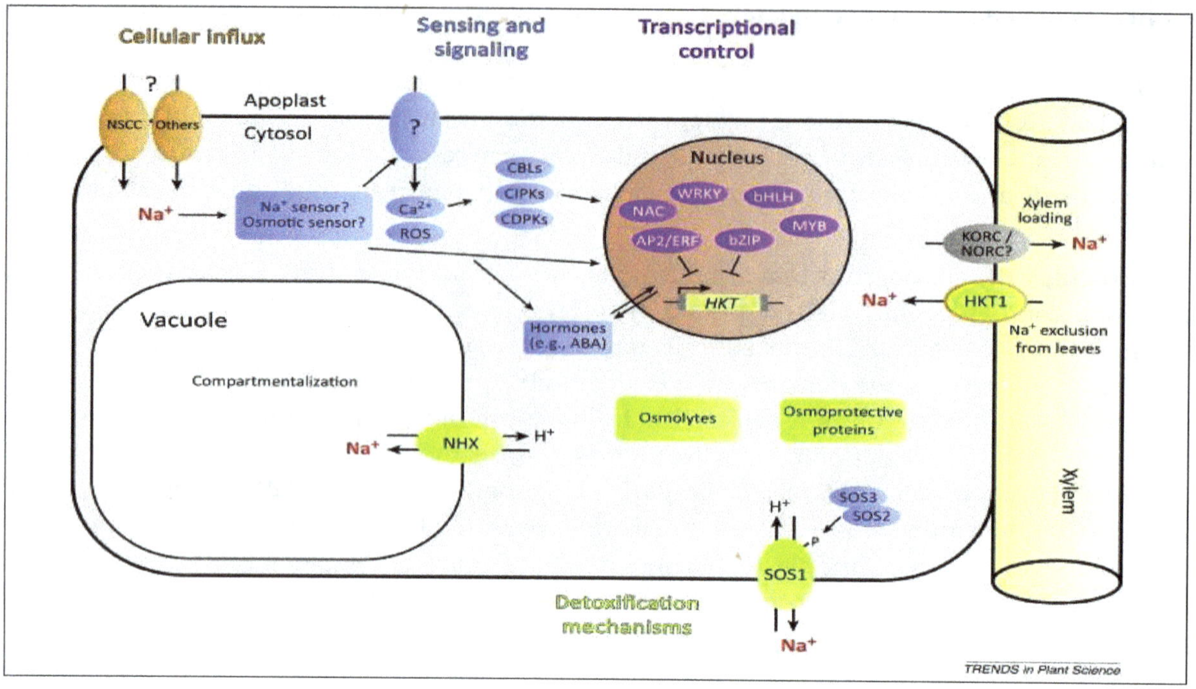

Figure 5. Overview of cellular Na^+ transport mechanisms and the salt stress response network in plant root cells. Na^+ (depicted in red) enters the cell via NSCCs and other membrane transporters (cellular Na^+-influx mechanisms highlighted in orange). Inside the cell, Na^+ is identified. This activates Ca^{2+}, ROS, and hormone signaling cascades. CBLs, CIPKs, and CDPKs are part of the Ca^{2+}-signaling pathway (sensing and signaling components highlighted in blue), which can alter the global transcriptional profile of the plant (transcription factor families in the nucleus depicted in purple; an AP2/ERF and a bZIP transcription factor that negatively regulate *HKT* gene expression are shown as an example). Ultimately, these early signaling pathways result in expression and activation of cellular detoxification mechanisms, including HKT, NHX, and the SOS Na^+ transport mechanisms, as well as osmotic protection strategies (cellular detoxification mechanisms highlighted in light green). Furthermore, the Na^+ distribution in the plant is regulated in a tissue-specific manner by unloading of Na^+ from the xylem. Abbreviations: NSCCs, nonselective cation channels; ROS, reactive oxygen species; CDPKs, calcium-dependent protein kinases; CBLs, calcineurin B-like proteins; CIPKs, CBL-interacting protein kinases; AP2/ERF, APETALA2/ETHYLENE RESPONSE FACTOR; bZIP, basic leucine zipper; NHX, Na^+/H^+ exchanger; SOS, salt overly sensitive [301].

Sodium is not considered an essential element for plant growth. Plants do not show any Na requirements, and they lack any particular transport system for Na. However, when the plant is exposed to high Na concentrations, Na finds its way into the plant cells following various pathways. It may gain entry into plant cell passively from soil solution osmolarity, or via voltage-dependent and independent cation channels [8]. Due to the similarity of hydrated ionic radii for Na^+ and K^+, Na^+ hampers K^+ absorption by the plant because transport proteins cannot distinguish between them. At higher a cystolic Na^+/K^+ concentration ratio, cellular processes are affected, which would otherwise be maintained by K^+, such as protein synthesis and enzyme activity [302]. Reduced K uptake can cause reductions in plant growth and productivity under saline conditions. Plants restrict the free cytosolic movement of Na in the cytosol by vacuole compartmentalization, in order to prevent Na from hampering the regular functioning of cytosolic enzymes. This mechanism is equally important for both glycophytes and halophytes [303]. Salt tolerance of a plant species is dependent on its ability to restrict translocation of toxic ions in shoots [304,305]. This ability is regulated by specific tissues [17], morphological features [306–308], and water use efficiency. These adaptive mechanisms alter the plant response to salinity at both cellular and entire plant levels [17,285,309].

8.2. Ionic Influx and Efflux

Sodium and chloride ions are the major ions which induce harmful effects on plant growth. The accumulation of these ions in leaves and roots determines the tolerance of a genotype to salt stress [310]. Roots are the primary organ which permits the entry of toxic ions into the plant under high saline conditions. Plant roots regulate Na and chloride contents via the extrusion mechanism. Plants respond to toxic salt levels either by exclusion to soil or by upward movement of ions through the xylem transpiration stream [311]. Phloem is a chief pathway for toxic ion transport from shoots to roots [17,278]. A passive mode of Na transport from external medium into the cell involves uniporters or ion channel type transporters like HKT, LCT1, and NSCC [45,312]. In roots, Na reaches the xylem using symplastic and apoplastic routes from the epidermis. The rate of Na transport is dependent on the barriers posed by the casparian strip and the Na extrusion from the cell. Sodium ion transport is also mediated by high-affinity potassium transporters which act as Na^+/K^+ symporters and also as Na^+-selective uniporters [313].

High salt concentration causes an increase in electrolyte leakage [106]. Sodium ion accumulation at the cellular level is a major consequence of saline environmental conditions. When Na ion levels are high, Na ions are either extruded or compartmentalized in vacuoles via Na^+/H^+ antiporters [45,314]. Ion transporters like Na^+/H^+ antiporters are involved in Na^+ ion extrusion or Na^+ ion compartmentalization in vacuoles. These antiporters may be plasma membrane-localized or tonoplast antiporters, which utilize the pH gradient, developed by P-type H^+-ATPases, respectively [7,8]. Electrochemical potential channels are the other source of ion transport across the cell membrane. These channels are ion specific and regulate ionic movement through gating, the period during which channels are open or closed [315].

Sodium extrusion and vacuolar compartmentalization are active (energy requiring) processes compared to sodium influx (passive transport). K^+ transporters are considered responsible for sodium influx into the cell. Movement of Na^+ ions from the cytosol to the vacuole is accomplished by Na^+/H^+ antiporters [316]. The plant's capacity to tolerate saline conditions depends on a high K^+/Na^+ cytosolic concentration. Plants adopt various strategies to cope with a high Na ion concentration, such as the restriction of Na^+ entry into the cell, Na^+ extrusion, and vacuole compartmentalization [279]. Sodium compartmentalized in the vacuole is an osmolytic process which provides a mechanism for water uptake. Sodium efflux is considered an adaptation of salt-sensitive plants, whereas Na compartmentalization is considered a feature of salt-tolerant species. The high salinity tolerance of halophytes can also be attributed to their ability to confine Na ions to their roots, making them Na accumulators [45,317]. Sodium exclusion from the plant body may also involve salt glands, present on the leaf surface [159].

Chloride also contributes to the undesirable effects of salt stress on plant growth. Salt-tolerant plant genotypes have the ability to inhibit chloride uptake. Chloride uptake is dependent on the shoot to root ratio and follows a passive transport system [286,318]. Higher chloride contents also induce succulence in salt-stressed plants [298], as well as a reduction in nitrate reductase activity [319]. Nitrate reductase activity inhibition is an indicator of poor nitrogen assimilation, leading to reduced protein synthesis and plant growth [320].

Adjusting the osmotic potential of the cell by K accumulation in vacuoles through different K channels and transporters appears to be a prominent plant strategy to cope with plant stress [321,322]. Potassium alters the membrane potential and turgor, maintains enzyme activities, and adjusts osmotic pressure and stomatal movement. It aids the plant in photosynthesis, protein synthesis, and oxidant metabolism [323–325].

9. Effects of Salinity on Potassium and Calcium

Potassium regulates protein synthesis, enzymatic metabolism, and photosynthesis. Low K levels were recorded in potato seedlings experiencing salt stress [31,73,326]. Potassium competes with Na under saline conditions. In saline soil, Na ions which reach the plasma membrane of a cell cause the membrane to depolarize and the K outward rectifier channels to open. This results in a loss of K from the cell [327]. Importantly, the K^+:Na^+ ratio determines the saline resistance of plants. Higher K^+ concentration is directly related to higher biomass production. Salt-tolerant plant species have the ability to retain a higher concentration of K [72,159].

Potassium and Ca occupy a critical position in the regulation of cell membrane integrity and function [328]. Calcium aids in various physiological processes, including solute movement, stomatal regulation, molecular signaling for cell defense systems, and cell repair under stress. Under saline conditions, Na^+/Ca^{2+} interactions are important because Na^+ has the ability to displace Ca^{2+} from its binding sites, thus causing a decline in Ca^{2+} availability. The cytosolic Ca concentration determines the salt sensitivity of the plant [8,329]. Calcium deficiency is evident in plants grown in saline soil conditions [330]. Calcium supplementation is beneficial in ameliorating the negative effects of salinity in beans [331,332]. The ability of plants to survive under osmotic stress is dependent on the plant's capacity to maintain high Ca^{2+}:Na^+ ratio and to exclude Na^+. Increasing Ca^{+2} (by addition of Ca^{2+} as gypsum $CaSO_4$) had an antagonistic effect on Mg^{2+} availability to the plant, as it removed Mg^{2+} from the soil complex [72,333].

Salinization management through Ca^{+2} supplementation is reported to be beneficial in enhancing the quality of celery by reducing the incidence of black heart. Calcium supplementation in the form of calcium sulfate improved the growth of tomato plants. This stabilized membrane permeability, increasing N and K concentrations in leaves [67]. Calcium sulfate has more pronounced effects on tomato growth parameters than calcium chloride, perhaps because calcium chloride can be a source of chloride ions [188]. The supplementation of K and Ca^{2+} to the salt-affected pepper plants increased vegetative growth and fruit production and decreased the incidence of blossom end rot, but caused a reduction in fresh fruit weight and marketable yield [4]. Calcium acetate positively stimulates photosynthesis and stomatal conductance. It also ameliorates the effects of low water content on osmotic potential from salt stress. However, Ca^{2+} is reported to have an inhibitory effect on proline accumulation, which affects the low osmotic potential [34,334].

Salinity and Nitrogen

Nitrogen is required in considerable quantities to satisfy the mineral needs of the plant. Moreover, nitrogen is a constituent of amino acids and nucleic acids [311]. Salt stress decreased protein contents in potatoes [335]. Nitrogen has a large effect on plant growth when there is ample water. Still, addition of nitrogen to plants, even under salt stress conditions, improves the yield in many crop plants, including tomato, millet, and wheat. Nevertheless, in a saline environment, plants have a decreased ability to uptake nitrogen [1,31]. Plants that are more susceptible to Cl^- toxicity can be managed through NO_3^- application since NO_3^- antagonizes Cl^- uptake [336]. Saline conditions reduce the regulation of nitrate uptake, metabolism, and utilization by plant species. The form of nitrogen applied affects the uptake of other nutrients like Na^+, Ca^{2+}, and K^+, and thus affects the plant's ability to tolerate saline conditions. Therefore, under saline conditions, a nitrogen fertilizer regime should be managed specifically considering the interactions between Na, NO_3, NH_4, Cl^-, etc. [337]. Potassium nitrate is a salt stress alleviating agent for melons [31].

Higher salt concentration in soil solutions interferes with the transport of nitrate in the shoots. For that reason, nitrate reductase, glutamine synthetase, and nitrite reductase enzymes all have reduced activity under stress conditions [246,338]. Nitrate reductase is responsible for the reduction of nitrate to ammonia. Nitrate reductase activity is generally less susceptible to salinity than nitrite reductase activity [339]. Increasing salinity levels decreased dry weight and protein contents of the leaves and roots of tomato seedlings. High Na and chloride ion levels suppressed the uptake of K and nitrate [340].

Nitrogen-containing compounds (NCC) accumulate in plants as a response to higher saline conditions. These compounds include amino acids, amides, quaternary ammonium compounds, and polyamines. Their reported functions are osmoprotection, osmotic adjustment, ROS scavenging, nitrogen provision, and maintenance of pH. Plant nitrogen metabolism can be regulated by phytohormone like cytokinin [29,341].

Phosphorus (P) is a prime constituent in nucleic acids, phospholipids, phosphoproteins, nucleotides, and ATP. Phosphorus uptake is reduced in plants growing under saline conditions. The availability of P is reduced due to strong ionic effects and low solubility of calcium phosphate minerals [302]. Navarro et al. [342] reported a decline in phosphorus mobility stored in vacuoles in salt-stressed melon plants. Plant growth promoting bacteria increased uptake of P and K in tomato plants grown under salt stress. P uptake may enhance the survivability of young plants in salt stress conditions [343–346].

10. Conclusions

Plants respond to salinity stress at physiological, cellular, genetic, and metabolic levels. Previous research demonstrates that among plant responses to salinity, mechanisms that control ion uptake, transport, and balance, as well as water potential, photosynthesis, cell division, osmotic adjustment, enzymatic activities, polyamine regulation, stress signaling, and regulation of root apoplastic barriers play critical roles in plant tolerance to salinity. In order to manage salt stress, significant work has been done on calcium biosensors to understand cation-sensing processes, ABA-dependent phosphorylation, changes in cell wall components, auxin and ABA associated modulations in root architecture, Na^+ exclusion mechanisms, signaling of phytohormones in roots and guard cells, and organ-specific expression of sodium transporter gene (*HKT1*). There is a need to integrate information from genomic, transcriptomic, proteomic, and metabolomics studies, as a collaborative approach for determining key pathways controlling salinity tolerance at the whole-plant level. Further studies on tissue-specific Na-sensing processes, association of Na and K biosensors with Na^+/K^+ homeostasis, ABA and other phytohormone signaling pathways, and interaction between phytohormones and ion transporters are required to illuminate the details of inter- and intracellular molecular interactions that are involved in plant stress tolerance. Further research in these areas will be helpful for mitigating salinity damage in commercially important crops.

Author Contributions: Conceived of the presented idea, M.A.S., A.S. and N.K.; developed the theory and performed the computations, M.A.S., N.K., R.M.B. and L.R.; writing—original draft preparation, M.A.S., A.S., N.K., S.A., C.G., N.M., W.N. and F.G.-S.; writing—review and editing, M.A.S., A.S., N.K. and S.A.; supervision, M.A.S. and N.K.; verified the numerical results, R.M.B., S.A., L.R., C.G., N.M., W.N. and F.G.-S. All authors have read and agreed to the published version of the manuscript.

References

1. Shrivastava, P.; Kumar, R. Soil salinity: A serious environmental issue and plant growth promoting bacteria as one of the tools for its alleviation. *Saudi J. Biol. Sci.* **2015**, *22*, 123–131. [CrossRef]
2. Rogel, J.A.; Ariza, F.A.; Silla, R.O. Soil salinity and moisture gradients and plant zonation in Mediterranean salt marshes of Southeast Spain. *Wetlands* **2000**, *20*, 357–372. [CrossRef]
3. Flowers, T.J.; Garcia, A.; Koyama, M.; Yeo, A.R. Breeding for salt tolerance in crop plants—The role of molecular biology. *Acta Physiol. Plant.* **1997**, *19*, 427–433. [CrossRef]
4. Rubio, J.S.; Garcia-Sanchez, F.; Rubio, F.; Martinez, V. Yield, blossom-end rot incidence and fruit quality in pepper plants under moderate salinity are affected by K^+ and Ca^{+2} fertilization. *Sci. Hortic.* **2009**, *119*, 79–87. [CrossRef]
5. Ghassemi, F.; Jakeman, A.J.; Nix, H.A. *Salinization of Land and Water Resources*; University of New south Wales Press Ltd.: Canberra, Australia, 1995.

6. Sonowal, H.; Pal, P.B.; Shukla, K.; Ramana, K.V. Ramana, Aspalatone Prevents VEGF-Induced Lipid Peroxidation, Migration, Tube Formation, and Dysfunction of Human Aortic Endothelial Cells. *Oxid. Med. Cell. Longev.* **2017**, *11*, 2017. [CrossRef]
7. Wang, H.; Tang, X.; Shao, C.; Shao, H.; Wang, L. Molecular cloning and bioinformatics analysis of a new plasma membrane Na+/H+ antiporter gene from the halophyte Kosteletzkya virginica. *Sci. World J.* **2014**, *2014*, 141675.
8. Xiong, L.; Zhu, J.K. Molecular and genetic aspects of plant responses to osmotic stress. *Plant Cell Environ.* **2002**, *25*, 131–139. [CrossRef] [PubMed]
9. Meng, X.; Zhou, J.; Sui, N. Mechanisms of salt tolerance in halophytes: Current understanding and recent advances. *Open Llife Sci.* **2018**, *13*, 149–154. [CrossRef]
10. Ashraf, M.H.P.J.C.; Harris, P.J. Photosynthesis under stressful environments: An overview. *Photosynthetica* **2013**, *51*, 163–190. [CrossRef]
11. Siew, R.; Klein, C.R. The effect of NaCl on some metabolic and fine structural changes during the greening of etiolated leaves. *J. Cell Biol.* **1969**, *37*, 590–596. [CrossRef]
12. Slama, I.; Abdelly, C.; Bouchereau, A.; Flowers, T.; Savoure, A. Diversity, distribution and roles of osmoprotective compounds accumulated in halophytes under abiotic stress. *Ann. Bot.* **2015**, *115*, 433–447. [CrossRef] [PubMed]
13. Cushman, J.C. Molecular cloning and expression of chloroplast NADP-malate dehydrogenase during crassulacean acid metabolism induction by salt stress. *Photosynth. Res.* **1990**, *35*, 15–27. [CrossRef]
14. Sévin, D.C.; Stählin, J.N.; Pollak, G.R.; Kuehne, A.; Sauer, U. Global Metabolic Responses to Salt Stress in Fifteen Species. *PLoS ONE* **2016**, *11*, e0148888. [CrossRef] [PubMed]
15. Dschida, W.J.; Platt-Aloia, K.A.; Thomson, W.W. Epidermal peels of *Avicennia germinans* (L.) Stearn: A useful system to study the function of salt glands. *Ann. Bot. Lond.* **1992**, *70*, 501–509. [CrossRef]
16. Robert, I.L.; Frank, W.J.; Summy, K.R.; Hudson, D.Y.; Richard, S. The Biological Flora of Coastal Dunes and Wetlands. *Avicennia Germinans J. Coast. Res.* **2017**, *33*, 191–207.
17. Tester, M.; Davenport, R. Na^+ tolerance and Na^+ transport in higher plants. *Ann. Bot.* **2003**, *91*, 503–527. [CrossRef] [PubMed]
18. Shahid, M.A.; Balal, R.M.; Khan, N.; Rossi, L.; Rathinasabapathi, B.; Liu, G.; Khan, J.; Cámara-Zapata, J.M.; Martínez-Nicolas, J.J.; Garcia-Sanchez, F. Polyamines provide new insights into the biochemical basis of Cr-tolerance in Kinnow mandarin grafted on diploid and double-diploid rootstocks. *Environ. Exp. Bot.* **2018**, *1*, 248–260. [CrossRef]
19. Zhu, J.; Hasegawa, P.M.; Bressan, R.A. Molecular aspects of osmotic stress in plants. *Citri. Rev. Plant Sci.* **1997**, *16*, 253–277. [CrossRef]
20. Serrano, R.; Mulet, J.M.; Rios, G.; Marquez, J.A.; de Larrinoa, I.F.; Leube, M.P.; Mendizabal, I.; Pascual-Ahuir, A.; Proft, M.; Ros, R.; et al. A glimpse of the mechanism of ion homeostasis during salt stress. *J. Exp. Bot.* **1999**, *50*, 1023–1036. [CrossRef]
21. Flowers, T.J.; Munns, R.; Colmer, T.D. Sodium chloride toxicity and the cellular basis of salt tolerance in halophytes. *Ann. Bot.* **2015**, *115*, 419–431. [CrossRef]
22. Khan, N.; Bano, A.; Babar, M.A. The stimulatory effects of plant growth promoting rhizobacteria and plant growth regulators on wheat physiology grown in sandy soil. *Arch. Microbiol.* **2019**, *201*, 769–785. [CrossRef] [PubMed]
23. Price, A.; Hendry, G. Iron catalysed oxygen radical formation and its possible contribution to drought damage in nine native grasses and three cereals. *Plant Cell Environ.* **1991**, *14*, 477–484. [CrossRef]
24. Khan, N.; Bano, A. Effects of exogenously applied salicylic acid and putrescine alone and in combination with rhizobacteria on the phytoremediation of heavy metals and chickpea growth in sandy soil. *Int. J. Phytoremed.* **2018**, *16*, 405–414. [CrossRef] [PubMed]
25. Stevens, R.M.; Pech, J.M.; Grigson, G.J. A Short Non-Saline Sprinkling Increases the Tuber Weights of Saline Sprinkler Irrigated Potatoes. *Agronomy* **2017**, *7*, 4. [CrossRef]
26. Nxele, X.; Klein, A.; Ndimba, B.K. Drought and salinity stress alters ROS accumulation, water retention, and osmolyte content in sorghum plants. *Afr. J. Bot.* **2017**, *1*, 261–266. [CrossRef]
27. Golldack, D.; Li, C.; Mohan, H.; Probst, N. Tolerance to drought and salt stress in plants: Unraveling the signaling networks. *Front. Plant Sci.* **2014**, *5*, 151. [CrossRef]

28. Naseem, H.; Ahsan, M.; Shahid, M.A.; Khan, N. Exopolysaccharides producing rhizobacteria and their role in plant growth and drought tolerance. *J. Basic Microbiol.* **2018**, *58*, 1009–1022. [CrossRef]
29. Zekri, M.; Parsons, L.R. Calcium influnces growth and leaf mineral concentration of citrus under saline conditions. *Hort. Sci.* **1990**, *25*, 784–786. [CrossRef]
30. Kumari, A.; Das, P.; Parida, A.K.; Agarwal, P.K. Proteomics, metabolomics, and ionomics perspectives of salinity tolerance in halophytes. *Front. Plant Sci.* **2015**, *6*, 537. [CrossRef]
31. Saha, J.; Brauer, E.K.; Sengupta, A.; Popescu, S.C.; Gupta, K.; Gupta, B. Polyamines as redox homeostasis regulators during salt stress in plants. *Front. Environ. Sci.* **2015**, *3*, 21. [CrossRef]
32. Mansour, M.M.F. Cell permeability under salt stress. In *Strategies for Improving Salt Tolerance in Higher Plants;* Jaiwal, P.K., Singh, R.P., Gulati, A., Eds.; Oxford and IBH: New Delhi, India, 1997; pp. 87–110.
33. Ali, S.; Xie, L. Plant Growth Promoting and Stress mitigating abilities of Soil Born Microorganisms. *Recent Pat. Food Nutr. Agric.* **2019**, *15*. [CrossRef] [PubMed]
34. Mansour, M.M.F.; Salama, K.H.A. Cellular basis of salinity tolerance in plants. *Environ. Exp. Bot.* **2004**, *52*, 113–122. [CrossRef]
35. Le Gall, H.; Philippe, F.; Domon, J.M.; Gillet, F.; Pelloux, J.; Rayon, C. Cell wall metabolism in response to abiotic stress. *Plants* **2015**, *4*, 112–166. [CrossRef] [PubMed]
36. Al-Amri, S.M. Improved growth, productivity and quality of tomato (*Solanum lycopersicum* L.) plants through application of shikimic acid. *Saudi J. Biol. Sci.* **2013**, *1*, 339–345. [CrossRef] [PubMed]
37. Mahlooji, M.; Seyed Sharifi, R.; Razmjoo, J.; Sabzalian, M.R.; Sedghi, M. Effect of salt stress on photosynthesis and physiological parameters of three contrasting barley genotypes. *Photosynthetica* **2017**, *56*, 549–556. [CrossRef]
38. Shekari, F.; Abbasi, A.; Mustafavi, S.H. Effect of silicon and selenium on enzymatic changes and productivity of dill in saline condition. *J. Saudi Soc. Agric. Sci.* **2017**, *1*, 367–374. [CrossRef]
39. Dolatabadian, A.; Sanavy, S.A.M.M.; Gholamhoseini, M.; Joghan, A.K.; Majdi, M.; Kashkooli, A. The role of calcium in improving photosynthesis and related physiological and biochemical attributes of spring wheat subjected to simulated acid rain. *Physiol. Mol. Biol. Plants.* **2013**, *19*, 189–198. [CrossRef]
40. Farooq, S.; Azam, F. The use of cell membrane stability (CMS) technique to screen for salt tolerant wheat varieties. *J. Plant Physiol.* **2006**, *163*, 629–637. [CrossRef]
41. Munns, R. Comparative physiology of salt and water stress. *Plant Cell Environ.* **2002**, *25*, 239–250. [CrossRef]
42. Dolatabadi, N.; Toorchi, M. Rapeseed (*Brassica napus* L.) genotypes response to NaCl salinity. *J. Biodiv. Environ. Sci.* **2017**, *10*, 265–270.
43. Wu, J.; Seliskar, D.M.; Gallagher, J.L. Stress tolerance in the march plant Spartina patens: Impact of NaCl on growth and root plasma membrane lipid composition. *Physiol. Plant* **1998**, *102*, 307–317. [CrossRef]
44. Balal, R.M.; Shahid, M.A.; Vincent, C.; Zotarelli, L.; Liu, G.; Mattson, N.S.; Garcia-Sanchez, F. (Kinnow mandarin plants grafted on tetraploid rootstocks are more tolerant to Cr-toxicity than those grafted on its diploids one. *Environ. Exp. Bot.* **2017**, *140*, 8–18. [CrossRef]
45. Plant, A.L.; Bray, E.A. *Regulation of Gene Expression by Abscissic Acid During Environmental Stresses;* Marcel Dekker: New York, NY, USA, 1999; pp. 303–331.
46. Delphine, S.; Alvino, A.; Zacchini, M.; Loreb, F. Consequences of salt stress on conductance to CO_2 diffusion, Rubisco characteristics and anatomy of spinach leaves. *Aust. J. Plant Physiol.* **1998**, *25*, 395–402. [CrossRef]
47. Mehmood, A.; Khan, N.; Irshad, M.; Hamayun, M. IAA Producing Endopytic Fungus Fusariun oxysporum wlw Colonize Maize Roots and Promoted Maize Growth Under Hydroponic Condition. *Eur. Exp. Biol.* **2018**, *8*, 24. [CrossRef]
48. Romero-Aranda, R.; Moya, J.L.; Tadeo, F.R.; Legaz, F.; Primo-Millo, E.; Talon, M. Physiological and anatomical disturbances induced by chloride salts in sensitive and tolerant citrus: Beneficial and detrimental effects of cations. *Plant Cell Environ.* **1998**, *21*, 1243–1253. [CrossRef]
49. Hasegawa, P.M.; Bressan, R.A.; Zhu, J.K.; Bohnert, H.J. Plant cellular and molecular responses to high salinity. *Annu. Rev. Plant Physiol.* **2000**, *51*, 463–499. [CrossRef]
50. Blumwald, E.; Aharon, G.S.; Apse, M.P. Sodium transport in plant cells. *Biochim. Biophys. Acta* **2000**, *1465*, 140–151. [CrossRef]
51. Maser, P.; Gierth, M.; Schroeder, J.I. Molecular mechanisms of potassium and sodium uptake in plants. *Plant Soil.* **2002**, *247*, 43–54. [CrossRef]

52. Khan, N.; Bano, A. Modulation of phytoremediation and plant growth by the treatment with PGPR, Ag nanoparticle and untreated municipal wastewater. *Int. J. Phytoremed.* **2016**, *18*, 1258–1269. [CrossRef]
53. Zribi, L.; Fatma, G.; Fatma, R.; Salwa, R.; Hassan, N.; Nejib, R.M. Application of chlorophyll fluorescence for the diagnosis of salt stress in tomato "*Solanum lycopersicum* (variety Rio Grande)". *Sci. Hortic.* **2009**, *120*, 367–372. [CrossRef]
54. Orlovsky, N.; Japakova, U.; Zhang, H.; Volis, S. Effect of salinity on seed germination, growth and ion content in dimorphic seeds of *Salicornia europaea* L. (Chenopodiaceae). *Plant Divers.* **2016**, *1*, 183–189. [CrossRef] [PubMed]
55. Shaterian, J.; Waterer, D.; De Jong, H.; Tanino, K.K. Differential stress responses to NaCl salt application in early and late maturing diploid potato (*Solanum* sp.) clones. *Environ. Exp. Bot* **2005**, *54*, 202–212. [CrossRef]
56. Chartzoulakis, K.; Klapaki, G. Response of two greenhouse pepper hybrids to NaCl salinity during different growth stages. *Sci. Hortic.* **2000**, *86*, 247–260. [CrossRef]
57. Kaya, C.; Tuna, A.L.; Ashrafa, M.; Altunlu, H. Improved salt tolerance of melon (*Cucumis melo* L.) by the addition of proline and potassium nitrate. *Environ. Exp. Bot* **2007**, *60*, 397–403. [CrossRef]
58. Giuffrida, F.; Graziani, G.; Fogliano, V.; Scuderi, D.; Romano, D.; Leonardi, C. Effects of nutrient and NaCl salinity on growth, yield, quality and composition of pepper grown in soilless closed system. *J. Plant Nutr.* **2014**, *37*, 1455–1474. [CrossRef]
59. Jiang, C.; Zu, C.; Lu, D.; Zheng, Q.; Shen, J.; Wang, H.; Li, D. Effect of exogenous selenium supply on photosynthesis, Na+ accumulation and antioxidative capacity of maize (*Zea mays* L.) under salinity stress. *Sci. Rep.* **2007**, *7*, 42039. [CrossRef]
60. Savvas, D.; Lenz, F. Effect of NaCl or nutrient-induced salinity on growth, yield and composition of egg plants grown in rockwool. *Sci. Hortic.* **2000**, *84*, 37–47. [CrossRef]
61. Abbas, W.; Ashraf, M.; Akram, N.A. Alleviation of salt-induced adverse effects in egg plant (*Solanum melongena* L.) by glycinebetaine and sugar beet extracts. *Sci. Hortic.* **2010**, *125*, 188–195. [CrossRef]
62. Mahjoor, F.; Ghaemi, A.A.; Golabi, M.H. Interaction effects of water salinity and hydroponic growth medium on eggplant yield, water-use efficiency, and evapotranspiration. *Int. Soil Water Conserv.* **2016**, *1*, 99–107. [CrossRef]
63. Balal, R.M.; Shahid, M.A.; Javaid, M.M.; Iqbal, Z.; Liu, G.D.; Zotarelli, L.; Khan, N. Chitosan alleviates phytotoxicity caused by boron through augmented polyamine metabolism and antioxidant activities and reduced boron concentration in *Cucumis sativus* L. *Acta Physiol. Plant.* **2017**, *1*, 31. [CrossRef]
64. Tabatabaie, S.J.; Nazari, J. Influence of nutrient concentrations and NaCl salinity on the growth, photosynthesis, and essential oil content of peppermint and lemon verbena. *Turk. J. Agric. For.* **2007**, *31*, 245–253.
65. Pardossi, A.; Bagnoli, G.; Malorgio, F.C.; Campiotti, A.; Tognoni, F. NaCl effects on celery (*Apium graveolens* L.) grown in NFT. *Sci. Hortic* **1999**, *81*, 229–242. [CrossRef]
66. Zheng, Y.; Jia, A.; Ning, T.; Xu, J.; Li, Z.; Jiang, G. Potassium nitrate application alleviates sodium chloride stress in winter wheat cultivars differing in salt tolerance. *J. Plant Physiol.* **2008**, *165*, 1455–1465. [CrossRef] [PubMed]
67. Saini, S.; Sharma, I.; Pati, P.K. Versatile roles of brassinosteroid in plants in the context of its homoeostasis, signaling and crosstalks. *Front. Plant Sci.* **2015**, *6*, 950. [CrossRef] [PubMed]
68. Kalhoro, N.A.; Rajpar, I.; Kalhoro, S.A.; Ali, A.; Raza, S.; Ahmed, M.; Kalhoro, F.A.; Ramzan, M.; Wahid, F. Effect of salts stress on the growth and yield of wheat (*Triticum aestivum* L.). *Am. J. Plant Sci.* **2016**, *7*, 2257. [CrossRef]
69. Xie, Z.; Duan, L.; Tian, X.; Wang, B.; Eneji, A.E.; Li, Z. Coronatine alleviates salinity stress in cotton by improving the antioxidation defence system and radical-scavenging activity. *J. Plant Physiol.* **2008**, *165*, 375–384. [CrossRef]
70. Sun, C.; Feng, D.; Mi, Z.; Li, C.; Zhang, J.; Gao, Y.; Sun, J. Impacts of Ridge-Furrow Planting on Salt Stress and Cotton Yield under Drip Irrigation. *Water* **2017**, *9*, 49. [CrossRef]
71. Caines, A.M.; Shennan, C. Interactive effects of Ca^{+2} and NaCl salinity on the growth of two tomato genotypes differing in Ca^{+2} use efficiency. *Plant Physiol. Biochem.* **1999**, *37*, 569–576. [CrossRef]
72. Evers, D.; Hemmer, K.; Hausman, J.F. Salt stress induced biometric and physiological changes in *Solanum tuberosum* L. cv. Bintje grown in vitro. *Acta Physiol. Plant* **1998**, *20*, 3–7. [CrossRef]
73. Gao, Y.; Lu, Y.; Wu, M.; Liang, E.; Li, Y.; Zhang, D.; Yin, Z.; Ren, X.; Dai, Y.; Deng, D.; et al. Ability to Remove Na^+ and Retain K^+ Correlates with Salt Tolerance in Two Maize Inbred Lines Seedlings. *Front. Plant Sci.* **2016**, *7*, 1716. [CrossRef]

74. Hu, Y.; Xia, S.; Su, Y.; Wang, H.; Luo, W.; Su, S.; Xiao, L. Brassinolide Increases Potato Root Growth In Vitro in a Dose-Dependent Way and Alleviates Salinity Stress. *Biomed. Res. Int.* **2016**, 8231873. [CrossRef]
75. Alom, R.; Hasan, M.A.; Islam, M.R.; Wang, Q.F. Germination characters and early seedling growth of wheat (*Triticum aestivum* L.) genotypes under salt stress conditions. *JCSB* **2016**, *1*, 383–392. [CrossRef]
76. Tuna, A.L.; Kaya, C.; Ashraf, M.; Altunlu, H.; Yokas, I.; Yagmur, B. The effect of calcium sulphate on growth, membrane stability and nutrient uptake of tomato plants grown under salt stress. *Environ. Exp. Bot.* **2007**, *59*, 173–178. [CrossRef]
77. De Pascale, S.; Barbieri, G. Effects of soil salinity and top removal on growth and yield of broad bean as a green vegetable. *Sci. Hortic* **1997**, *71*, 147–165. [CrossRef]
78. Verma, S.; Mishra, S.N. Putrescine alleviation of growth in salt stressed *Brassica juncea* by inducing antioxidative defense system. *J. Plant Physiol.* **2005**, *162*, 669–677. [CrossRef] [PubMed]
79. Hu, Y.; Schmidhalter, U. Drought and Salinity: A comparison of their effects on mineral nutrition of plants. *J. Plant Nutr. Soil. Sci.* **2005**, *168*, 541–549. [CrossRef]
80. Rahnama, H.; Ebrahimzadeh, H. The effect of NaCl on antioxidant activities in potato seedlings. *Biol. Plant.* **2005**, *49*, 93–97. [CrossRef]
81. Korkmaz, A.; Şirikçi, R.; Kocaçınar, F.; Değer, Ö.; Demirkırıan, A.R. Alleviation of salt-induced adverse effects in pepper seedlings by seed application of glycinebetaine. *Sci. Hortic.* **2012**, *4*, 197–205. [CrossRef]
82. Roy, T.S.; Chakraborty, R.; Parvez, M.N.; Mostofa, M.; Ferdous, J.; Ahmed, S. Yield, dry matter and specific gravity of exportable potato: Response to salt. *Univ. J. Agric. Res.* **2017**, *5*, 98–103.
83. Mulholland, B.J.; Taylor, I.B.; Jackson, A.C.; Thornpson, A.J. Can ABA mediate responses of salinity have stressed tomato. *Environ. Exp. Bot.* **2003**, *50*, 17–28. [CrossRef]
84. Mayak, S.; Tirosh, T.; Glick, B.R. Plant growth promoting bacteria confer resistance in tomato plants to salt stress. *Plant Sci. Plant Sci.* **2004**, *42*, 565–572.
85. Smalle, J.; Van Der Straeten, D. Ethylene and vegetative development. *Physiol. Plant.* **1997**, *100*, 593–605. [CrossRef]
86. Akhtar, S.S.; AkhtarAktas, H.; Abak, K.; Cakmak, I. Genotypic variation in the response of pepper to salinity. *Sci. Hortic.* **2006**, *110*, 260–266.
87. Delgado, M.J.; Garrido, J.M.; Ligero, F.; Lluch, C. Nitrogen fixation and carbon metabolism by nodules and bacteroids of pea plants under sodium chloride stress. *Physiol. Plant.* **1993**, *89*, 824–829. [CrossRef]
88. Borucki, W.; Sujkowska, M. The effects of sodium chloride-salinity upon growth, nodulation, and root nodule structure of pea (*Pisum sativum* L.) plants. *Acta Physiol. Plant.* **2008**, *30*, 293–301. [CrossRef]
89. Andrzej, T.; Philip, P. Role of root microbiota in plant productivity. *J. Exp. Bot.* **2015**, *66*, 2167–2175.
90. Zahran, H.H. Rhizobium-Legume symbiosis and nitrogen fixation under severe conditions and in an arid climate. *Microbiol. Mol. Biol. R.* **1999**, *63*, 968–989. [CrossRef]
91. Batista-Santos, P.; Duro, N.; Rodrigues, A.P.; Semedo, J.N.; Alves, P.; da Costa, M.; Graça, I.; Pais, I.P.; Scotti-Campos, P.; Lidon, F.C.; et al. Is salt stress tolerance in *Casuarina glauca* Sieb. ex Spreng. associated with its nitrogen-fixing root-nodule symbiosis? An analysis at the photosynthetic level. *Plant Physiol. Biochem.* **2015**, *1*, 97–109. [CrossRef]
92. Othman, Y.; Al-karaki, G.; Al-Tawaha, A.R.; Al-Horani, A. Variation in germination and ion uptake in barley genotypes under salinity conditions. *World J. Agric. Res.* **2006**, *2*, 11–15.
93. Kandil, A.A.; Shareif, E.; Gad, M.A. Effect of Salinity on Germination and Seeding Parameters of Forage Cowpea Seed. *Res. J. Seed Sci.* **2017**, *10*, 17–26.
94. Khan, M.A.; Rizvi, Y. Effect of salinity, temperature and growth regulators on the germination and early seedling growth of *Atriplex griffithi* Var. Stocksii. *Can J. Bot.* **1994**, *72*, 475–479. [CrossRef]
95. Rajjou, L.; Duval, M.; Gallardo, K.; Catusse, J.; Bally, J.; Job, C.; Job, D. Seed germination and vigor. *Annu. Rev. Plant Biol.* **2012**, *2*, 507–533. [CrossRef]
96. Meena, K.K.; Sorty, A.M.; Bitla, U.M.; Choudhary, K.; Gupta, P.; Pareek, A.; Minhas, P.S. Abiotic Stress Responses and Microbe-Mediated Mitigation in Plants: The Omics Strategies. *Front. Plant Sci.* **2017**, *8*, 172. [CrossRef]
97. Stepien, P.; Klobus, G. Water relations and photosynthesis in *Cucumus sativus* L. leaves under salt stress. *Biol. Plant.* **2006**, *50*, 610–616. [CrossRef]
98. Dobrota, C. Energy dependant plant stress acclimation. *Rev. Environ. Sci.* **2006**, *5*, 243–251. [CrossRef]

99. Yeo, A.R.; Capron, S.J.M.; Flowers, T.J. The effect of salinity upon photosynthesis in rice (*Oryza sativa* L.) Gas exchange by individual leaves relation to their salt content. *J. Exp. Bot.* **1985**, *36*, 1240–1248. [CrossRef]
100. Muradoglu, F.; Gundogdu, M.; Ercisli, S.; Encu, T.; Balta, F.; Jaafar, H.Z.; Zia-Ul-Haq, M. Cadmium toxicity affects chlorophyll a and b content, antioxidant enzyme activities and mineral nutrient accumulation in strawberry. *Biol. Res.* **2015**, *48*, 1–7. [CrossRef]
101. El-Shintinawy, F.; El-Ansary, A. Differential Effect of Cd2+ and Ni2+ on Amino Acid Metabolism in Soybean Seedlings. *Biol. Plant.* **2000**, *43*, 79–84. [CrossRef]
102. Puniran-Hartley, N.; Hartley, J.; Shabala, L.; Shabala, S. Salinity-induced accumulation of organic osmolytes in barley and wheat leaves correlates with increased oxidative stress tolerance: In planta evidence for cross-tolerance. *Plant Physiol. Biochem.* **2014**, *1*, 32–39. [CrossRef] [PubMed]
103. Garcia-Legaz, M.F.; Ortiz, J.M.; Garcia-Lindon, A.; Cerda, A. Effect of salinity on growth, ion content and CO_2 assimilation rate in lemon varieties on different rootstocks. *Physiol. Plant.* **1993**, *89*, 427–432. [CrossRef]
104. Acosta-Motos, J.R.; Ortuño, M.F.; Bernal-Vicente, A.; Diaz-Vivancos, P.; Sanchez-Blanco, M.J.; Hernandez, J.A. Plant responses to salt stress: Adaptive mechanisms. *Agronomy* **2017**, *7*, 18. [CrossRef]
105. Hanachi, S.; Van Labeke, M.C.; Mehouachi, T. Application of chlorophyll fluorescence to screen eggplant (*Solanum melongena* L.) cultivars for salt tolerance. *Photosynthetica* **2014**, *52*, 57–62. [CrossRef]
106. Tiwari, J.K.; Munshi, A.D.; Kumar, R.; Pandey, R.N.; Arora, A.; Bhat, J.S.; Sureja, A.K. Effect of salt stress on cucumber: Na+–K+ ratio, osmolyte concentration, phenols and chlorophyll content. *Acta Physiol. Plant.* **2010**, *1*, 103–114. [CrossRef]
107. Lycoskoufis, I.H.; Savvas, D.; Mavrogianopoulos, G. Growth, gas exchange, and nutrient status in pepper (*Capsicum annuum* L.) grown in recirculating nutrient solution as affected by salinity imposed to half of the root system. *Sci. Hortic.* **2005**, *106*, 147–161. [CrossRef]
108. Zhang, T.; Gong, H.; Wen, X.; Lu, C. Salt stress induces a decrease in oxidation energy transfer from phycobilisomes to photosystem II but an increase to photosystem I in the cyanobacterium Spirulina platensis. *J. Plant Physiol.* **2010**, *12*, 20.
109. Wu, X.; He, J.; Chen, J.; Yang, S.; Zha, D. Alleviation of exogenous 6-benzyladenine on two genotypes of eggplant (*Solanum melongena* Mill.) growth under salt stress. *Protoplasma* **2014**, *251*, 169–176. [CrossRef] [PubMed]
110. Yan, K.; Shao, H.; Shao, C.; Chen, P.; Zhao, S.; Brestic, M.; Chen, X. Physiological adaptive mechanisms of plants grown in saline soil and implications for sustainable saline agriculture in coastal zone. *Acta Physiol. Plant.* **2013**, *1*, 2867–2878. [CrossRef]
111. Hayat, S.; Hasan, S.A.; Yusuf, M.; Hayat, Q.; Ahmad, A. Effect of 28- homobrassinolide on photosynthesis, fluorescence and antioxidant system in the presence or absence of salinity and temperature in *Vigna radiata*. *Environ. Exp. Bot.* **2010**, *69*, 105–112. [CrossRef]
112. Misra, A.N.; Sahu, S.M.; Misra, M.; Singh, P.; Meera, I.; Das, N.; Kar, M.; Shau, P. Sodium chloride induced changes in leaf growth, and pigment and protein contents in two rice cultivars. *Biol. Plant.* **1997**, *39*, 257–262. [CrossRef]
113. Neelesh, K.; Veena, P. Effect of Salt Stress on Growth Parameters, Moisture Content, Relative Water Content and Photosynthetic Pigments of Fenugreek Variety RMt-1. *J. Plant Sci.* **2015**, *10*, 210–221. [CrossRef]
114. Bethke, P.C.; Drew, M.C. Stomatal and non-stomatal components to inhibition of photosynthesis in leaves of *Capsicum annum* during progressive exposure to NaCl salinity. *Plant Physiol.* **1992**, *99*, 219–226. [CrossRef] [PubMed]
115. Reddy, M.P.; Vora, A.B. Changes in pigment composition, Hill reaction activity and saccharides metabolism in Bajra (*Pennisetum typhoides*) leaves under NaCl salinity. *Photosynthetica* **1986**, *20*, 50–55.
116. Gul, B.; Ansari, R.; Flowers, T.J.; Khan, M.A. Germination strategies of halophyte seeds under salinity. *Environ. Exp. Bot.* **2013**, *1*, 4–18. [CrossRef]
117. Parida, A.K.; Das, A.B. Salt tolerance and salinity affects on plant. *Ecotoxicol. Environ.* **2005**, *60*, 324–349. [CrossRef] [PubMed]
118. Cao, Y.Y.; Duan, H.; Yang, L.N.; Wang, Z.Q.; Zhou, S.C.; Yang, J.C. Effect of heat stress during meiosis on grain yield of rice cultivars differing in heat tolerance and its physiological mechanism. *Acta Agron. Sin.* **2008**, *34*, 2134–2142. [CrossRef]
119. Elhamid, E.; Sadak, M.; Tawfik, M. Alleviation of Adverse Effects of Salt Stress in Wheat Cultivars by Foliar

Treatment with Antioxidant 2—Changes in Some Biochemical Aspects, Lipid Peroxidation, Antioxidant Enzymes and Amino Acid Contents. *Agric. Sci.* **2014**, *5*, 1269–1280. [CrossRef]

120. Khan, N.; Zandi, P.; Ali, S.; Mehmood, A.; Adnan Shahid, M.; Yang, J. Impact of salicylic acid and PGPR on the drought tolerance and phytoremediation potential of Helianthus annus. *Front. Microbiol.* **2018**, *9*, 2507. [CrossRef]
121. Kartashov, A.V.; Radyukina, N.L.; Ivanov, Y.V.; Pashkovskii, P.P.; Shevyakova, N.I.; Kuznetsov, V.V. Role of antioxidants systems in wild plant adaptation to salt stress. *Russ. J. Plant Physiol.* **2008**, *55*, 463–468. [CrossRef]
122. Katsumi, M. Physiological modes of brassinolide action in cucumber hypocotyl growth. In *Brassinosteriods: Chemistry, Bioactivity and Applications*; Cutler, H.G., Yokota, T., Adam, G., Eds.; American Chemical Society: Washington, DC, USA, 1991; pp. 246–254.
123. Faried, H.N.; Ayyub, C.M.; Amjad, M.; Ahmed, R. Salinity impairs ionic, physiological and biochemical attributes in potato. *Pak. J. Agric. Sci.* **2016**, *53*, 17–25.
124. Sofo, A.; Scopa, A.; Nuzzaci, M.; Vitti, A. Ascorbate Peroxidase and Catalase Activities and Their Genetic Regulation in Plants Subjected to Drought and Salinity Stresses. *Int. J. Mol. Sci.* **2015**, *16*, 13561–13578. [CrossRef] [PubMed]
125. Gupta, B.; Huang, B. Mechanism of salinity tolerance in plants: Physiological, biochemical, and molecular characterization. *Int. J. Genom.* **2014**, *2014*. [CrossRef]
126. Bhattacharyya, A.; Chattopadhyay, R.; Mitra, S.; Crowe, S.E. Oxidative Stress: An Essential Factor in the Pathogenesis of Gastrointestinal Mucosal Diseases. *Physiol. Rev.* **2014**, *94*, 329–354. [CrossRef] [PubMed]
127. Ashraf, M.; Bhatti, B. Physiological responses of some salt tolerance and salt sensitive lines of wheat (*Crthamus tinctorius* L.). *Acta Physiol. Plant.* **1999**, *117*, 23–28.
128. Adnan, A.; Erkan, Y. Effect of salinity stress on chlorophyll, carotenoid content, and proline in *Salicornia prostrata* Pall. and *Suaeda prostrata* Pall. subsp. *prostrata* (Amaranthaceae). *Braz. J. Bot.* **2016**, *39*, 101–106.
129. Kamal, A.; Qureshi, M.S.; Ashraf, M.Y.; Hussain, M. Salinity induced some changes in growth and physio-chemical aspects of two soybean [*Glycine max* (L.) merr] genotype. *Pak. J. Bot.* **2003**, *35*, 93–97.
130. Parvin, K.; Haque, M.N. Protective Role of Salicylic Acid on Salt Affected Broccoli Plants. *J. Agric. Ecol.* **2017**, *10*, 1–10. [CrossRef]
131. Stepien, P.; Klobus, G. Antioxidant defense in the leaves of C_3 and C_4 plants under salinity stress. *Physiol. Plant.* **2005**, *125*, 31–40. [CrossRef]
132. Chandrasekaran, M.; Kim, K.; Krishnamoorthy, R.; Walitang, D.; Sundaram, S.; Joe, M.M.; Selvakumar, G.; Hu, S.; Oh, S.H.; Sa, T. Mycorrhizal symbiotic efficiency on C3 and C4 plants under salinity stress—A meta-analysis. *Front. Microbiol.* **2016**, *11*, 1246. [CrossRef] [PubMed]
133. Akbar, S.; Sultan, S. Soil bacteria showing a potential of chlorpyrifos degradation and plant growth enhancement. *Braz. J. Microbiol.* **2016**, *47*, 563–570. [CrossRef]
134. Bhantana, P.; Lazarovitch, N. Evapotranspiration, Crop coefficient and growth of two young pomegranate (*Punica granatum* L.) varieties under salt stress. *Agric. Water Manag.* **2010**, *97*, 715–722. [CrossRef]
135. Gorai, M.; Ennajeh, M.; Khemira, H.; Neffati, M. Combined effect of NaCl-salinity and hypoxia on growth, photosynthesis, water relations and solute accumulation in *Phragmite saustralis* plants. *Flora* **2010**, *205*, 462–470. [CrossRef]
136. Baby, T.; Collins, C.; Tyerman, S.D.; Gilliham, M. Salinity negatively affects pollen tube growth and fruit set in grapevines and is not mitigated by silicon. *Am. J. Enol. Vitic.* **2016**, *67*, 218–228. [CrossRef]
137. Van Zelm, E.; Zhang, Y.; Testerink, C. Salt tolerance mechanisms of plants. *Ann. Rev. Plant Biol.* **2020**, *71*, 403–433. [CrossRef] [PubMed]
138. Aldesuquy, H.; Baka, Z.; Mickky, B. Kinetin and spermine mediated induction of salt tolerance in wheat plants: Leaf area, photosynthesis and chloroplast ultrastructure of flag leaf at ear emergence. *Egypt. J. Basic Appl. Sci.* **2014**, *1*, 77–87. [CrossRef]
139. Szalai, G.; Janda, K.; Darkó, É.; Janda, T.; Peeva, V.; Pál, M. Comparative analysis of polyamine metabolism in wheat and maize plants. *Plant Physiol. Biochem.* **2017**, *1*, 239–250. [CrossRef] [PubMed]
140. Syvertsen, J.; Levy, Y. *Salinity Interactions with Other Abiotic and Biotic Stresses in Citrus*; University of Florida, Indian River Research and Education Center: Ft. Pierce, FL, USA, 2005.
141. Negrão, S.; Schmöckel, S.M.; Tester, M. Evaluating physiological responses of plants to salinity stress. *Ann. Bot.* **2017**, *119*, 1–11. [CrossRef] [PubMed]

142. Moir-Barnetson, L.; Veneklaas, E.J.; Colmer, T.D. Salinity tolerances of three succulent halophytes (*Tecticornia* spp.) differentially distributed along a salinity gradient. *Funct. Plant Biol.* **2016**, *4*, 739–750. [CrossRef]
143. Khan, N.; Bano, A.; Babar, M.A. The root growth of wheat plants, the water conservation and fertility status of sandy soils influenced by plant growth promoting rhizobacteria. *Symbiosis* **2017**, *72*, 195–205. [CrossRef]
144. Bohnert, H.J.; Shen, B. Transformation and compatible solutes. *Sci. Hortic.* **1999**, *78*, 237–260. [CrossRef]
145. Shuyskaya, E.V.; Rakhamkulova, Z.F.; Lebedeva, M.P.; Kolesnikov, A.V.; Safarova, A.; Borisochkina, T.I.; Toderich, K.N. Different mechanisms of ion homeostasis are dominant in the recretohalophyte Tamarix ramosissima under different soil salinity. *Acta Physiol. Plant.* **2017**, *1*, 81. [CrossRef]
146. Hernandez, J.A.; Almansa, M.S. Short term effects of salt stress on antioxidant systems and leaf water relations of pea leaves. *Physiol. Plant.* **2002**, *115*, 251–257. [CrossRef] [PubMed]
147. Khan, N.; Ali, S.; Zandi, P.; Mehmood, A.; Ullah, S.; Ikram, M.; Ismail, M.; Babar, M.A. Role of sugars, amino acids and organic acids in improving plant abiotic stress tolerance. *Pak. J. Bot.* **1997**, *52*, 2. [CrossRef]
148. Hong, B.; Barg, R.; Ho, T.D. Developmental and organ specific expression of an ABA and stress induced protein in barley. *Plant Mol. Biol.* **1992**, *18*, 663–674. [CrossRef]
149. Hare, P.D.; Cress, W.A.; Staden, J.V. Dissecting the roles of osmolyte accumulation during stress. *Plant Cell Environ.* **1998**, *21*, 535–554. [CrossRef]
150. Galinski, E.A. Compatible solutes of halophilic eubacteria-eubacteria-molecular principles, water-solute interactions, stress protection. *Cell Mol. Life Sci.* **1993**, *49*, 487–496. [CrossRef]
151. Lutts, S.; Lefèvre, I. How can we take advantage of halophyte properties to cope with heavy metal toxicity in salt-affected areas? *Ann. Bot.* **2015**, *1*, 509–528. [CrossRef]
152. Brown, A.D. *Microbial Water Stress Physiology, Principles and Perspectives*; Wiley: Hoboken, NJ, USA, 1990.
153. Solomon, A.; Beer, S.; Waisel, Y.; Jones, G.P.; Paleg, L.G. Effects of NaCl on the Carboxylating activity of Rubisco from *Tamarix jordanis* in the presence and absence of proline-related compatible solutes. *Plant Physiol.* **1994**, *90*, 198–204. [CrossRef]
154. Papageorgiou, G.; Murata, N. The unusually strong stability effects of glycine betaine on the structure and function of the oxygen evolving photosystem II complex. *Photosynth. Res.* **1995**, *44*, 243–252. [CrossRef]
155. Bose, J.; Rodrigo-Moreno, A.; Shabala, S. ROS homeostasis in halophytes in the context of salinity stress tolerance. *J. Exp. Bot.* **2014**, *1*, 1241–1257. [CrossRef]
156. Stoop, J.M.H.; Williamson, J.D.; Pharr, D.M. Mannitol metabolism in plants: A method for coping with stress. *Trends Plant Sci.* **1996**, *1*, 139–144. [CrossRef]
157. Moran, J.F.; Becana, M.; Iturbe-Ormaetxe, I.; Frechilla, S.; Klucas, R.V.; Aparicio-Tejo, P. Drought induces oxidative stress in pea plants. *Planta* **1994**, *94*, 346–352. [CrossRef]
158. Managbanag, J.R.; Torzilli, A.P. An analysis of trehalose, glycerol, and mannitol accumulation during heat and salt stress in a salt marsh isolate of *Aureobasidium pullulans*. *Mycologia* **2002**, *1*, 384–391. [CrossRef]
159. Khan, N.; Bano, A.; Zandi, P. Effects of exogenously applied plant growth regulators in combination with PGPR on the physiology and root growth of chickpea (*Cicer arietinum*) and their role in drought tolerance. *J. Plant Interact.* **2018**, *1*, 239–247. [CrossRef]
160. Tipirdamaz, R.; Gagneul, D.; Duhazé, C.; Aïnouche, A.; Monnier, C.; Özkum, D.; Larher, F. Clustering of halophytes from an inland salt marsh in Turkey according to their ability to accumulate sodium and nitrogenous osmolytes. *Environ. Exp. Bot.* **2006**, *57*, 139–153. [CrossRef]
161. Ashraf, M.; Harris, P.J.C. Potential biochemical indicators of salinity tolerance in plants. *Plant Sci.* **2004**, *166*, 3–16. [CrossRef]
162. Sandhu, D.; Cornacchione, M.V.; Ferreira, J.F.; Suarez, D.L. Variable salinity responses of 12 alfalfa genotypes and comparative expression analyses of salt-response genes. *Sci. Rep.* **2017**, *22*, 42958. [CrossRef]
163. Liang, X.; Zhang, L.; Natarajan, S.K.; Becker, D.F. Proline Mechanisms of Stress Survival. *Antioxid. Redox Signal.* **2013**, *19*, 998–1011. [CrossRef]
164. Dar, M.I.; Naikoo, M.I.; Rehman, F.; Naushin, F.; Khan, F.A. Proline accumulation in plants: Roles in stress tolerance and plant development. In *Osmolytes and Plants Acclimation to Changing Environment: Emerging Omics Technologies*; Springer: New Delhi, India, 2016; pp. 155–166.
165. Ali, B.; Hayat, S.; Ahmad, A. 28-homobrassinolide ameliorates the saline stress in *Cicer arietinum* L. *Environ. Exp. Bot.* **2007**, *59*, 217–223. [CrossRef]

166. Mantilla, B.S.; Marchese, L.; Casas-Sánchez, A.; Dyer, N.A.; Ejeh, N.; Biran, M.; Silber, A.M. Proline Metabolism is Essential for Trypanosoma brucei brucei Survival in the Tsetse Vector. *PLoS Pathog.* **2017**, *13*, e1006158. [CrossRef]
167. Kahlaoui, B.; Hachicha, M.; Misle, E.; Fidalgo, F.; Teixeira, J. Physiological and biochemical responses to the exogenous application of proline of tomato plants irrigated with saline water. *J. Saudi Soc. Agric. Sci.* **2018**, *1*, 17–23. [CrossRef]
168. Nemoto, Y.; Sasakuma, T. Differential stress responses of early salt-stress responding genes in common wheat. *Phytochemistry* **2002**, *61*, 129–133. [CrossRef]
169. Jaarsma, R.; de Vries, R.S.; de Boer, A.H. Effect of salt stress on growth, Na+ accumulation and proline metabolism in potato (*Solanum tuberosum*) cultivars. *PLoS ONE* **2013**, *8*, e60183. [CrossRef] [PubMed]
170. Joseph, E.A.; Radhakrishnan, V.V.; Mohanan, K.V. A Study on the Accumulation of Proline—An Osmoprotectant Amino Acid under Salt Stress in Some Native Rice Cultivars of North Kerala, India. *Univ. J. Agric. Res.* **2015**, *3*, 15–22. [CrossRef]
171. Knight, H.; Anthony, J.; Knight, M.R. Calcium signaling in *Arabidopsis thaliana* responding to drought and salinity. *Plant J.* **1997**, *12*, 1067–1078. [CrossRef]
172. Misra, N.; Gupta, A.K. Interactive effects of sodium and calcium on proline metabolism in salt tolerant green gram cultivar. *Am. J. Plant Physiol.* **2006**, *1*, 1–2.
173. Ashraf, M.; Foolad, M.R. Roles of glycinebetaine and proline in improving plant abiotic stress resistance. *Environ. Exp. Bot.* **2007**, *59*, 206–216. [CrossRef]
174. Huang, Z.; Zhao, L.; Chen, D.; Liang, M.; Liu, Z.; Shao, H.; Long, X. Salt Stress Encourages Proline Accumulation by Regulating Proline Biosynthesis and Degradation in Jerusalem Artichoke Plantlets. *PLoS ONE* **2013**, *8*, e62085. [CrossRef]
175. Lutts, S.; Majerus, V.; Kinet, J.M. NaCl effects on proline metabolism in rice (*Oryza sativa*) seedlings. *Int. Rice Res.* **1999**, *25*, 39–40. [CrossRef]
176. Devnarain, N.; Crampton, B.G.; Chikwamba, R.K.; Becker, J.V.W.; O'Kennedy, M.M. Physiological responses of selected African sorghum landraces to progressive water stress and re-watering. *S. Afr. J.* **2016**, *103*, 61–69. [CrossRef]
177. Yan, H.; Gang, L.Z.; Zhao, C.Y.; Guo, W.Y. Effects of exogenous proline on the physiology of soybean plantlets regenerated from embryo in vitro and on the ultra-structure of their mitochondria under NaCl stress. *Soybean Sci.* **2000**, *19*, 314–319.
178. Hare, P.D.; Cress, W.A.; Staden, J.V. Disruptive effects of exogenous proline on chloroplast and mitochondrial ultrastructure in *Arabidopsis* leaves. *S. Afr. J.* **2002**, *68*, 393–396. [CrossRef]
179. Zeyner, A.; Romanowski, K.; Vernunft, A.; Harris, P.; Müller, A.M.; Wolf, C.; Kienzle, E. Effects of Different Oral Doses of Sodium Chloride on the Basal Acid-Base and Mineral Status of Exercising Horses Fed Low Amounts of Hay. *PLoS ONE* **2017**, *12*, e0168325. [CrossRef] [PubMed]
180. Jimenez-Bremont, J.F.; Becerra-Flora, A.; Hernandez-Lucero, E.; Rodriguez-Kessler, M.; Acosta-Gallegos, J.A.; Ramirez-Pimental, J.G. Proline accumulation in two bean cultivars under salt stress and the effect of polyamines and ornithine. *Biol. Plant.* **2006**, *50*, 763–766. [CrossRef]
181. Gill, S.S.; Tuteja, N. Polyamines and abiotic stress tolerance in plants. *Plant Signal Behav.* **2010**, *5*, 26–33. [CrossRef] [PubMed]
182. Griffith, S.M.; Banowetz, G.M. Nitrogen nutrition and flowering. In *Nitrogen Nutrition in Higher Plants*; Srivastava, H.S., Singh, R.P., Eds.; Associated Publ. Co.: New Delhi, India, 1995; pp. 385–400.
183. Kumar, R.; Khurana, A.; Sharma, A.K. Role of plant hormones and their interplay in development and ripening of fleshy fruits. *J. Exp. Bot.* **2013**, *65*, 4561–4575. [CrossRef]
184. Diao, Q.; Song, Y.; Shi, D.; Qi, H. Interaction of Polyamines, Abscisic Acid, Nitric Oxide, and Hydrogen Peroxide under Chilling Stress in Tomato (Lycopersicon esculentum Mill.) Seedlings. *Front. Plant Sci.* **2017**, *8*, 203. [CrossRef] [PubMed]
185. Shabala, S.; Cuin, T.A.; Pottosin, I. Polyamines prevent NaCl-induced K^+ efflux from pea mesophyll by blocking non-selective cation channels. *FEBS Lett.* **2007**, *581*, 1993–1999. [CrossRef]
186. Rodríguez, A.A.; Maiale, S.J.; Menéndez, A.B.; Ruiz, O.A. Polyamine oxidase activity contributes to sustain maize leaf elongation under saline stress. *J. Exp. Bot.* **2009**, *1*, 4249–4262. [CrossRef]
187. Mishra, S.N.; Sharma, I. Putrescine as a growth inducer and as a source of nitrogen for mustard seedlings under sodium chloride salinity. *Indian J. Exp. Biol.* **1994**, *32*, 916–918.

188. Lakra, N.; Tomar, P.C.; Mishra, S.N. Growth response modulation by putrescine in Indian mustard *Brassica juncea* L. under multiple stress. *Indian J. Exp. Biol.* **2016**, *54*, 262–270.
189. Maruri-López, I.; Jiménez-Bremont, J. Hetero- and homodimerization of Arabidopsis thaliana arginine decarboxylase AtADC1 and AtADC2. *Biochem. Biophys. Res. Commun.* **2017**, *11*, 508–513. [CrossRef] [PubMed]
190. Wani, S.H.; Singh, N.B.; Haribhushan, A.; Mir, J.I. Compatible solute engineering in plants for abiotic stress tolerance-role of glycine betaine. *Curr. Genom.* **2013**, *14*, 157. [CrossRef] [PubMed]
191. Baloda, A.; Madanpotra, S.; Jaiwal, P.K. Transformation of mung bean plants for abiotic stress tolerance by introducing codA gene, for an osmoprotectant glycine betaine. *J. Plant Stress Physiol.* **2017**, *3*, 5–11. [CrossRef]
192. Wei, D.; Zhang, W.; Wang, C.; Meng, Q.; Li, G.; Chen, T.H.; Yang, X. Genetic engineering of the biosynthesis of glycinebetaine leads to alleviate salt-induced potassium efflux and enhances salt tolerance in tomato plants. *Plant Sci.* **2017**, *257*, 74–83. [CrossRef] [PubMed]
193. Park, E.J.; Jeknic, Z.; Chen, T.H.H. Exogenous application of glycine betaine increases chilling tolerance in tomato plants. *Plant Cell Physiol.* **2006**, *47*, 706–714. [CrossRef] [PubMed]
194. Park, H.J.; Kim, W.Y.; Yun, D.J. A New Insight of Salt Stress Signaling in Plant. *Mol. Cells* **2016**, *39*, 447–459. [CrossRef] [PubMed]
195. Khan, N.; Bano, A. Rhizobacteria and Abiotic Stress Management. In *Plant Growth Promoting Rhizobacteria for Sustainable Stress Management*; Springer: Singapore, 2019; pp. 65–80.
196. Mohammad, N.; Naeem, K.; Zakia, A.; Abdul, G. Genetic diversity and disease response of rust in bread wheat collected from Waziristan Agency, Pakistan. *Int. J. Biodivers. Conserv.* **2011**, *3*, 10–18.
197. Shanker, A.; Venkateswarlu, B. *Abiotic Stress in Plants: Mechanisms and Adaptations. BoD–Books on Demand*; IntechOpen: Norderstedt, Germany, 2011; p. 22.
198. Nawaz, K.; Ashraf, M. Exogenous application of glycine betaine modulates activities of antioxidants in maize plants subjected to salt stress. *J. Agron. Crop Sci.* **2009**, *196*, 28–37. [CrossRef]
199. Annunziata, M.G.; Ciarmiello, L.F.; Woodrow, P.; Maximova, E.; Fuggi, A.; Carillo, P. Durum Wheat Roots Adapt to Salinity Remodeling the Cellular Content of Nitrogen Metabolites and Sucrose. *Front. Plant Sci.* **2016**, *7*, 2035. [CrossRef]
200. Wyn-Jones, R.G.; Gorham, J.; McDonnell, E. Organic and inorganic solute contents as selection criteria for salt tolerance in the Triticeae. In *Salinity Tolerance in Plants: Strategies for Crop Improvement*; Staples, R., Toennissen, G.H., Eds.; Wiley and Sons: New York, NY, USA, 1984; pp. 189–203.
201. Woodrow, P.; Ciarmiello, L.F.; Annunziata, M.G.; Pacifico, S.; Iannuzzi, F.; Mirto, A.; Carillo, P. Durum wheat seedling responses to simultaneous high light and salinity involve a fine reconfiguration of amino acids and carbohydrate metabolism. *Plant Physiol.* **2016**. [CrossRef]
202. Majeed, K.; Al-Hamzawi, A. Effect of Calcium Nitrate, Potassium Nitrate and Anfaton on Growth and Storability of Plastic Houses Cucumber (*Cucumis sativus* L. cv. Al-Hytham). *Am. J. Plant Sci.* **2010**, *5*, 278–290.
203. Kong, X.; Luo, Z.; Dong, H.; Eneji, A.E.; Li, W. Effects of non-uniform root zone salinity on water use, Na+ recirculation, and Na+ and H+ flux in cotton. *J. Exp. Bot.* **2012**, *63*, 2105–2116. [CrossRef] [PubMed]
204. Wu, H.; Shabala, L.; Liu, X.; Azzarello, E.; Zhou, M.; Pandolfi, C.; Shabala, S. Linking salinity stress tolerance with tissue-specific Na+ sequestration in wheat roots. *Front. Plant Sci.* **2015**, *6*. [CrossRef]
205. Maleka, P.; Kontturi, M.; Pehu, B.; Somersalo, S. Photosynthesis response of drought and salt stressed tomato and turnip plants to foliar applied glycinebetaine. *Physiol. Plant.* **1999**, *105*, 45–50.
206. Athar, H.U.R.; Zafar, Z.U.; Ashraf, M. Glycinebetaine Improved Photosynthesis in Canola under Salt Stress: Evaluation of Chlorophyll Fluorescence Parameters as Potential Indicators. *J. Agron. Crop Sci.* **2015**, *201*, 428–442. [CrossRef]
207. Luo, J.Y.; Zhang, S.; Peng, J.; Zhu, X.Z.; Lv, L.M.; Wang, C.Y.; Cui, J.J. Effects of Soil Salinity on the Expression of Bt Toxin (Cry1Ac) and the Control Efficiency of Helicoverpa armigera in Field- Grown Transgenic Bt Cotton. *PLoS ONE* **2017**, *12*, e0170379. [CrossRef]
208. Raza, M.A.S.; Shahid, A.M.; Saleem, M.F.; Khan, I.H.; Ahmad, S.; Ali, M.; Iqbal, R. Effects and management strategies to mitigate drought stress in oilseed rape (*Brassica napus* L.): A review. *Zemdirbyste-Agriculture* **2017**, *104*, 85–94. [CrossRef]
209. Akram, R.; Fahad, S.; Hashmi, M.Z.; Wahid, A.; Adnan, M.; Mubeen, M.; Khan, N.; Rehmani, M.I.; Awais, M.; Abbas, M.; et al. Trends of electronic waste pollution and its impact on the global environment and ecosystem. *Environ. Sci. Pollut. Res. Int.* **2019**, *1*, 1–6. [CrossRef]

210. Bhuiyan, M.S.; Maynard, G.; Raman, A.; Hodgkins, D.; Mitchell, D.; Nicol, H. Salt effects on proline and glycine betaine levels and photosynthetic performance in Melilotus siculus, Tecticornia pergranulata and Thinopyrum ponticum measured in simulated saline conditions. *Funct. Plant Biol.* **2016**, *4*, 254–265. [CrossRef]
211. Demiral, T.; Türkan, I. Does exogenous glycinebetaine affect antioxidative system of rice seedlings under NaCl treatment? *J. Plant Physiol.* **2004**, *18*, 1089–1100. [CrossRef] [PubMed]
212. Talaat, N.B.; Shawky, B.T. Influence of arbuscular mycorrhizae on yield, nutrients, organic solutes, and antioxidant enzymes of two wheat cultivars under salt stress. *J. Soil Sci. Plant Nutr.* **2011**, *174*, 283–291. [CrossRef]
213. Ruiz-Lozano, J.M.; Porcel, R.; Azcón, C.; Aroca, R. Regulation by arbuscular mycorrhizae of the integrated physiological response to salinity in plants: New challenges in physiological and molecular studies. *J. Exp. Bot.* **2012**, *28*, 4033–4044. [CrossRef]
214. Ghosh, B.; Ali, M.N.; Saikat, G. Response of Rice under Salinity Stress: A Review Update. *J. Res. Rice* **2016**, *4*, 1–8. [CrossRef]
215. Mack, G.; Hoffmann, C.M.; Marlander, B. Nitrogen compounds in organs of two sugar beet genotypes (*Beta vulgaris* L.) during the season. *Field Crop. Res.* **2007**, *102*, 210–218. [CrossRef]
216. Khan, N.; Bano, A.; Rahman, M.A.; Guo, J.; Kang, Z.; Babar, M.A. Comparative physiological and metabolic analysis reveals a complex mechanism involved in drought tolerance in chickpea (*Cicer arietinum* L.) induced by PGPR and PGRs. *Sci. Rep.* **2019**, *9*, 1–19. [CrossRef]
217. Ashraf, M. Biotechnological approach of improving plant salt tolerance using antioxidants as markers. *Biotechnol. Adv.* **2009**, *27*, 84–93. [CrossRef] [PubMed]
218. Turan, S.; Tripathy, B.C. Salt and genotype impact on antioxidative enzymes and lipid peroxidation in two rice cultivars during de-etiolation. *Protoplasma* **2013**, *1*, 209–222. [CrossRef]
219. Das, M.K.; Roychoudhury, A. ROS and responses of antioxidant as ROS-scavengers during environmental stress in plants. *Front. Environ. Sci.* **2014**, *2*, 1–3. [CrossRef]
220. Elstner, E.F. Mechanisms of oxygen activation in different compartments of plant cell. In *Active Oxygen/ Oxidative Stress and Plant Metabolism*; Pell, E.J., Stefen, K.L., Eds.; American society of Plant Physiol: Rockville, MD, USA, 1991; pp. 13–25.
221. Vardhini, B.V.; Anjum, N.A. Brassinosteroids make plant life easier under abiotic stresses mainly by modulating major components of antioxidant defense system. *Front. Environ. Sci.* **2015**, *12*, 67. [CrossRef]
222. Rich, P.R.; Bonner, W.D., Jr. The sites of superoxide anion generation in higher plant mitochondria. *Arch Biochem. Biophys.* **1978**, *188*, 206–213. [CrossRef]
223. Song, X.S.; Hu, W.H.; Mao, W.H.; Ogweno, J.O.; Zhou, Y.H.; Yu, J.Q. Response of ascorbate peroxidase isoenzymes and ascorbate regeneration system to abiotic stresses in *Cucumis sativus* L. *Plant Physiol. Biochem.* **2005**, *1*, 1082–1088. [CrossRef] [PubMed]
224. Mehmood, A.; Hussain, A.; Irshad, M.; Hamayun, M.; Iqbal, A.; Khan, N. In vitro production of IAA by endophytic fungus Aspergillus awamori and its growth promoting activities in Zea mays. *Symbiosis* **2019**, *77*, 225–235. [CrossRef]
225. Xia, X.J.; Gao, C.J.; Song, L.X.; Zhou, Y.H.; Shi, K.A.I.; Yu, J.Q. Role of H_2O_2 dynamics in brassinosteroid-induced stomatal closure and opening in S olanum lycopersicum. *Plant Cell Environ.* **2014**, *37*, 2036–2050. [CrossRef] [PubMed]
226. Blokhina, O.; Virolainen, E.; Fagerstedt, K.V. Antioxidants, oxidative damage and oxygen deprivation stress: A review. *Ann. Bot.* **2003**, *91*, 79–194. [CrossRef]
227. Das, S.K.; Patra, J.K.; Thatoi, H. Antioxidative response to abiotic and biotic stresses in mangrove plants: A review. *Intern. Rev. Hydrobiol.* **2016**, *101*, 3–19. [CrossRef]
228. Sairam, R.K.; Tyagi, A. Physiology and molecular biology of salinity stress tolerance in plants. *Curr. Sci.* **2004**, *86*, 407–421.
229. Zhang, M.; Fang, Y.M.; Ji, Y.H.; Jiang, Z.P.; Wang, L. Effects of salt stress on ion content, antioxidant enzymes and protein profile in different tissues of Broussonetia Papyrifera. *S. Afr. J. Bot.* **2013**, *85*, 1–9. [CrossRef]
230. Wahid, A.; Parveen, M.; Gelani, S.; Basra, S.M.A. Pretreatment of seed with H_2O_2 improves salt tolerance of wheat seedlings by alleviation of oxidative damage and expression of stress proteins. *J. Plant Physiol.* **2007**, *164*, 283–294. [CrossRef]

231. Shalata, A.; Neumann, P.M. Exogenous ascorbic acid increases resistance to salt stress and reduces lipid peroxidation. *J. Exp. Bot.* **2001**, *52*, 2207–2211. [CrossRef]
232. Khan, N.; Ali, S.; Shahid, M.A.; Kharabian-Masouleh, A. Advances in detection of stress tolerance in plants through metabolomics approaches. *Plant Omics* **2017**, *10*, 153. [CrossRef]
233. Aliniaeifard, S.; Hajilou, J.; Tabatabaei, S.J.; Sifi-Kalhor, M. Effects of Ascorbic Acid and Reduced Glutathione on the Alleviation of Salinity Stress in Olive Plants. *Int. J. Fruit Sci.* **2016**, *16*, 395–409. [CrossRef]
234. Ding, X.; Jiang, Y.; He, L.; Zhou, Q.; Yu, J.; Hui, D.; Huang, D. Exogenous glutathione improves high root-zone temperature tolerance by modulating photosynthesis, antioxidant and osmolytes systems in cucumber seedlings. *Sci. Rep.* **2016**, *6*, 35424. [CrossRef] [PubMed]
235. Ahmad, B. Interactive Effects of Silicon and Potassium Nitrate in Improving Salt Tolerance of Wheat. *J. Integr. Agric.* **2013**. [CrossRef]
236. Parihar, P.; Singh, S.; Singh, R.; Singh, V.P.; Prasad, S.M. Effect of salinity stress on plants and its tolerance strategies: A review. *Environ. Sci. Pollut. Res.* **2015**, *22*, 4056–4075. [CrossRef] [PubMed]
237. Wang, N.; Qi, H.; Qiao, W.; Shi, J.; Xu, Q.; Zhou, H.; Yan, G.; Huang, Q. Cotton (*Gossypium hirsutum* L.) genotypes with contrasting K+/Na+ ion homeostasis: Implications for salinity tolerance. *Acta Physiol. Plant.* **2017**, *1*, 77. [CrossRef]
238. Walia, H.; Wilson, C.; Condamine, P.; Liu, X.; Ismail, A.M.; Close, T.J. Large scale expression profiling and physiological characterization of Jasmonic acid-mediated adaptation of barley to salinity stress. *Plant Cell Environ.* **2007**, *30*, 410–421. [CrossRef] [PubMed]
239. Qiu, Z.; Guo, J.; Zhu, A.; Zhang, L.; Zhang, M. Exogenous jasmonic acid can enhance tolerance of wheat seedlings to salt stress. *Ecotoxicol. Environ. Saf.* **2014**, *1*, 202–208. [CrossRef] [PubMed]
240. Ahmad, P.; Rasool, S.; Gul, A.; Sheikh, S.A.; Akram, N.A.; Ashraf, M.; Kazi, A.M.; Gucel, S. Jasmonates: Multifunctional Roles in Stress Tolerance. *Front. Plant Sci.* **2016**, *7*, 813. [CrossRef]
241. Khan, N.; Bano, A.; Babar, M.A. Metabolic and physiological changes induced by plant growth regulators and plant growth promoting rhizobacteria and their impact on drought tolerance in *Cicer arietinum* L. *PLoS ONE* **2019**, *14*, e0213040. [CrossRef] [PubMed]
242. Wani, S.H.; Kumar, V.; Shriram, V.; Sah, S.K. Phytohormones and their metabolic engineering for abiotic stress tolerance in crop plants. *Crop J.* **2016**, *4*, 162–176. [CrossRef]
243. Seif El-Yazal, S.A.; Seif El-Yazal, M.A.; Dwidar, E.F.; Rady, M.M. Phytohormone crosstalk research: Cytokinin and its crosstalk with other phytohormones. *Curr. Protein Pept. Sci.* **2015**, *16*, 395–405. [CrossRef] [PubMed]
244. Khan, N.; Bano, A.; Ali, S.; Babar, M.A. Crosstalk amongst phytohormones from planta and PGPR under biotic and abiotic stresses. *Plant Growth Regul.* **2020**, *90*, 189–203. [CrossRef]
245. Nafisi, M.; Fimognari, L.; Sakuragi, Y. Interplays between the cell wall and phytohormones in interaction between plants and necrotrophic pathogens. *Phytochemistry* **2015**, *1*, 63–71. [CrossRef] [PubMed]
246. Aldesuquy, H.S.; Gaber, A.M. Effect of growth regulators on leaf area, pigment content and photosynthetic activity on *Vicia faba* plants irrigated by sea water. *Biol. Plant.* **1993**, *35*, 519–527. [CrossRef]
247. Babar, S.; Siddiqi, E.H.; Hussain, I.; Hayat Bhatti, K.; Rasheed, R. Mitigating the effects of salinity by foliar application of salicylic acid in fenugreek. *Physiol. J.* **2014**, *2014*. [CrossRef]
248. Vishwakarma, K.; Upadhyay, N.; Kumar, N.; Yadav, G.; Singh, J.; Mishra, R.K.; Sharma, S. Abscisic Acid Signaling and Abiotic Stress Tolerance in Plants: A Review on Current Knowledge and Future Prospects. *Front. Plant Sci.* **2017**, *8*, 161. [CrossRef]
249. Mehmood, A.; Hussain, A.; Irshad, M.; Khan, N.; Hamayun, M.; Ismail; Afridi, S.G.; Lee, I.J. IAA and flavonoids modulates the association between maize roots and phytostimulant endophytic Aspergillus fumigatus greenish. *J. Plant Interact.* **2018**, *1*, 532–542. [CrossRef]
250. Barnawal, D.; Bharti, N.; Tripathi, A.; Pandey, S.S.; Chanotiya, C.S.; Kalra, A. ACC-deaminase-producing endophyte Brachybacterium paraconglomeratum strain SMR20 ameliorates Chlorophytum salinity stress via altering phytohormone generation. *J. Plant Growth Reg.* **2016**, *35*, 553–564. [CrossRef]
251. Beaudoin, N.; Serizet, C.; Gosti, F.; Giraudat, J. Interaction between abscisic acid and ethylene signalling cascades. *Plant Cell.* **2000**, *2*, 1103–1115. [CrossRef]
252. Khan, N.; Bano, A.; Rahman, M.A.; Rathinasabapathi, B.; Babar, M.A. UPLC-HRMS-based untargeted metabolic profiling reveals changes in chickpea (Cicer arietinum) metabolome following long-term drought stress. *Plant Cell Environ.* **2019**, *42*, 115–132. [CrossRef] [PubMed]

253. Anuradha, S.; Rao, S.S.R. Effect of brassinosteroids on salinity stress induced inhibition of seed germination and seedling growth of rice (*Oryza sativa* L.). *Plant Growth Regul.* **2001**, *33*, 151–153. [CrossRef]
254. Sharma, I.; Ching, E.; Saini, S.; Bhardwaj, R.; Pati, P.K. Exogenous application of brassinosteroid offers tolerance to salinity by altering stress responses in rice variety Pusa Basmati-1. *Plant Physiol. Biochem.* **2013**, *69*, 17–26. [CrossRef]
255. Anuradha, S.; Rao, S.S.R. Application of brassinosteroids to rice seeds (*Oryza sativa* L.) reduced the impact of salt stress on growth, prevented photosynthetic pigments loss and increased nitrate reductase activity. *Plant Growth Regul.* **2003**, *40*, 29–32. [CrossRef]
256. Shahid, M.A.; Balal, R.M.; Khan, N.; Simón-Grao, S.; Alfosea-Simón, M.; Cámara-Zapata, J.M.; Mattson, N.S.; Garcia-Sanchez, F. Rootstocks influence the salt tolerance of Kinnow mandarin trees by altering the antioxidant defense system, osmolyte concentration, and toxic ion accumulation. *Sci. Hortic.* **2019**, *10*, 1. [CrossRef]
257. Shahid, M.A.; Balal, R.M.; Khan, N.; Zotarelli, L.; Liu, G.D.; Sarkhosh, A.; Fernández-Zapata, J.C.; Nicolás, J.J.; Garcia-Sanchez, F. Selenium impedes cadmium and arsenic toxicity in potato by modulating carbohydrate and nitrogen metabolism. *Ecotoxicol. Environ. Saf.* **2019**, *30*, 588–599. [CrossRef] [PubMed]
258. Houimli, S.I.M.; Denden, M.; El-Hadj, S.B. Induction of salt tolerance in pepper (*Capsicum annum*) by 24-epibrassinolide. *Eurasian J. Biosci.* **2008**, *2*, 83–90.
259. Thussagunpanit, J.; Jutamanee, K.; Sonjaroon, W.; Kaveeta, L.; Chai-Arree, W.; Pankean, P.; Suksamrarn, A. Effects of brassinosteroid and brassinosteroid mimic on photosynthetic efficiency and rice yield under heat stress. *Photosynth* **2015**, *1*, 312–320. [CrossRef]
260. Qayyum, B.; Shabbaz, M.; Akram, N.A. Interachive effect of foliar application of 24—Epibrassinolide and root zone salinity on morpho – physiological attributes of wheat (*triticum aestivum* L.). *Int. J. Agric. Biol.* **2007**, *9*, 584–589.
261. Hairat, S.; Khurana, P. Improving Photosynthetic Responses during Recovery from Heat Treatments with Brassinosteroid and Calcium Chloride in Indian Bread Wheat Cultivars. *Am. J. Plant Sci.* **2015**, *6*, 1827–1849. [CrossRef]
262. Amzallag, G.N. Brassinosteroids as metahormones evidences from specific influence during critical period in sorghum development. *Plant Biol.* **2002**, *4*, 656–663. [CrossRef]
263. Gruszka, D.; Janeczko, A.; Dziurka, M.; Pociecha, E.; Oklestkova, J.; Szarejko, I. Barley Brassinosteroid Mutants Provide an Insight into Phytohormonal Homeostasis in Plant Reaction to Drought Stress. *Front. Plant Sci.* **2016**, *7*, 1824. [CrossRef]
264. Kamuro, Y.; Takatsuto, S. Practical application of brassinosteroids in agricultural fields. In *Brassinosteroids: Steroidal Plant Hormones*; Sakurai, A., Yokota, T., Clouse, S.D., Eds.; Springer: Tokyo, Japan, 1999; pp. 223–241.
265. Ali, B.; Hasan, S.A.; Hayat, S.; Hayat, Q.; Yadav, S.; Fariduddin, Q.; Ahmad, A. A role for brassinosteroids in the amelioration of aluminium stress through antioxidant system in mung bean (*Vigna radiata* L. Wilczek). *Environ. Exp. Bot.* **2008**, *62*, 153–159. [CrossRef]
266. Hasan, S.A.; Hayat, S.; Ali, B.B.; Ahmad, A. 28-Homobrassinolide protects chickpea (*Cicer arietinum*) from cadmium toxicity by stimulating antioxidants. *Environ. Pollut.* **2008**, *151*, 60–66. [CrossRef] [PubMed]
267. Rady, M.M.; Osman, A.S. Response of growth and antioxidant system of heavymetal-contaminated tomato plants to 24-epibrassinolide. *Afr. J. Agric. Res.* **2012**, *7*, 3249–3254.
268. Jain, M.; Mathur, G.; Koul, S.; Sarin, N. Ameliorative effects of proline on salt induced lipid peroxidation in Cell lines of groundnut (*Arachis hypogea* L.). *Plant Cell Rep.* **2001**, *20*, 463–468. [CrossRef]
269. Abdullahi, B.A.; Gu, X.; Gan, Q.; Yang, Y. Brassinolide amelioration of aluminum toxicity in mung bean seedling growth. *J. Plant Nutr.* **2003**, *26*, 1725–1734. [CrossRef]
270. Martínez-Cuenca, M.R.; Primo-Capella, A.; Forner-Giner, M.A. Influence of Rootstock on Citrus Tree Growth: Effects on Photosynthesis and Carbohydrate Distribution, Plant Size, Yield, Fruit Quality, and Dwarfing Genotypes. *Plant Growth* **2016**, *16*, 107.
271. Khelil, A.; Menu, T.; Ricard, B. Adaptive response to salt involving carbohydrate metabolism in leaves of a salt-sensitive tomato cultivars. *Plant Physiol. Biochem.* **2007**, *45*, 551–559. [CrossRef]
272. Bartels, D.; Sunkar, R. Drought and salt tolerance in plants. *Crit. Rev. Plant Sci.* **2005**, *24*, 23–58. [CrossRef]
273. Hoekstra, F.A.; Golovina, E.A.; Buitkink, J. Mechanisms of plant desiccation tolerance. *Trends Plant Sci.* **2001**, *6*, 431–438. [CrossRef]
274. Pierik, R.; Christa, T. The Art of Being Flexible: How to Escape from Shade, Salt, and Drought. *Top. Rev. Plast.* **2014**, *166*, 5–22. [CrossRef] [PubMed]

275. Libal-Wekseler, J.; Fernak, B.; Michael, S.T.; Zhu, J.K. Carbohydrate metabolism is affected in response to salt stress. *Horticulture* **1997**, *20*, 1043–1045.
276. Khan, N.; Bano, A. Role of plant growth promoting rhizobacteria and Ag-nano particle in the bioremediation of heavy metals and maize growth under municipal wastewater irrigation. *Int. J. Phytorem.* **2016**, *18*, 211–221. [CrossRef]
277. Schreiber, L. Transport barriers made of cutin, suberin and associated waxes. *Trends Plant Sci.* **2010**, *15*, 546–553. [CrossRef] [PubMed]
278. Lux, A.; Šottníková, A.; Opatrná, J.; Greger, M. Differences in structure of adventitious roots in Salix clones with contrasting characteristics of cadmium accumulation and sensitivity. *Physiol. Plant.* **2004**, *120*, 537–545. [CrossRef] [PubMed]
279. Shabala, S. Learning from halophytes: Physiological basis and strategies to improve abiotic stress tolerance in crops. *Ann. Bot.* **2013**, *112*, 1209–1221. [CrossRef] [PubMed]
280. White, P.J.; Broadley, M.R. Chloride in soil and its uptake and movement within the plant. A review. *Ann. Bot.* **2001**, *88*, 967–988. [CrossRef]
281. Grattan, S.R.; Grieve, C.M. Mineral nutrient acquisition and response by plants grown in saline environments. In *Hand Book of Plant and Crop Stress*; Pessarakli, M., Ed.; Marcel Dekker: New York, NY, USA, 1999; pp. 203–299.
282. Ikram, M.; Ali, N.; Jan, G.; Guljan, F.; Khan, N. Endophytic fungal diversity and their interaction with plants for agriculture sustainability under stressful condition. *Recent Pat. Food Nutr. Agric.* **2019**, *12*. [CrossRef]
283. Marzec, M.; Muszynska, A.; Gruszka, D. The Role of Strigolactones in Nutrient-Stress Responses in Plants. *Int. J. Mol. Sci.* **2013**, *14*, 9286–9304. [CrossRef] [PubMed]
284. Ali, S.; Khan, N.; Nouroz, F.; Erum, S.; Nasim, W.; Shahid, M.A. In vitro effects of GA 3 on morphogenesis of CIP potato explants and acclimatization of plantlets in field. *In Vitro Cell. Dev. Biol.* **2018**, *54*, 104–111. [CrossRef]
285. Wang, M.; Zheng, Q.; Shen, Q.; Guo, S. The critical role of potassium in plant stress response. *Int. J. Mol. Sci.* **2013**, *14*, 7370–7390. [CrossRef]
286. Ali, S.; Khan, N.; Nouroz, F.; Erum, S.; Nasim, W.A. Effects of sucrose and growth regulators on the microtuberization of cip potato (*Solanum tuberosum*) germplasm. *Pak. J. Bot.* **2018**, *1*, 763–768.
287. Maathuis, F.J.M.; Ahmad, I.; Patishtan, J. Regulation of Na^+ fluxes in plants. *Front. Plant Sci.* **2014**, *5*, 467. [CrossRef] [PubMed]
288. Conde, A.; Chaves, M.M.; Gerós, H. Membrane transport, sensing and signaling in plant adaptation to environmental stress. *Plant Cell Physiol.* **2011**, *1*, 1583–1602. [CrossRef]
289. Wyn Jones, R.G.; Pollard, A. Proteins, enzymes and inorganic ions. In *Encyclopedia of Plant Physiology*; New Series; Lauchli, A., Person, A., Eds.; Springer: New York, NY, USA, 1983; Volume 15B, pp. 528–562.
290. Wang, H.; Wu, Z.; Chen, Y.; Yang, C.; Shi, D. Effects of salt and alkali stresses on growth and ion balance in rice (*Oryza sativa* L.). *Plant Soil Environ.* **2011**, *8*, 286–294. [CrossRef]
291. Rossi, L.; Francini, A.; Minnocci, A.; Sebastiani, L. Salt stress modifies apoplastic barriers in olive (*Olea europaea* L.): A comparison between a salt-tolerant and a salt-sensitive cultivar. *Sci. Hortic.* **2015**, *192*, 38–46. [CrossRef]
292. Krishnamurthy, P.; Ranathunge, K.; Nayak, S.; Schreiber, L.; Mathew, M.K. Root apoplastic barriers block Na+ transport to shoots in rice (*Oryza sativa* L.). *J. Exp. Bot.* **2011**, *62*, 4215–4228. [CrossRef]
293. Kosová, K.; Vítámvás, P.; Urban, M.O.; Prášil, I.T. Plant proteome responses to salinity stress–comparison of glycophytes and halophytes. *Funct. Plant Biol.* **2013**, *23*, 775–786. [CrossRef] [PubMed]
294. Moya, J.L.; Primo-Millo, E.; Talon, M. Morphological factors determining salt tolerance in citrus seedlings: The shoot to root ratio modulates passive root uptake of chloride ions and their accumulation in leaves. *Plant Cell Environ.* **1999**, *22*, 1425–1433. [CrossRef]
295. Hussain, M.; Park, H.W.; Farooq, M.; Jabran, K.; Lee, D.J. Morphological and physiological basis of salt resistance in different rice genotypes. *Int. J. Agric. Biol.* **2013**, *15*, 113–118.
296. Carden, D.E.; Walker, J.D.; Flowers, T.J.; Miller, A.J. Single cell measurement of the contributions of cytosolic Na^+ and K^+ to salt tolerance. *Plant Physiol.* **2003**, *131*, 676–683. [CrossRef] [PubMed]
297. Mahmud, S.; Sharmin, S.; Chowdhury, B.L.; Hossain, M.A. Research Article Effect of Salinity and Alleviating Role of Methyl Jasmonate in Some Rice Varieties. *Asian J. Plant Sci.* **2017**, *16*, 87–93. [CrossRef]

298. Manaa, A.; Ben Ahmed, H.; Valot, B.; Bouchet, J.P.; Aschi-Smiti, S.; Causse, M.; Faurobert, M. Salt and genotype impact on plant physiology and root proteome variations in tomato. *J. Exp. Bot.* **2011**, *62*, 2797–2813. [CrossRef] [PubMed]
299. Akca, Y.; Samsunlu, E. The effect of salt stress on growth, chlorophyll content, proline and nutrient accumulation and K/Na ratio in walnut. *Pak. J. Bot.* **2012**, *44*, 1513–1520.
300. Hakim, M.A.; Juraimi, A.S.; Hanafi, M.M.; Ismail, M.R.; Rafii, M.Y.; Islam, M.M.; Selamat, A. The effect of salinity on growth, ion accumulation and yield of rice varieties. *J. Anim. Plant Sci.* **2014**, *24*, 874–885.
301. Deinlein, U.; Stephan, A.B.; Horie, T.; Luo, W.; Xu, G.; Schroeder, J.I. Plant salt-tolerance mechanisms. *Trends Plant. Sci.* **2014**, *19*, 371–379. [CrossRef]
302. Adabnejad, H.; Kavousi, H.R.; Hamidi, H.; Tavassolian, I. Assessment of the vacuolar Na+/H+ antiporter (*NHX1*) transcriptional changes in *Leptochloa fusca* L. in response to salt and cadmium stresses. *Mol. Biol. Res. Commun.* **2015**, *4*, 133–142. [PubMed]
303. Guyton, A.C.; Hall, J.E. *Textbook of Medical Physiology*; Saunders: Philadelphia, PA, USA, 1986; p. 20.
304. Kumari, P.H.; Kumar, S.A.; Sivan, P.; Katam, R.; Suravajhala, P.; Rao, K.S.; Kishor, P.B.K. Overexpression of a Plasma Membrane Bound Na^+/H^+ Antiporter-Like Protein (*SbNHXLP*) Confers Salt Tolerance and Improves Fruit Yield in Tomato by Maintaining Ion Homeostasis. *Front. Plant Sci.* **2016**, *7*, 2027. [CrossRef] [PubMed]
305. Volkov, V. Salinity tolerance in plants. Quantitative approach to ion transport starting from halophytes and stepping to genetic and protein engineering for manipulating ion fluxes. *Front. Plant Sci.* **2015**, *6*, 873. [CrossRef]
306. Veraplakorn, V.; Nanakorn, M.; Kaveeta, L.; Srisom, S.; Bennett, I.J. Variation in ion accumulation as a measure of salt tolerance in seedling and callus of Stylosanthes guianensis. *Exp. Plant Physiol.* **2013**, *25*, 106–115. [CrossRef]
307. Mukhtar, I.; Shahid, M.A.; Khan, M.W.; Balal, R.M.; Iqbal, M.M.; Naz, T.; Zubair, M.; Ali, H.H. Improving salinity tolerance in chili by exogenous application of calcium and sulphur. *Soil Environ.* **2016**, *1*, 35.
308. Bao, A.; Zhao, Z.; Ding, G.; Shi, L.; Xu, F.; Cai, H. The Stable Level of *Glutamine synthetase 2* Plays an Important Role in Rice Growth and in Carbon-Nitrogen Metabolic Balance. *Int. J. Mol. Sci.* **2015**, *16*, 12713–12736. [CrossRef]
309. Bohnert, H.J.; Sheveleva, E. Plant stress adaptations-making metabolism move. *Curr. Opin. Biol.* **1998**, *1*, 267–274. [CrossRef]
310. Bonales-Alatorre, E.; Pottosin, I.; Shabala, L.; Chen, Z.H.; Zeng, F.; Jacobsen, S.E.; Shabala, S. Differential Activity of Plasma and Vacuolar Membrane Transporters Contributes to Genotypic Differences in Salinity Tolerance in a Halophyte Species, *Chenopodium quinoa*. *Int. J. Mol. Sci.* **2013**, *14*, 9267–9285. [CrossRef] [PubMed]
311. Cherel, L. Regulation of K^+ channel activities in plants: From physiological to molecular aspects. *J. Exp. Bot.* **2004**, *55*, 337–351. [CrossRef] [PubMed]
312. Shabala, S.; Pottosin, I. Regulation of potassium transport in plants under hostile conditions: Implications for abiotic and biotic stress tolerance. *Physiol. Plant.* **2014**, *151*, 257–279. [CrossRef] [PubMed]
313. Humphries, E.S.A.; Dart, C. Neuronal and Cardiovascular Potassium Channels as Therapeutic Drug Targets: Promise and Pitfalls. *J. Biomol. Screen.* **2015**, *20*, 1055–1073. [CrossRef] [PubMed]
314. Machnicka, B.; Czogalla, A.; Hryniewicz-Jankowska, A.; Bogusławska, D.M.; Grochowalska, R.; Heger, E.; Sikorski, A.F. Spectrins: A structural platform for stabilization and activation of membrane channels, receptors and transporters. *Biochim. Biophys. Acta* **2014**, *1*, 620–634. [CrossRef]
315. Tavakoli, M.; Poustini, K.; Besharati, H.; Ali, S. Variable salinity responses of 25 alfalfa genotypes and comparative salt-response ion distribution. *Russ. J. Plant Physiol.* **2019**, *66*, 231–239. [CrossRef]
316. Carillo, P.; Parisi, D.; Woodrow, P.; Pontecorvo, G.; Massaro, G.; Annunziata, M.G.; Fuggi, A.; Sulpice, R. Salt-induced accumulation of glycine betaine is inhibited by high light in durum wheat. *Funct. Plant Biol.* **2011**, *22*, 139–150. [CrossRef]
317. Awada, S.; Campbell, W.F.L.; Dudle, M.; Jurinak, J.J. Interactive effects of sodium chloride, sodium sulfate, calcium sulfate and calcium chloride on snapbean growth, photosynthesis and ion uptake. *J. Plant Nutr.* **1995**, *18*, 889–900. [CrossRef]
318. Multari, S.; Stewart, D.; Russell, W.R. Potential of Fava Bean as Future Protein Supply to Partially Replace Meat Intake in the Human Diet. *Compr. Rev. Food Sci.* **2015**, *14*, 511–522. [CrossRef]

319. Hura, T.; Szewczyk-Taranek, B.; Hura, K.; Nowak, K.; Pawłowska, B. Physiological Responses of *Rosa rubiginosa* to Saline Environment. *Water Air Soil Pollut.* **2017**, *228*, 81. [CrossRef] [PubMed]
320. Gowayed, M.H.; Al-Zahrani, H.S.; Metwali, E.M. Improving the salinity tolerance in potato (*Solanum tuberosum*) by exogenous application of silicon dioxide nanoparticles. *Int. J. Agric. Biol.* **2017**, *1*, 19.
321. Pessarakli, M.; Haghighi, M.; Sheibanirad, A. Plant Responses under Environmental Stress Conditions. *Adv. Plants Agric. Res.* **2015**, 2, 00073. [CrossRef]
322. Debouba, M.; Gouia, A.; Suzuki, M.H.; Ghorbel, M. NaCl stress effects on enzymes involved in nitrogen assimilation pathway in tomato "*Lycopersicon esculentum*" seedlings. *J. Plant Physiol.* **2006**, *163*, 1247–1258. [CrossRef]
323. Wang, H.; Zhang, M.; Guo, R.; Shi, D.; Liu, B.; Lin, X.; Yang, C. Effects of slat stress on ion balance and nitrogen metabolism of old and young leaves in rice (*Oryza sativa* L.). *BMC Plant Biol.* **2012**, *12*, 194. [CrossRef]
324. Pandey, M.; Srivastava, A.K.; D'Souza, S.F.; Penna, S. Thiourea, a ROS Scavenger, Regulates Source-to-Sink Relationship to Enhance Crop Yield and Oil Content in *Brassica juncea* (L.). *PLoS ONE* **2013**, *8*, e73921. [CrossRef]
325. Koksal, N.; Alkan-Torun, A.; Kulahlioglu, I.; Ertargin, E.; Karalar, E. Ion uptake of marigold under saline growth conditions. *Springer Plus.* **2016**, *5*, 139. [CrossRef]
326. Khan, N.; Bano, A. Role of PGPR in the Phytoremediation of Heavy Metals and Crop Growth under Municipal Wastewater Irrigation. In *Phytoremediation*; Springer: Cham, Switzerland, 2018; pp. 135–149.
327. Pravin, V.; Rosazlin, A.; Tumirah, K.; Salmah, I.; Amru, N.B. Role of Plant Growth Promoting Rhizobacteria in Agricultural Sustainability—A Review. *Molecules* **2016**, *21*, 573. [CrossRef]
328. Pal, D.; Khozin-Goldberg, I.; Cohen, Z.; Boussiba, S. The effect of light, salinity, and nitrogen availability on lipid production by *Nannochloropsis* sp. *Appl. Microbiol. Biotech.* **2011**, *90*, 1429–1441. [CrossRef] [PubMed]
329. Chen, W.; Hou, Z.; Wu, L.; Liang, Y.; Wei, C. Effects of salinity and nitrogen on cotton growth in arid environment. *Plant Soil.* **2010**, *326*, 61–73. [CrossRef]
330. Seemann, J.R.; Thomas, D.S. Salinity and nitrogen effects on photosynthesis, ribulose-1, 5-bisphosphate carboxylase and metabolite pool sizes in *Phaseolus vulgaris* L. *Plant Physiol.* **1986**, *82*, 555–560. [CrossRef] [PubMed]
331. Giblin, A.E.; Weston, N.B.; Banta, G.T.; Tucker, J.; Hopkinson, C.S. The effects of salinity on nitrogen losses from an oligohaline estuarine sediment. *Estuaries Coasts* **2010**, *33*, 1054–1068. [CrossRef]
332. Laura, R.D. Salinity and nitrogen mineralization in soil. *Soil Biol. Bioch.* **1977**, *9*, 333–336. [CrossRef]
333. Villa-Castorena, M.; Ulery, A.L.; Catalán-Valencia, E.A.; Remmenga, M.D. Salinity and nitrogen rate effects on the growth and yield of chile pepper plants. *Soil Sci. Soc. Am. J.* **2003**, *67*, 1781–1789. [CrossRef]
334. Kahn, A.E.; Durako, M.J. Thalassia testudinum seedling responses to changes in salinity and nitrogen levels. *J. Exp. Mar. Biol. Ecol.* **2006**, *335*, 1–12. [CrossRef]
335. Zhang, D.; Li, W.; Xin, C.; Tang, W.; Eneji, A.E.; Dong, H. Lint yield and nitrogen use efficiency of field-grown cotton vary with soil salinity and nitrogen application rate. *Field Crops Res.* **2012**, *138*, 63–70. [CrossRef]
336. Al-Harbi, A.R.; Wahb-Allah, M.A.; Abu-Muriefah, S.S. Salinity and nitrogen level affects germination, emergence, and seedling growth of tomato. *Int. J. Veg. Sci.* **2008**, *14*, 380–392. [CrossRef]
337. Akhtar, M.; Hussain, F.; Ashraf, M.Y.; Qureshi, T.M.; Akhter, J.; Awan, A.R. Influence of salinity on nitrogen transformations in soil. *Commun. Soil Sci. Plant Anal.* **2012**, *43*, 1674–1683. [CrossRef]
338. Esmaili, E.; Kapourchal, S.A.; Malakouti, M.J.; Homaee, M. Interactive effect of salinity and two nitrogen fertilizers on growth and composition of sorghum. *Plant Soil Environ.* **2008**, *54*, 537–546. [CrossRef]
339. Azizian, A.; Sepaskhah, A.R. Maize response to different water, salinity and nitrogen levels: Agronomic behavior. *Int. J. Plant Prod.* **2014**, *8*, 107–130.
340. Tang, Z.; Liu, Y.; Guo, X.; Zu, Y. The combined effects of salinity and nitrogen forms on Catharanthus roseus: The role of internal ammonium and free amino acids during salt stress. *J. Plant Nutr. Soil Sci.* **2011**, *174*, 135–144. [CrossRef]
341. Garg, N.; Chandel, S. The effects of salinity on nitrogen fixation and trehalose metabolism in mycorrhizal *Cajanus cajan* (L.) millsp. plants. *J. Plant Growth Reg.* **2011**, *30*, 490–503. [CrossRef]
342. Navarro, J.M.; Botella, M.A.; Cerda, A.; Martinez, V. Phosphorus uptake and translocation in salt stressed melon plants. *J. Plant Physiol.* **2001**, *158*, 375–381. [CrossRef]
343. Zakery-Asl, M.A.; Bolandnazar, S.; Oustan, S. Effect of salinity and nitrogen on growth, sodium, potassium

accumulation, and osmotic adjustment of halophyte Suaeda aegyptiaca (Hasselq.) Zoh. *Arch. Agron. Soil Sci.* **2014**, *60*, 785–792. [CrossRef]

344. Zeng, W.Z.; Xu, C.; Wu, J.W.; Huang, J.S.; Zhao, Q.; Wu, M.S. Impacts of salinity and nitrogen on the photosynthetic rate and growth of sunflowers (*Helianthus annuus* L.). *Pedosphere* **2014**, *24*, 635–644. [CrossRef]
345. Cardeñosa, V.; Medrano, E.; Lorenzo, P.; Sánchez-Guerrero, M.C.; Cuevas, F.; Pradas, I.; Moreno-Rojas, J.M. Effects of salinity and nitrogen supply on the quality and health-related compounds of strawberry fruits (Fragaria× ananassa cv. Primoris). *J. Sci. Food Agric.* **2015**, *95*, 2924–2930. [CrossRef] [PubMed]
346. Min, W.; Guo, H.; Zhang, W.; Zhou, G.; Ma, L.; Ye, J.; Liang, Y.; Hou, Z. Response of soil microbial community and diversity to increasing water salinity and nitrogen fertilization rate in an arid soil. *Acta Agric. Scand. Sect. B Soil Plant. Sci.* **2016**, *66*, 117–126. [CrossRef]

8

Identification and Quantification of Plant Growth Regulators and Antioxidant Compounds in Aqueous Extracts of *Padina durvillaei* and *Ulva lactuca*

Israel Benítez García [1], Ana Karen Dueñas Ledezma [1], Emmanuel Martínez Montaño [1,2], Jesús Aarón Salazar Leyva [1], Esther Carrera [3] and Idalia Osuna Ruiz [1,*]

1 Maestría en Ciencias Aplicadas, Unidad Académica de Ingeniería en Biotecnología, Universidad Politécnica de Sinaloa (UPSIN), Carretera Municipal Libre Mazatlán Higueras Km 3., Mazatlán, SIN 82199, Mexico; ibenitez@upsin.edu.mx (I.B.G.); 2017031067@upsin.edu.mx (A.K.D.L.); emartinez@upsin.edu.mx (E.M.M.); jsalazar@upsin.edu.mx (J.A.S.L.)

2 Cátedras CONACYT, Consejo Nacional de Ciencia y Tecnología, Ciudad de México 03940, Mexico

3 Instituto de Biología Molecular y Celular de Plantas (UPV-CSIC), Instituto de Biología Molecular y Celular de Plantas (IBMCP), Universidad Politécnica de Valencia-Consejo Superior de Investigaciones Científicas (CSIC), 46022 Valencia, Spain; ecarrera@ibmcp.upv.es

* Correspondence: iosuna@upsin.edu.mx

Abstract: Aqueous seaweed extracts have diverse compounds such as Plant-Growth Regulators (PGRs) which have been utilized in agricultural practices for increasing crop productivity. Algal biomass of *Padina durvillaei* and *Ulva lactuca* have been suggested for use as biofertilizers because of plant growth-enhancing properties. This work aimed to identify the main PGRs and antioxidant properties in *P. durvillaei* and *U. lactuca* extracts, such as abscisic acid, auxins, cytokinins, gibberellins, jasmonates, and salicylates, to assess their potential use as biofertilizers that improve plant growth and crop yield. Phytochemical analyses of two seaweed extracts showed a significantly higher content of sulfates, flavonoids, and phenolic compounds in *P. durvillaei* extract, which could be linked to its higher antioxidant activity (DPPH, ABTS, and FRAP) compared to *U. lactuca* extract. The identification and quantification of PGRs showed two gibberellins (GA1 and GA4), abscisic acid (ABA), indoleacetic acid (IAA), three cytokinins (tZ, IP, and DHZ), jasmonic acid (JA), and salicylic acid (SA) in two seaweed extracts. However, GA4, tZ, and DHZ contents were significantly higher in *P. durvillaei* compared to *U. lactuca* extracts. These findings evidence that *P. durvillaei* and *U. lactuca* extracts are suitable candidates for use as biofertilizers.

Keywords: seaweed extract; phytohormone profiling; fertilizers; antioxidant; plant growth regulators; brown seaweed; green algae

1. Introduction

Seaweed extracts have been used in agricultural activities due to their content of macroelements (alginate, agar, carrageenan, etc.), which activate the synthesis of endogenous hormones in plants [1,2] and contribute microelements (N, Ca, Mg, Mn, B, Br, I, Zn, Cu, and Co), amino acids, and vitamins that enrich soil in plant crops [3]. Besides, seaweed extracts contain biochemical compounds such as chlorophylls, carotenoids, and phenolics that confer antioxidant protection [4,5]. The antioxidant properties of algae extracts have been widely evaluated and attributed to sulfated polysaccharides, pigments, and phenolic compounds [4–8], which provide desirable characteristics for their potential use in crops, since, in addition to conferring antioxidant protection, compounds such as polysaccharides have been linked to growth promoting activities [8].

Some environmentally-friendly extraction methods generally include boiling or soaking with distilled water, which have been used as biostimulants for plant growth [9]. Phytohormone-like Plant-Growth Regulators (PGRs) have been identified in algal extracts, such as abscisic acid, auxins, cytokinins, gibberellins, jasmonates, or salicylates, all of which regulate plant cell metabolism and boost production and growth [10–12]. For this reason, marine algae have been used in agriculture as organic fertilizers to achieve sustainable crop production [13,14] and counter the excessive use of fertilizers and synthetic hormones (e.g., 2.4-dichlorophenoxyacetic acid and naphthaleneacetic acid) that may potentially affect both the environment and humans [12,15].

Several authors have reported that algal extracts induce physiological processes in treated plants, such as germination, emergence, root growth, nutrient mobilization, maturation, tolerance to stress, and disease resistance; these responses are similar to those observed in plant crops treated with synthetic hormones [3,10,16–18]. Some seaweed extracts are marketed as liquid biofertilizers or biostimulants [3,16,19], mostly enriched with biomass of *Ascophyllum nodosum*, *Sargassum* spp., and *Macrocystis pyrifera* [3,20,21]. However, the different species of algae show variations regarding PGR biosynthesis [22]; thus, algal extracts exert variable physiological effects on different crops.

In Latin America, algal extracts have recently been used as biostimulants. Some reports demonstrate the benefits of the application of seaweed extracts harvested in coastal areas on various crops. The macroalgae *M. pyrifera*, *Gelidium robustum*, *Chondracanthus canaliculatus*, *Sargassum* spp., *Ulva lactuca*, and *Padina gymnospora*, have been used as biostimulants, fertilizers, and root promoters, as well as to stimulate growth and increase antifungal protection in tomato plants (*Solanum lycopersicum*) [20,23]. However, the exploitation of marine algae, mainly those involved in massive arrivals, is still incipient; moreover, it is not well known if improvements in yield and production of crops fertilized with algal extracts are due to the presence of PGRs in algal organic matter and/or if a possible contribution of other metabolites contribute to the biostimulant effects. Therefore, it is necessary to study the chemical and bioactive composition of a new algal extract when it is prepared to consider its potential use as plant grow stimulant.

In addition, PGRs in seaweeds have been insufficiently studied. Therefore, information is needed to support the use of marine sources, such as algae, to achieve sustainable agriculture practices in the future. This will reduce the environmental impact associated with the excessive use of chemical fertilizers, and also the potential risks to consumers resulting from the indiscriminate application of synthetic PGRs, along with the fact that algae provide other bioactive compounds that enhance protection against stress oxidative, improving plant health.

The objective of the present investigation was to characterize the chemical and bioactive composition (antioxidant activity, PGR identification and content) of aqueous extracts of the macroalgae *Ulva lactuca* and *Padina durvillaei*, and evaluate the use of such aqueous extracts as a potential biofertilizer.

2. Material and Methods

2.1. Seaweed Collection and Reagents

Specimens of the seaweeds *Padina durvillaei* (Bory Saint-Vicent, 1957) and *Ulva lactuca* (Linnaeus 1753) were collected in Mazatlan Bay, Sinaloa, Mexico (23°1′2 9.1′′ LN, 106°25′29.7′ LW), in March 2017. Fresh samples were rinsed with distilled water, lyophilized, ground with a commercial grinder, and stored at −20 °C until used. All chemicals used in this research were analytical grade and supplied from Sigma (Sigma-Aldrich Co., St. Louis, MO, USA), unless otherwise specified.

2.2. Seaweed Extracts

Seaweed extracts were obtained using distilled water according to Tierney et al. [24], modified as follows: dried algal material was mixed with water at 21 °C (1:10, *w:v*) with stirring for 3 h; then, the extract was filtered through a fiber glass filter (1.2 µm pore size) and the algal residue extracted again (twice). Filtrates were pooled and centrifuged at 12,000× g and 4 °C for 20 min; then, the supernatant was collected. Finally, the aqueous extract was lyophilized and stored at −20 °C until analyzed.

The extraction yield was calculated according to Equation (1):

$$\textit{Extraction yield (\%)} = (\text{grams of dry aqueous extract/grams of dry seaweed}) \times 100 \quad (1)$$

2.3. Chemical Composition

Carbohydrate content was measured using the phenol–sulfuric acid method [25] using D-glucose as standard. Soluble protein content was determined with Bradford's method using Bovine Serum Albumin (BSA) as standard [26].

Sulfate content was measured with the barium chloride-gelatin assay using potassium sulfate as standard [27]. The uronic acid content was determined with the sulfuric acid-carbazole colorimetric method using D-glucuronic acid as standard [28].

2.3.1. Total Phenolic Content (TPC)

Total soluble phenolic content was determined using the Folin–Ciocalteu method [29]. Dry samples were reconstituted with acetone (1 mg/mL); then, a 100 mL of each sample was mixed with 150 mL of Folin solution (previously diluted 1:1 with deionized water) followed by the addition of 1 mL of 2% sodium carbonate in 0.4% sodium hydroxide. The mixture was incubated in the dark at room temperature for 20 min. The resulting blue complex was read in a spectrophotometer at 750 nm. Phenolic content was expressed as mg of gallic acid equivalent (GAE) per g of sample (dry weight). A gallic acid standard curve was constructed at the concentration range of 0–0.25 mg/mL.

2.3.2. Total Flavonoids Content (TFC)

Total flavonoid content was assessed according to Luximon-Ramma et al. [30]. Samples of solutions (1 mL) were diluted in equal volumes of a 2% aluminum chloride solution (2 g of $AlCl_3 \cdot 6H_2O$ in 100 mL of methanol). The mixture was incubated at room temperature for 10 min. Absorbance was read at 367 nm. The results were expressed in mg of quercetin equivalents (QE) per gram of sample (dry weight). A quercetin standard curve was constructed at the concentration range of 0–0.5 mg/mL.

2.4. Antioxidant Evaluation

2.4.1. DPPH Free-Radical Scavenging Activity

The free-radical scavenging potential of the seaweed extracts was analyzed according to the method proposed by Mensor et al. [31], modified as follows: a 100 mL aliquot of each extract (at concentrations of 0.0015 to 1.5 mg/mL) was mixed with 900 mL of an ethanol solution of 0.3 mM 2,2-diphenyl-1-picrylhydrazyl (DPPH); the mixture was incubated for 30 min in the dark at room temperature. Then, absorbance was measured at 518 nm. Trolox was used as standard, whereas the DPPH solution served as control to calculate the degree of radical scavenging by samples as well as the reference standard.

The percentage of DPPH scavenging was calculated with Equation (2):

$$\textit{\% DPPH scavenging} = [(1 - \text{Absorbance of sample})/\text{Absorbance of Control}] \times 100 \quad (2)$$

2.4.2. ABTS Free-Radical Scavenging Activity

The scavenging activity of 2,2′-azinobis [3-ethylbenzthiazoline]-6-sulphonic acid (ABTS) was determined according to the method by Przygodzka et al. [32], modified as follows: the ABTS radical was previously activated for 12–16 h at room temperature in the dark; the resulting ABTS radical solution was diluted with ethanol and its absorbance read at 734 nm, yielding a value of 0.80. A 100 μL aliquot of each sample (at a concentration of 0.0015 to 1.5 mg/mL) was mixed with 2.9 mL of ABTS solution and the absorbance was read 10 min after mixing. Trolox was used as reference standard, whereas the ABTS radical solution served as control to calculate the degree of radical scavenging by samples as well as the reference standard.

The percentage of ABTS scavenging was calculated using Equation (3):

$$\% \, ABTS \, scavenging = [(\mathrm{A} - \mathrm{B})/\mathrm{A}] \times 100 \tag{3}$$

where A is absorbance of the *ABTS* control solution and B is absorbance of the test solution.

2.4.3. Ferric Reducing Antioxidant Power (FRAP) Assay

The FRAP assay was performed according to the methods of Benzie and Strain [33], with the minor modification reported by Szôllôsi and Varga [34]. The FRAP reagent was made from three different solutions: Solution A: 300 mM acetate buffer, pH 3.6; Solution B: 10 mM TPTZ dissolved in 40 mM HCl; and Solution C: 20 mM $FeCl_3{\cdot}6H_2O$. The work solution was prepared by mixing A, B, and C in a 10:1:1 ratio (by volume). For the assay, 100 µL of sample were mixed with 1400 µL of FRAP, and then incubated at room temperature for 30 min in the dark. Finally, absorbance was read at 593 nm. Trolox was used as reference standard.

2.5. Identification and Quantification of Plant Growth Regulator Profiles

PGRs for acid hormones (gibberellins, GAs; indolacetic acid, IAA; jasmonic acid, JA; abscisic acid, ABA; and salicylic acid, SA), and cytokinins or basic hormones (dihydrozeatine, DHZ; isopentyladenine, iP; and t-zeatine, tZ) were identified and quantified by ultra-high performance liquid chromatography–mass spectrometry (UHPLC-MS) using a Thermo Scientific™ Q Exactive™ Hybrid Quadrupole-Orbitrap mass spectrometer at Institute for Plant Molecular and Cell Biology (IBMCP), Spain. Extraction and separation of plant hormone profiles were performed as described by Seo et al. [35]. The lyophilized extract was suspended in 80% methanol (MeOH) containing 1% acetic acid, mixed by stirring for 1 h at 4 °C and centrifuged at 14 000× *g* at 4 °C for 4 min. The supernatant extract was stored at −20 °C overnight and then centrifuged at 14 000× *g* at 4 °C for 4 min. Then, the supernatant was dried in a vacuum evaporator. The dry residue was dissolved in 1% (*v*/*v*) acetic acid and passed consecutively through an Oasis HLB reverse-phase column (30 mg; Waters) and an Oasis MCX cation exchanger. Acid hormones were eluted with MeOH and basic hormones with 60% MeOH containing 5% aqueous ammonia [35]. The final residues were dried and dissolved in 5% (*v*/*v*) acetonitrile, 1% (*v*/*v*) MeOH, and 1% (*v*/*v*) acetic acid. Then, hormones were separated by UHPLC with a reverse Accucore C18 column (2.6 mm inner diameter, 100 mm length; Thermo Fisher Scientific) with a 2% to 55% (*v*/*v*) acetonitrile gradient containing 0.05% (*v*/*v*) acetic acid at 400 mL for 21 min. The plant hormones were analyzed with a Q-Exactive mass spectrometer (Orbitrap detector; Thermo Fisher Scientific) by targeted selected ion monitoring (capillary temperature, 300 °C; S-lens RF level, 70; resolution, 70,000) and electrospray ionization (spray voltage, 3 kV; heater temperature, 150 °C; sheath gas-flow rate, 1.90 mL/min; auxiliary gas-flow rate, 0.42 mL/min). The concentrations of plant hormones in extracts were determined using embedded calibration curves and the Xcalibur 2.2 SP1 build 48 and TraceFinder programs. The internal standards for quantification of each plant hormone were D6-ABA, D2-GA1, D2-GA4, D5-tZ, D3-DHZ, D6-iP, D2-IAA, D6-SA, and D2-JA (Olchemim Ltd., Olomouc, Czech).

2.6. Statistical Analyses

The results of all assays are reported as mean ± standard deviation. Comparisons between groups were performed using the Student's *t*-test. Statistical significance was set at $p < 0.05$.

All statistical analyses were performed using the statistical program Sigma Plot version 11.0 (2018 Systat Software, Inc.; Erkrath, Germany).

3. Results

3.1. Chemical Composition and Antioxidant Capacity Evaluation of Seaweed Extracts

The highest extraction yield was achieved in the *Ulva lactuca* extract (Table 1), which was nearly twice the yield for *Padina durvillaei* (7.55 ± 4.05% and 3.34 ± 1.20%, respectively). With regard to the chemical composition of the soluble components in both extracts, similar contents of carbohydrates and uronic acids were found; however, the percentage of sulfates and the content of polyphenols and flavonoids varied, with higher values in the *P. durvillaei* extract (6.63 ± 0.7%, 34.26 ± 1.39 GAE/g, and 16.16 ± 2.87 QE/g, respectively). Soluble protein was found in the *P. durvillaei* extract (1.28 ± 0.5%), but not in the *U. lactuca* extract.

Table 1. Chemical composition of seaweed extracts.

	Padina Durvillaei	*Ulva Lactuca*
Extraction Yield (%)	3.34 ± 1.20	7.55 ± 4.05
Chemical Composition (dwt.)		
Soluble Protein (%)	1.28 ± 0.56	N. D.
Carbohydrates (%)	16.36 ± 0.08	16.19 ± 0.07
Uronic Acids (%)	8.79 ± 0.60	8.01 ± 0.21
Sulfates (%)	6.63 ± 0.76 a	4.05 ± 1.13 b
Total Polyphenols Content (mg GAE/g)	34.26 ± 1.39 a	27.29 ± 1.57 b
Total Flavonoids Content (mg QE/g)	16.16 ± 2.87 a	10.22 ± 0.96 b

All chemicals components are expressed on dry weight basis (dwt.). Values represented are mean of triplicates; values followed by different letters are significantly different at $p < 0.05$. N.D., not detected; GAE/g, mg of Gallic acid equivalent per gram of seaweed extract; QE/g, mg of Quercetin equivalent per gram of seaweed extract.

Regards to the antioxidant activity in extracts (Figure 1), greater activity was recorded in the *P. durvillaei* extract for the DPPH and ABTS tests, with 0.37 and 0.07 TEAC per gram of extract, respectively (Figure 1A,B). The antioxidant capacity, as measured by DPPH, was three-fold for the *P. durvillaei* extract versus the *U. lactuca* extract (0.13 TEAC/g). The ferric reducing antioxidant power (FRAP) was only recorded in *P. durvillaei* extracts, with TEAC values of 0.02/g (Figure 1C).

3.2. PGR Contents in Seaweed Extracts

Two groups of PGRs (acidic and basic) were identified by UHPLC-MS (Figure 2); their identification and quantification in aqueous algal extracts correspond to abcisic acid (ABA), an auxin (IAA), three cytokinins (dihydrozeatine, DHZ; isopentyladenine, IP; and t-zeatine, tZ), two gibberellins (GA1 and GA4), jasmonic acid (JA), and salicylic acid (SA). The quantification of acidic PGRs showed that SA and IAA are the main chemical groups in both algal extracts. SA concentration was 88.8 ± 52.9 ng/g in the *P. durvillaei* extract and 67.6 ± 4.2 ng/g in the *U. lactuca* extract (Figure 2A). IAA concentration was slightly higher in *U. lactuca* (49.3 ± 5.2 ng/g) relative to *P. durvillaei* (39.0 ± 12.8 ng/g), but these differences were not statistically significant (Figure 2A). On the other hand, the *U. lactuca* extract contained the highest ABA concentration (17.8 ± 5.2 ng/g) compared to the *P. durvillaei* extract (1.5 ± 0.5 ng/g); JA concentrations were similar in both extracts (Figure 2A). With regard to gibberellins, the highest GA4 concentration (1.51 ± 0.64 ng/g) was noted in the *U. lactuca* extract, compared to the *P. durvillaei* extract (0.37 ± 0.08 ng/g), while GA1 showed similar levels in both extracts (Figure 2B).

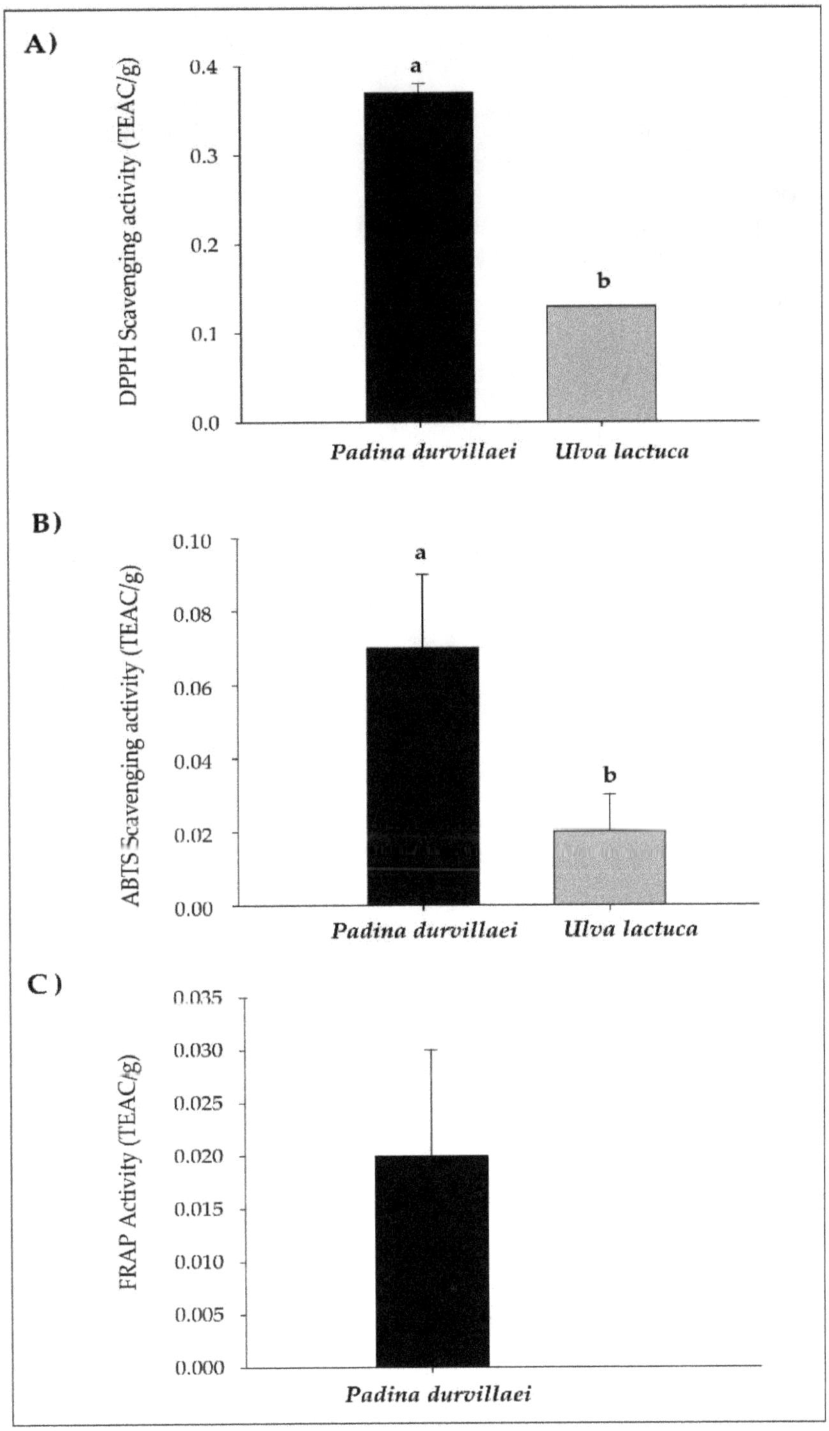

Figure 1. In vitro antioxidant activity of seaweed extracts of *Padina durvillaei* and *Ulva lactuca* assessed by different methods: (**A**) DPPH scavenging activity; (**B**) ABTS scavenging activity; and (**C**) FRAP activity. Bars represent mean of triplicates ± SD; different letters are significantly different at $p < 0.05$. TEAC/g, mmol of Trolox equivalent of antioxidant activity per gram of seaweed extract.

The quantification of basic PGRs (Figure 2C) showed that tZ is the most abundant hormone in the *U. lactuca* extract (10.95± 4.31 ng/g). On the other hand, IP concentrations were similar in both extracts (4.41 1.85 ng/g in *P. durvillaei* and 4.57 ± 0.38 ng/g in *U. lactuca*). DHZ concentrations were low in both extracts, with 1.09 ± 0.44 ng/g in *P. durvillaei* and 0.77 ± 0.09 ng/g in *U. lactuca*.

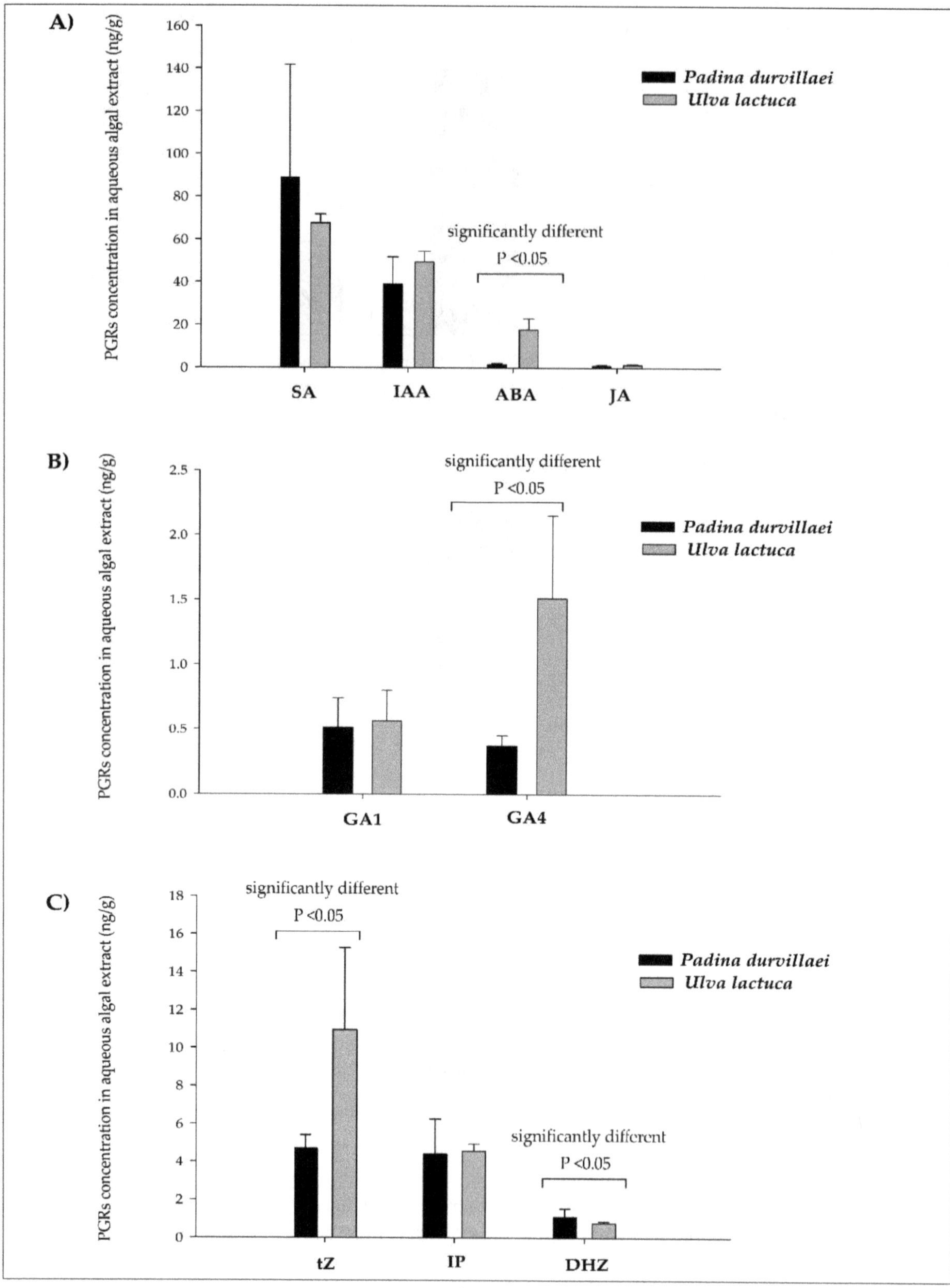

Figure 2. PGRs quantification by UHPLC-MS in aqueous algal extracts of *Padina durvillaei* and *Ulva lactuca*: (**A**) acid PGRs (salicylic acid, SA; indolacetic acid, IAA; abscisic acid, ABA; and jasmonic acid, JA); (**B**) gibberellins (GA1 and GA4); and (**C**) cytokinins or basic hormones (t-zeatine, tZ; isopentyladenine, iP; and dihydrozeatine, DHZ). Bars represent mean of triplicates ± SD. Significant differences at $p < 0.05$ of some type of PGR between algal extracts have been indicated.

4. Discussion

In this study, we used two species of algae to obtain aqueous extracts. Both species are frequently observed as floating seaweed mats that reach the northeast Pacific coasts; they belong to different taxonomic groups: *P. durvillaei* to the class Phaeophyceae (brown algae) and *U. lactuca* to the class Chlorophyceae (green algae). This leads to potential chemo-taxonomic differences, coupled with spatiotemporal variations associated with growth and environmental adaptation. Ultimately, these differences result in the variability in chemical composition, as observed in the aqueous extracts analyzed in the present study. Such variability has also been observed in other studies reporting taxonomic compositional differences in marine algae thriving in temperate waters (e.g., *Caulerpa sertularioides, Rhizoclonium riparium, Gracilaria vermoculophylla*, and *Spyridia filamentosa*), including the presence of biochemical compounds associated with bioactive characteristics, such as fucosterol, β-sitosterol, omega-3 fatty acids, and various photosynthetic pigments [36]. An example is the case of temperate macroalgae in Denmark, where aqueous extracts were analyzed from 16 species of macroalgae in different taxonomic families and whose total polyphenolics content showed no statistically significant differences [6].

From the compositional differences observed in macroalgae, their extracts exhibit different levels of biological activity, particularly antioxidant activity; the present study recorded a higher activity of the *P. durvillaei* extract. This extract has a higher content of polyphenols and flavonoids, compounds with a known ability to scavenge synthetic radicals in in vitro tests (DPPH and ABTS), as well as the presence of sulfated polysaccharides. Note that no ferric reducing antioxidant power (FRAP) was detected in *Ulva* extracts, while it was very weak in the *Padina* extract; this may indicate the influence of compositional characteristics of sulfated polysaccharides in extracts. Besides, substances that interfere in the antioxidant tests used were found, such as proteins and uronic acids. High contents of sulfates, protein, and uronic acids were also observed, which have been associated with the low antioxidant activity observed in polysaccharide fractions isolated from aqueous extracts of green algae such as *Ulva fasciata* [37].

It was also observed that the presence of fucose in sulfated polysaccharides from brown algae conferred a higher ferric reducing activity relative to polysaccharides from green algae that do not contain this sugar (e.g., *Sargassum wightii* vs. *Ulva lactuca*) [7]; a similar effect is likely to influence the behavior observed in our study. In general, secondary metabolites such as polyphenols are more abundant in brown algae, with some, e.g., 2,4,6-trihydroxybenzoate (a benzoic acid derivative), being unique to this group of macroalgae [38]. It should be stressed that polyphenols are compounds with electrons that can be donated, thus conferring a higher antioxidant response when classified methods are used, including electron transfer such as ABTS and FRAP [39].

On the other hand, a higher antioxidant activity was observed in both extracts with the DPPH method. This indicates that the antioxidant chemicals in extracts function mainly through a reaction mechanism involving the transfer of a hydrogen atom from these compounds [6]; this sort of antioxidant capacity was about three times higher for the *P. durvillaei* extract relative to the *U. lactuca* extract. Some studies have shown that aqueous extracts from brown algae have a higher polyphenolics content relative to red and green algae, which confer on them greater antioxidant activity. For example, the antioxidant activity (assessed with the DPPH test) of aqueous extracts from the brown algae *Fucus serratus* and *F. vesiculus* was 42 times higher than in extracts from the green algae *Enteromorpha intestinalis* and 89 times higher than in *Ulva lactuca* [6]. The same behavior was observed for the FRAP test, since the reducing power from brown algae extracts was 3–5 times lower relative to green algae; total polyphenols content and type of polyphenols (ferulic, vanillic, coumaric, and gallic acids) contribute highly to the antioxidant and reducing activities [6]. Likewise, Wang et al. [8] reported higher total polyphenols contents in aqueous extracts of brown algae relative to red and green algae, which led to a high in vitro antioxidant activity measured with DPPH for extracts obtained from *Fucus vesiculus, F. serratus, Ascophyllum nodosum*, and *Laminaria hyperborea*, with antioxidant values representing up to 200 times the antioxidant activity observed for *Ulva lactuca*.

Originally, PGRs in plants have been found in trace amounts (fmol to pmol per gram, wet weight), with gibberellins, abscisic acid, cytokinins, indoleacetic acid, ethylene, jasmonates, and salicylates as those most studied PGRs [40]. Unlike plants, algae accumulate higher concentrations of PGRs (pmol to μmol per gram fresh weight), mainly auxins and cytokinins; some authors hypothesize that bioactive gibberellins inducing germination and growth may also accumulate, as well as jasmonates and salicylates. However, few species of macroalgae contain gibberellins, jasmonate, and salicylates [3].

The presence and number of PGRs in algae and terrestrial plants differ according to the species or variety [12,22]. For example, Mori et al. [41] reported the identification of auxins, cytokinins, and salicylic acid (SA), but not AG3 and jasmonates, in extracts of two red algae (*Pyropia yezoensis* and *Bangia fuscopurpurea*). Another study involving fourteen seaweeds in the Turkey coast reported the presence of five PGRs (t-zeatine (t-Z), IAA, GA3, ABA, and 6-benzyl amino purine (BAP) in two algae, namely *Petalonia fascia* (brown algae) and *Caulerpa racemosa* var. cylindracea (green algae), and the absence of GA3 and ABA in the eleven remaining seaweeds, including *Sargassum vulgare* and *Ulva rigida*, which are used to produce biofertilizers [22]. By contrast, this work identified nine PGRs (SA, IAA, ABA, JA, t-Z, IP, DHZ, GA4, and GA1) in extracts from two algae (*P. durvillaei* and *U. lactuca*) distributed along the Mexican Pacific coasts. The presence of SA and JA is worth noting, as these have not been identified in any other algae; these PGRs strengthen the defense capabilities of plants by inducing acquired systemic resistance (SAR).Gibberellins GA4 and GA1 are also present, which show biological activity in plants.

Mexico produces and markets fertilizers based on extracts of seaweeds such as *Macrocystis pyrifera*, *Sargassum* spp, *Ascophyllum nodosum*, *Laminaria* spp, *Egregia menziesii*, and *Gelidium robustum*, using methods that involve hydrothermal treatment under acidic, neutral, and alkaline conditions [20]. These extracts boost germination, rooting, and plant growth, likely resulting from the content of polysaccharides, macro- and microelements [20,42–44]. However, although studies on algal extracts show evident effects on plant growth, few reports demonstrate the presence of PGRs in such extracts, and these can only be inferred from the physiological effects shown [42]. Unlike marketed products based on seaweed extracts containing polysaccharides obtained using methods under alkaline, acidic, and neutral conditions, this study used aqueous extraction, which ensured the extraction of nine PGRs in high concentrations for use in plant crops.

Studies related to the identification and quantification of PGRs in algae, such as *Ascophyllum nodosum* and *Sargassum muticum* (Phaeophyceae) used in the development of biofertilizers [45], report results differing from those obtained in this work. For example, *A. nodosum*, whose extracts are used in products such as Phylgreen®, contains concentrations of IAA, ABA, and IP of 7.53, 17.63, and 16.11 pmol/g dry weight (DW), respectively [46], while extracts of *Sargassum heterophyllum* accumulate IP and t-Z at 48.2 and 2.4 pmol/g DW, respectively [47]. In contrast, this study reports concentrations in extracts of *P. durvillaei* (IAA 39.0 ng/g, ABA 1.5 ng/g, IP 4.41 ng/g DW, and t-Z 4.7 ng/g DW) and *U. lactuca* (IAA 49.3 ng/g, ABA 17.8 ng/g, IP 4.57 ng/g, and t-Z 10.95 ng/g DW) that are higher than those reported by Jannin et al. [46] and Stirk et al. [47]. This finding suggests that the algae studied here are potential alternatives suitable for use as biofertilizers based on their high PGR content.

On the other hand, the identification of GA1 and GA4 in this study contrasts with reports by Dumale et al. (2018) [48], who identified GA3 in extracts of *Caulerpa racemosa* (green algae), and by Shoubaky et al. [49] in *U. lactuca* extracts, who identified two groups of PGRs, namely ABA and eight gibberellins (GA7 methyl ester, GA8, GA13, GA19, GA23, GA44, and GA75). However, the gibberellins identified in extracts of *P. durvillaei* and *U. lactuca* are considered to be biologically active in plants, contrary to those reported by Shoubaky et al. [49]; this difference is relevant, as PGRs should be in the active form for use in plant crops [50] Note that the identification of PGRs in *P. durvillaei* extracts has not been reported in the literature previously; thus, the data in this paper support the potential use of brown algae or extracts thereof as growth promoters in plants of agricultural importance.

5. Conclusions

The present study focused on the analysis of the main PGRs and the biochemical characterization of their aqueous extracts in two macroalgae involved in massive arrivals. The chromatographic analysis revealed the presence of abcisic acid (AB), auxins (IAA), cytokinins (tZ, IP, and DHZ), gibberellins (AG1 and AG4), jasmonates (AJ), and salicylates (AS); of these, AS attained the highest levels in both extracts. On the other hand, the phytochemical analysis revealed the presence of soluble compounds such as carbohydrates and uronic acids, as well as bioactive compounds such as polyphenols and flavonoids that confer antioxidant activity to extracts. The identification of PGRs in algal extracts opens the possibilities for use of algae involved in massive arrivals as potential environmentally-friendly organic biofertilizers that serve as growth promoters in agricultural crops.

Author Contributions: Conceptualization, I.B.G. and I.O.R.; methodology, A.K.D.L., E.C., I.O.R., and I.B.G.; validation, J.A.S.L., E.C., I.B.G., and I.O.R.; formal analysis, E.M.M., E.C., and I.O.R.; investigation, A.K.D.L., I.B.G., and I.O.R.; resources, E.C., I.O.R., and I.B.G.; data curation, I.B.G. and I.O.R.; writing—original draft preparation, I.B.G. and I.O.R.; writing—review and editing, I.B.G., E.C., I.O.R., E.M.M., and J.A.S.L.; visualization, E.M.M. and J.A.S.L.; supervision, I.O.R., I.B.G., E.M.M., and J.A.S.L.; project administration, I.B.G. and I.O.R.; and funding acquisition, I.O.R. and I.B.G. All authors have read and agreed to the published version of the manuscript.

Acknowledgments: The authors are grateful to the Universidad Politécnica de Sinaloa and the Instituto de Biología Molecular y Celular de Plantas (IBMCP) CSIC-Universidad Politécnica de Valencia for providing the necessary facilities to carry out the research.

References

1. Ben Salah, I.; Aghrouss, S.; Douira, A.; Aissam, S.; El Alaoui-Talibi, Z.; Filali-Maltouf, A.; El Modafar, C. Seaweed polysaccharides as bio-elicitors of natural defenses in olive trees against verticillium wilt of olive. *J. Plant Interact.* **2018**, *13*, 248–255. [CrossRef]
2. Vera, J.; Castro, J.; Gonzalez, A.; Moenne, A. Seaweed Polysaccharides and Derived Oligosaccharides Stimulate Defense Responses and Protection Against Pathogens in Plants. *Mar. Drugs* **2011**, *9*, 2514–2525. [CrossRef]
3. Khan, W.; Rayirath, U.P.; Subramanian, S.; Jithesh, M.N.; Rayorath, P.; Hodges, D.M.; Critchley, A.T.; Craigie, J.S.; Norrie, J.; Prithiviraj, B. Seaweed Extracts as Biostimulants of Plant Growth and Development. *J. Plant Growth Regul.* **2009**, *28*, 386–399. [CrossRef]
4. Osuna-Ruiz, I.; López-Saiz, C.-M.; Burgos-Hernández, A.; Velázquez, C.; Nieves-Soto, M.; Hurtado-Oliva, M.A. Antioxidant, antimutagenic and antiproliferative activities in selected seaweed species from Sinaloa, Mexico. *Pharm. Biol.* **2016**, *54*, 2196–2210. [CrossRef] [PubMed]
5. Osuna-Ruíz, I.; Salazar-Leyva, J.A.; López-Saiz, C.M.; Burgos-Hernández, A.; Hernández-Garibay, E.; Lizardi-Mendoza, J.; Hurtado-Oliva, M.A. Enhancing antioxidant and antimutagenic activity of the green seaweed Rhizoclonium riparium by bioassay-guided solvent partitioning. *J. Appl. Phycol.* **2019**, *31*, 3871–3881. [CrossRef]
6. Farvin, K.S.; Jacobsen, C. Phenolic compounds and antioxidant activities of selected species of seaweeds from Danish coast. *Food Chem.* **2013**, *138*, 1670–1681. [CrossRef] [PubMed]
7. Meenakshi, S.; Umayaparvathi, S.; Arumugam, M.; Balasubramanian, T. In vitro antioxidant properties and FTIR analysis of two seaweeds of Gulf of Mannar. *Asian Pac. J. Trop. Biomed.* **2011**, *1*, 66–70. [CrossRef]
8. Wang, T.; Jónsdóttir, R.; Olafsdottir, G. Total phenolic compounds, radical scavenging and metal chelation of extracts from Icelandic seaweeds. *Food Chem.* **2009**, *116*, 240–248. [CrossRef]
9. Godlewska, K.; Michalak, I.; Tuhy, Ł.; Chojnacka, K. Plant Growth Biostimulants Based on Different Methods of Seaweed Extraction with Water. *BioMed Res. Int.* **2016**, *2016*, 1–11. [CrossRef]
10. Ghaderiardakani, F.; Collas, E.; Damiano, D.K.; Tagg, K.; Graham, N.S.; Coates, J.C. Effects of green seaweed extract on Arabidopsis early development suggest roles for hormone signalling in plant responses to algal fertilisers. *Sci. Rep.* **2019**, *9*, 1983. [CrossRef]

11. Prasad, K.; Das, A.K.; Oza, M.D.; Brahmbhatt, H.; Siddhanta, A.K.; Meena, R.; Eswaran, K.; Rajyaguru, M.R.; Ghosh, P.K. Detection and Quantification of Some Plant Growth Regulators in a Seaweed-Based Foliar Spray Employing a Mass Spectrometric Technique sans Chromatographic Separation. *J. Agric. Food Chem.* **2010**, *58*, 4594–4601. [CrossRef] [PubMed]
12. Yalçın, S.; Şükran Okudan, E.; Karakaş, Ö.; Önem, A.N.; Sözgen Başkan, K. Identification and quantification of some phytohormones in seaweeds using UPLC-MS/MS. *J. Liq. Chromatogr. Relat. Technol.* **2019**, *42*, 475–484. [CrossRef]
13. Hurtado, A.Q.; Neish, I.C.; Critchley, A.T. Phyconomy: The extensive cultivation of seaweeds, their sustainability and economic value, with particular reference to important lessons to be learned and transferred from the practice of eucheumatoid farming. *Phycologia* **2019**, *58*, 472–483. [CrossRef]
14. Kumar, R.; Trivedi, K.; Anand, K.G.V.; Ghosh, A. Science behind biostimulant action of seaweed extract on growth and crop yield: Insights into transcriptional changes in roots of maize treated with Kappaphycus alvarezii seaweed extract under soil moisture stressed conditions. *J. Appl. Phycol.* **2019**, *32*, 599–613. [CrossRef]
15. Hąc-Wydro, K.; Flasiński, M. The studies on the toxicity mechanism of environmentally hazardous natural (IAA) and synthetic (NAA) auxin—The experiments on model Arabidopsis thaliana and rat liver plasma membranes. *Colloids Surf. B Biointerfaces* **2015**, *130*, 53–60. [CrossRef]
16. Arioli, T.; Mattner, S.W.; Winberg, P. Applications of seaweed extracts in Australian agriculture: Past, present and future. *J. Appl. Phycol.* **2015**, *27*, 2007–2015. [CrossRef]
17. Hayashi, L.; Reis, R.P.; Dos Santos, A.A.; Castelar, B.; Robledo, D.; De Vega, G.B.; Msuya, F.E.; Eswaran, K.; Yasir, S.M.; Ali, M.K.M.; et al. The cultivation of Kappaphycus and Eucheuma in tropical and sub-tropical waters. In *Tropical Seaweed Farming Trends, Problems and Opportunities*; Springer: Cham, Switzerland, 2017; pp. 55–90.
18. Stirk, W.A.; Van Staden, J. Plant growth regulators in seaweeds: Occurrence, regulation and functions. In *Advances in Botanical Research*; Elsevier: Amsterdam, The Netherlands, 2014; Volume 71, pp. 125–159.
19. Uthirapandi, V.; Suriya, S.; Boomibalagan, P.; Eswaran, S.; Ramya, S.S.; Vijayanand, N.; Kathiresan, D. Bio-Fertilizer potential of seaweed liquid extracts of marine macro algae on growth and biochemical parameters of Ocimum sanctum. *J. Pharm. Phytochem.* **2018**, *7*, 3528–3532.
20. Hernández-Herrera, R.M.; Santacruz-Ruvalcaba, F.; Briceño-Domínguez, D.R.; Di Filippo-Herrera, D.A.; Hernández-Carmona, G. Seaweed as potential plant growth stimulants for agriculture in Mexico. *Hidrobiológica* **2018**, *28*, 129–140. [CrossRef]
21. Sharma, H.S.; Fleming, C.; Selby, C.; Rao, J.R.; Martin, T. Plant biostimulants: A review on the processing of macroalgae and use of extracts for crop management to reduce abiotic and biotic stresses. *J. Appl. Phycol.* **2013**, *26*, 465–490. [CrossRef]
22. Yalçın, S.; Okudan, E.Ş.; Karakaş, Ö.; Önem, A.N. Determination of Major Phytohormones in Fourteen Different Seaweeds Utilizing SPE–LC–MS/MS. *J. Chromatogr. Sci.* **2019**, *58*, 98–108. [CrossRef]
23. Hernández-Herrera, R.M.; Santacruz-Ruvalcaba, F.; Ruiz-López, M.A.; Norrie, J.; Hernández-Carmona, G. Effect of liquid seaweed extracts on growth of tomato seedlings (Solanum lycopersicum L.). *J. Appl. Phycol.* **2013**, *26*, 619–628. [CrossRef]
24. Tierney, M.S.; Smyth, T.; Rai, D.K.; Soler-Vila, A.; Croft, A.K.; Brunton, N. Enrichment of polyphenol contents and antioxidant activities of Irish brown macroalgae using food-friendly techniques based on polarity and molecular size. *Food Chem.* **2013**, *139*, 753–761. [CrossRef] [PubMed]
25. Dubois, M.; Gilles, K.A.; Hamilton, J.K.; Rebers, P.A.; Smith, F. Colorimetric Method for Determination of Sugars and Related Substances. *Anal. Chem.* **1956**, *28*, 350–356. [CrossRef]
26. Bradford, M.M. A rapid and sensitive method for the quantitation of microgram quantities of protein utilizing the principle of protein-dye binding. *Anal. Biochem.* **1976**, *72*, 248–254. [CrossRef]
27. Dodgson, K.; Price, R.; Lash, J.W.; Whitehouse, M.W.; Moretti, A.; Harborne, J. A note on the determination of the ester sulphate content of sulphated polysaccharides. *Biochem. J.* **1962**, *84*, 106–110. [CrossRef] [PubMed]
28. Bitter, T.; Muir, H. A modified uronic acid carbazole reaction. *Anal. Biochem.* **1962**, *4*, 330–334. [CrossRef]
29. Marigo, G. Sur une méthode de fractionnement et d'estimation des composés phénoliques chez les végétaux. *Annalusis* **1973**, *2*, 106–110.
30. Luximon-Ramma, A.; Bahorun, T.; Soobrattee, M.A.; Aruoma, O.I. Antioxidant Activities of Phenolic, Proanthocyanidin, and Flavonoid Components in Extracts of Cassia fistula. *J. Agric. Food Chem.* **2002**, *50*, 5042–5047. [CrossRef]

31. Mensor, L.L.; Menezes, F.S.; Leitão, G.G.; Reis, A.S.; Dos Santos, T.C.; Coube, C.S.; Leitão, S.G. Screening of Brazilian plant extracts for antioxidant activity by the use of DPPH free radical method. *Phytother. Res.* **2001**, *15*, 127–130. [CrossRef]
32. Przygodzka, M.; Zielińska, D.; Ciesarová, Z.; Kukurová, K.; Zieliński, H. Comparison of methods for evaluation of the antioxidant capacity and phenolic compounds in common spices. *LWT-Food Sci. Technol.* **2014**, *58*, 321–326. [CrossRef]
33. Benzie, I.; Strain, J. The Ferric Reducing Ability of Plasma (FRAP) as a Measure of "Antioxidant Power": The FRAP Assay. *Anal. Biochem.* **1996**, *239*, 70–76. [CrossRef] [PubMed]
34. Szőllősi, R.; Varga, I.S.I. Total antioxidant power in some species of Labiatae: Adaptation of FRAP method. *Acta Biol. Szeged.* **2002**, *46*, 125–127.
35. Seo, M.; Jikumaru, Y.; Kamiya, Y. Profiling of hormones and related metabolites in seed dormancy and germination studies. In *Seed Dormancy*; Springer: Boston, MA, USA, 2011; pp. 99–111.
36. Osuna-Ruiz, I.; Nieves-Soto, M.; Manzano-Sarabia, M.M.; Hernández-Garibay, E.; Lizardi-Mendoza, J.; Burgos-Hernández, A.; Hurtado-Oliva, M.Á. Gross chemical composition, fatty acids, sterols, and pigments in tropical seaweed species off Sinaloa, Mexico. *Cienc. Mar.* **2019**, *45*, 101–120. [CrossRef]
37. Shao, P.; Chen, M.; Pei, Y.; Sun, P. In intro antioxidant activities of different sulfated polysaccharides from chlorophytan seaweeds Ulva fasciata. *Int. J. Biol. Macromol.* **2013**, *59*, 295–300. [CrossRef] [PubMed]
38. Belghit, I.; Rasinger, J.D.; Heesch, S.; Biancarosa, I.; Liland, N.S.; Torstensen, B.; Waagbø, R.; Lock, E.J.; Bruckner, C.G. In-depth metabolic profiling of marine macroalgae confirms strong biochemical differences between brown, red and green algae. *Algal Res.* **2017**, *26*, 240–249. [CrossRef]
39. Prior, R.L.; Wu, X.; Schaich, K. Standardized Methods for the Determination of Antioxidant Capacity and Phenolics in Foods and Dietary Supplements. *J. Agric. Food Chem.* **2005**, *53*, 4290–4302. [CrossRef]
40. Šimura, J.; Antoniadi, I.; Široká, J.; Tarkowská, D.; Strnad, M.; Ljung, K.; Novak, O. Plant Hormonomics: Multiple Phytohormone Profiling by Targeted Metabolomics. *Plant Physiol.* **2018**, *177*, 476–489. [CrossRef]
41. Mori, I.C.; Ikeda, Y.; Matsuura, T.; Hirayama, T.; Mikami, K. Phytohormones in red seaweeds: A technical review of methods for analysis and a consideration of genomic data. *Bot. Mar.* **2017**, *60*, 153–170. [CrossRef]
42. Briceño-Domínguez, D.; Hernández-Carmona, G.; Moyo, M.; Stirk, W.; van Staden, J. Plant growth promoting activity of seaweed liquid extracts produced from Macrocystis pyrifera under different pH and temperature conditions. *J. Appl. Phycol.* **2014**, *26*, 2203–2210. [CrossRef]
43. Castellanos-Barriga, L.G.; Santacruz-Ruvalcaba, F.; Hernández-Carmona, G.; Ramírez-Briones, E.; Hernández-Herrera, R.M. Effect of seaweed liquid extracts from Ulva lactuca on seedling growth of mung bean (Vigna radiata). *J. Appl. Phycol.* **2017**, *29*, 2479–2488. [CrossRef]
44. Hernández-Herrera, R.M.; Santacruz-Ruvalcaba, F.; Zañudo-Hernández, J.; Hernández-Carmona, G. Activity of seaweed extracts and polysaccharide-Enriched extracts from Ulva lactuca and Padina gymnospora as growth promoters of tomato and mung bean plants. *J. Appl. Phycol.* **2016**, *28*, 2549–2560. [CrossRef]
45. Silva, L.D.; Bahcevandziev, K.; Pereira, L. Production of bio-Fertilizer from Ascophyllum nodosum and Sargassum muticum (Phaeophyceae). *J. Oceanol. Limnol.* **2019**, *37*, 918–927. [CrossRef]
46. Jannin, L.; Arkoun, M.; Etienne, P.; Laîné, P.; Goux, D.; Garnica, M.; Fuentes, M.; Francisco, S.S.; Baigorri, R.; Cruz, F.; et al. Brassica napus Growth is Promoted by Ascophyllum nodosum (L.) Le Jol. Seaweed Extract: Microarray Analysis and Physiological Characterization of N, C, and S Metabolisms. *J. Plant Growth Regul.* **2012**, *32*, 31–52. [CrossRef]
47. Stirk, W.; Novak, O.; Strnad, M.; Van Staden, J. Cytokinins in macroalgae. *Plant Growth Regul.* **2003**, *41*, 13–24. [CrossRef]
48. Dumale, J.; Gamoso, G.; Manangkil, J.; Divina, C. Detection and Quantification of Auxin and Gibberellic Acid in Caulerpa racemosa. *IJAT* **2018**, *14*, 653–660.
49. El Shoubaky, G.A.; Salem, E.A. Effect of abiotic stress on endogenous phytohormones profile in some seaweeds. *IJPPR* **2016**, *8*, 124–134.
50. Hedden, P.; Sponsel, V. A century of gibberellin research. *J. Plant Growth Regul.* **2015**, *34*, 740–760. [CrossRef]

Influence of Metal-Resistant *Staphylococcus aureus* Strain K1 on the Alleviation of Chromium Stress in Wheat

Fanrong Zeng [1], Munazza Zahoor [2], Muhammad Waseem [3], Alia Anayat [4], Muhammad Rizwan [2], Awais Ahmad [5], Tahira Yasmeen [2,*], Shafaqat Ali [2,6,*], Mohamed A. El-Sheikh [7], Mohammed Nasser Alyemeni [7] and Leonard Wijaya [7]

1 School of Agriculture, Yangtze University, 88 Jingmi Road, Jingzhou 434025, China; fanrong.zeng@yangtzeu.edu.cn
2 Department of Environmental Sciences and Engineering, Government College University, Faisalabad 38000, Pakistan; anglicmind5@gmail.com (M.Z.); mrazi1532@yahoo.com (M.R.)
3 Department of Microbiology, Government College University, Faisalabad 38000, Pakistan; muhammad.waseem@gcuf.edu.pk
4 Soil & Water Testing Lab., Ayub Agricultural Research Institute, Jhang Road, Faisalabad 38000, Pakistan; aliaanayat@gmail.com
5 Department of Chemistry, the University of Lahore, Lahore 54590, Pakistan; awaisahmed@gcuf.edu.pk
6 Department of Biological Sciences and Technology, China Medical University, Taichung 40402, Taiwan
7 Department of Botany and Microbiology, College of Science, King Saud University, Riyadh 11451, Saudi Arabia; melsheikh@ksu.edu.sa (M.A.E.-S.); mnyemeni@ksu.edu (M.N.A.); leon077@gmail.com (L.W.)
* Correspondence: tahirayasmeen@gcuf.edu.pk (T.Y.); shafaqataligill@yahoo.com or shafaqataligill@gcuf.edu.pk (S.A.)

Abstract: Chromium (Cr) is recognized as a toxic metal that has detrimental effects on living organisms; notably, it is discharged into soil by various industries as a result of anthropogenic activities. Microbe-assisted phytoremediation is one of the most emergent and environmentally friendly methods used for the detoxification of pollutants. In this study, the alleviative role of *Staphylococcus aureus* strain K1 was evaluated in wheat (*Triticum aestivum* L.) under Cr stress. For this, various Cr concentrations (0, 25, 50 and 100 mg·kg^{-1}) with and without peat-moss-based bacterial inoculum were applied in the soil. Results depicted that Cr stress reduced the plants' growth by causing oxidative stress in the absence of *S. aureus* K1 inoculation. However, the application of *S. aureus* K1 regulated the plants' growth and antioxidant enzymatic activities by reducing oxidative stress and Cr toxicity through conversion of Cr^{6+} to Cr^{3+}. The Cr^{6+} uptake by wheat was significantly reduced in the *S. aureus* K1 inoculated plants. It can be concluded that the application of *S. aureus* K1 could be an effective approach to alleviate the Cr toxicity in wheat and probably in other cereals grown under Cr stress.

Keywords: chromium; *Staphylococcus aureus*; wheat; oxidative stress; antioxidants

---

1. Introduction

Environmental pollution by toxic metals has dramatically increased because of various man-made actions taken while revolutionizing industries and urban life. Although these activities have substantially improved the living standards of humans, they have, at same time, deteriorated the environment [1]. Direct or indirect discharge of sewage and industrial effluent into surface water bodies has resulted in augmentation of chromium (Cr) and other toxic metals in soils [2], causing toxicity to plants [3], animals and humans [4]. In agricultural systems, Cr can easily move to different parts of crops and accumulate there to be later consumed by the animals and humans [5]. Soil contamination by Cr and other heavy metals impacts biodiversity negatively and badly disturbs the living entities in the soil [6].

Major origins of Cr contamination are the leather industry [7,8], mining [1], steel industry, paint industry, wood preservatives, volcanic eruption and weathering [9]. Chromium exerts negative effects on plants by reducing the plant height and root growth, interrupting the germination process, causing disproportion in nutrient levels, exerting harmful effects on photosynthesis, retarding soil microbial activities, inhibiting enzyme activity and stimulating the formation of reactive oxygen species (ROS) which result in induction of oxidative stress in plants [1,10,11]. Chromium can cause different malfunctions in human biological systems that may lead to the death of affected persons [12–14]. Wastewater effluents from the industries are discharged directly into water bodies that are utilized mostly for irrigation purposes. Farmers have to rely on this untreated contaminated water due to limited resources and inadequate sanitation facilities [15].

A major staple food across the world is wheat (*Triticum aestivum* L.), which fulfils food requirements of about 50% of the worldwide population [16]. Amongst wheat-producing countries, Pakistan comes ninth in the world. Wheat subsidizes necessary amino acids, vitamins and minerals, dietary fibers and phytochemicals in our diet [17]. Wheat can accumulate higher Cr concentration in stems followed by leaves and grains [18]. According to the literature, increased heavy metal accumulation in wheat tissues has become a potential source of food chain contamination that can cause serious abnormalities to human biological systems [19,20]. Crops may also have the ability to reduce the Cr from Cr^{6+} to Cr^{3}. This reduction process is likely to happen in roots as a detoxification mechanism [21]. There are a number of remediation methods used to treat sites contaminated with toxic metals. Presently, scientists have made rampant use of biologically centered techniques to deal with such toxic contaminants in order to remove them from environmental entities, including water, air and soil, or at least make them less damaging to the ecosystem [22]. The phytoremediation technique is a modernized method with a lower budget and environmentally sustainable system [23]; it destroys contaminants by using plants along with their rhizospheric microorganisms. Microbial-assisted phytoremediation helps to deal with toxic heavy metals by stabilizing or transforming them to less toxic forms in carrier materials such as soil, shallow water, sediments or groundwater [24]. Microbes have the capability to modify their genetic sequences in response to variation in environmental factors [25]. In soil polluted with heavy metals, microbes assist the plants by producing various growth-regulating substances, such as organic acids, hormones, siderophores and enzymes, that help in plant growth promotion by involving diverse mechanisms, namely acidification, precipitation, redox reactions and chelation [26]. Likewise, roots excrete beneficial nutrients to support the successful colonization and growth of microbes [26]. Chromium-reducing bacteria have the capability to remediate Cr toxicity by reducing Cr^{6+} into Cr^{3+} in the rhizosphere through bioaccumulation and biosorption mechanisms [27]. *Staphylococcus aureus* is a Gram-positive, ubiquitous and round-shaped facultative anaerobe that grows in clusters, forming a biofilm on surfaces. It can grow in a range of growth temperature from 7 to 48 °C, with 37 °C as the optimal temperature for growth [28]. It was isolated from tannery effluent and characterized as a chromium-reducing bacterium. The application of phytoremediation, along with Cr-resistant bacteria for detoxification of Cr^{6+}, has been considered a safe, effective and economical technique over customary techniques [29,30]. In this study, the alleviative role of *Staphylococcus aureus* strain K1 under Cr stress was evaluated in wheat plants. It was hypothesized that microbes (such as *Staphylococcus aureus* strain K1) may alleviate Cr toxicity in wheat by enhancing antioxidant enzymatic activities of wheat while reducing oxidative stress through biotransformation (Cr^{6+} into Cr^{3+}) and biosorption of Cr.

2. Materials and Methods

2.1. Soil Preparation

Sandy clay loam soil was brought from nursery and was air-dried without direct sunlight. After air-drying, soil sieving was done by a mesh with a pore size of 2 mm. Soil was then sterilized at a temperature of 121 °C for 20–30 min for the purpose of removing any kind of contaminant or bacteria that can cause hindrance in further findings [31]. Chromium solutions of different concentrations were prepared from stock solution of $K_2Cr_2O_7$ in the laboratory, and soil was spiked with final Cr concentrations of 0, 25, 50 and 100 mg·kg^{-1} of soil.

These different concentrations of Cr were taken to determine the maximum concentration of hexavalent Cr tolerable by strain K1. However, in case of Cr reduction, the lower concentration of Cr was used due to the fact that Cr is found in lower concentrations in the natural environment, especially in industrial effluents [31]. The concentrations of Cr used were similar to those used in the literature and were chosen considering the fact that, in field conditions, we had to establish the reduction ability of this particular strain rather than its maximum potential to survive in response to metal stress [15]. The soil was added in the pots (5 kg soil per pot) with proper mixing following the treatment plan. Electrical conductivity and pH from saturated soil were determined by making a soil-to-water ratio of 1:25. Soil was extracted with ammonium bicarbonate diethylenetriaminepentaacetic acid (AB-DTPA) solution for the measurement of bioavailable trace elements in the soil [32]. Soil organic matter was determined following the prescribed method [33]. Soil physicochemical characteristics are given in Table 1.

Table 1. Soil characterization of pot experiment.

Soil Properties	Unit	Values
Texture Properties		**Sandy Clay Loam**
pH	-	7.71
Sand	%	63.7
Clay	%	21.9
Silt	%	14.4
Electrical Conductivity (EC)	dSm^{-1}	4.77
Soluble Ion Values		
Cl^-	mmol$_c$L^{-1}	7.15
$Ca^{2+} + Mg^{2+}$	mmol$_c$L^{-1}	14.92
CO_3^{2-}	mmol$_c$L^{-1}	0.85
OM	%	0.90
CEC	cmol$_c$kg^{-1}	13.2
HCO_3^-	mmol$_c$L^{-1}	3.84
Metal Concentration		
Available Cr^{6+}	mg·kg^{-1}	0.04
Available Zn^{2+}	mg·kg^{-1}	0.72
Available Cu^{2+}	mg·kg^{-1}	0.23

2.2. Segregation of Cr-Resistant Bacteria

A modified method of serial dilution was adopted to isolate the Cr-tolerant bacteria from metal-contaminated industrial effluent [34]. For this, ten-fold serial dilutions (10^{-1}, 10^{-2}, 10^{-3} and 10^{-4}) were prepared from samples of collected wastewater using sterilized distilled water [34]. Then, 0.1 mL from each dilution was added to petri plates having Tryptic Soy Agar complemented with 0.5 mM Cr^{6+}. Morphologically different colonies were picked and transferred to petri plates supplemented with gradually elevated levels (0.0, 0.5, 2.5, 5.0, 10.0, 15.0, 20, 22 and 23 mM) of Cr^{6+} [35]. The bacteria that showed maximum resistance to the highest concentration of hexavalent Cr were selected for use in further studies.

2.3. Bacterial Identification

Molecular characterization was carried out through the amplification of 16S rDNA gene via polymerase chain reaction (PCR) using the following universal primers: 27F (5′-AAACTCAAATGAATTGACGG-3′) and 1492R (5′-ACGGGCGGTGTGTAC-3′) [36]. For genomic DNA extraction, Favorgen DNA extraction kit was used following the manufacturer's guideline. The initial denaturation temperature was set at 94 °C for a period of 5 min, and this was followed by 40 recurring cycles of denaturizing DNA at 94 °C for 45 s, annealing at 53 °C for 45 s and elongation at 72 °C for 60 s. Final extension was set at 72 °C for 10 min, and this was followed by temperature being held at 4 °C [37]. PCR product (5 μL) was loaded in gel wells, and the reaction was allowed to complete; the product was then visualized using Gel Documentation System (Slite 200 W) under ultraviolet light [37]. After validation, 30 μL PCR product was delivered to Macrogen (Seoul, Korea) for the purpose of sequencing. ChormasPro (v1.7.1) software was used for correction of sequences that were submitted to GenBank for accession number. A phylogenetic tree was constructed by downloading similar partial 16S rDNA gene sequences from the NCBI BLAST database with the help of computer software MEGA (v7.0.) [38].

2.4. Bacterial Inoculum Preparation

In order to obtain pure inoculum of *S. aureus* strain K1, an individual isolated colony was inoculated in 250 mL sterilized nutrient broth and incubated at 150 rpm on orbital rotary shaker for 48 h (at 37 °C). The pure culture was harvested via centrifugation at 6000× *g* for 10 min, and the supernatant was discarded. The pellet was washed with sterilized distilled water and resuspended in 100 mL of normal saline (0.85% NaCl) solution. Overall, cell density for the inoculum was maintained at 1×10^8 CFU mL^{-1} [39].

2.5. Seed Coating and Pot Experiment

For this study, seeds of wheat variety Sehar were taken from Ayub Agriculture Research Institute, Faisalabad, Pakistan. Seeds were first washed thoroughly with distilled water, and this was followed by surface sterilization using 10% hydrogen peroxide (H_2O_2) for 30 min [40]. The sterilized seeds were immersed in double volume of bacterial suspension (1×10^8 CFU mL^{-1}) and kept at 37 ± 2 °C on a rotary shaker (90 rpm) for 2 h. To facilitate the attachment of bacterial inoculum to the seeds, carboxymethyl cellulose (CMC) (2%) was added to the suspension as a sticking agent. Seeds were dried under shade after 2 h of inoculation for further experimental use. Uninoculated sterilized seeds were used as control. Clay and peat moss in equal parts (1:1) were mixed and the seeds were added to this mixture, which was shaken well for proper coating and incubated overnight in the dark. The completely randomized design had a total of eight treatments, with three replicates for each treatment. A total of eight seeds per pot were sown, and thinning was performed to result in four seedlings per pot after 3 weeks of seed germination.

2.6. Treatments

The experiment was conducted in plastic pots using different concentrations of Cr (0, 25, 50 and 100 $mg{\cdot}kg^{-1}$) in the presence and absence of bacterial inoculation. Different treatments were as follows: T1 (Control), 0 $mg{\cdot}kg^{-1}$ Cr; T2, 25 $mg{\cdot}kg^{-1}$ Cr; T3, 50 $mg{\cdot}kg^{-1}$ Cr; T4, 100 $mg{\cdot}kg^{-1}$ Cr; T5, 0 $mg{\cdot}kg^{-1}$ Cr + *S. aureus* K1; T6, 25 $mg{\cdot}kg^{-1}$ Cr + *S. aureus* K1; T7, 50 $mg{\cdot}kg^{-1}$ Cr + *S. aureus* K1; T8, 100 $mg{\cdot}kg^{-1}$ Cr + *S. aureus* K1.

2.7. Plant Harvesting

At 135 days after seed sowing, plants were harvested at maturity. The height and spike lengths of plants were measured with a meter rod. Shoots, roots, spikes and grains were separated properly.

Then, 0.1 M HCl was used to remove the metals from the root surface, and the roots were washed with distilled water. Samples of roots and shoots were kept in a hot air oven (70 °C) for a period of 72 h. Afterwards, dry weight was recorded and samples were crushed to small pieces and processed for further analyses.

2.8. Determination of Chlorophyll Contents and Gas Exchange Parameters

At 8 weeks after seed germination, fresh leaf samples were taken to determine chlorophyll contents using acetone (85% *v*/*v*) for pigment extraction. These leaf samples were kept in the dark at 4 °C for 24 h. Centrifugation of samples was done to get the supernatant. Absorbance was recorded by spectrophotometer at three different wavelengths (470, 647 and 664.5 nm), and final chlorophyll contents were calculated by following the prescribed method [41]. Photosynthetic rate, transpiration rate and stomatal conductance of samples were recorded 8 weeks after seed germination on a fully sunny day using an infrared gas analyzer (IRGA, LCA-4, Analytical Development Company, Hoddesdon, UK).

2.9. Determination of Reactive Oxygen Species and Antioxidant Enzyme Activities

At 2 months after seed sowing, fresh leaves of plants were sampled for the estimation of reactive oxygen species (ROS) through the assessment of electrolyte leakage (EL) and the contents of malondialdehyde (MDA) and hydrogen peroxide (H_2O_2). Additionally, the activities of enzymes such as superoxide dismutase (SOD), peroxidase (POD), catalase (CAT) and ascorbate peroxidase (APX) were assessed. For the EL estimation, distilled water tubes were used to place leaf samples. Samples were autoclaved at 32 °C for period of 2 h, and the observed EC of the solution was termed as EC_1. Afterwards, this solution was autoclaved at 121 °C for 20 min to measure EC_2, and finally EL was calculated using the following equation as described by Dionisio-Sese and Tobita [42]:

$$EL = (EC_1/EC_2) \times 100$$

The concentration of MDA was measured using the method of Heath and Packer (1968) as modified by Dhindsa et al. [43] and Zhang and Kirham [44]. Hydrogen peroxide was recorded through homogenization of samples in phosphate buffer 50 mM (pH 6.5) and centrifugation followed by addition of 20% H_2SO_4 (*v*/*v*). Samples were centrifuged once more for 15 min, and readings were taken by spectrophotometer at 410 nm absorbance [45]. A spectrophotometer was utilized to record the activities of antioxidant enzymes such as SOD, POD, CAT and APX. Fresh leaf samples were crushed in liquid nitrogen (N_2), and 0.05 M phosphate buffer (pH 7.8) was utilized for the purpose of standardization. This was followed by centrifugation at 4 °C on 12,000× *g* for a period of 10 min. Supernatant was collected for the sake of antioxidant enzyme activity measurements. The method of Zhang [46] was employed to measure SOD and POD activities, while the Aebi method [47] was used for CAT activity. APX contents were estimated using the method of Nakano and Asada [48].

2.10. Estimation of Cr Contents in Plants

Digestion of dry shoot and root samples was performed for 1 g of each sample in 4:1 (*v*/*v*) ratio of HNO_3:$HClO_4$ as described by Rehman et al. [49]. Finally, the digested samples were run on an atomic absorption spectrophotometer for the estimation of Cr concentrations in the processed samples.

2.11. Statistical Analysis

The IBM SPSS Statistics for Windows, Version 21.0, was used for the data analyses, using the analysis of variance (ANOVA) tool at a 5% probability level. Tukey's HSD post hoc test was performed for multiple comparison of triplicates.

3. Results

The current study was envisaged to assess the capability of metal-resistant *Staphylococcus aureus* strain K1 to ameliorate the Cr stress in wheat plants.

3.1. Growth Characteristics of Isolate K1

The bacterial strain K1, capable of tolerating a Cr concentration of up to 22 mM, was selected for further studies. Morphologically, it is characterized by Gram-positive cocci (≈1 μm) with yellowish golden color. Chemically, it is oxidase- and coagulase-negative and catalase-positive (Table 2). The BLASTn investigation showed that it has a close resemblance (99%) to *Staphylococcus aureus* strain ATCC 12600 (NR_115606.1) and *Staphylococcus aureus* strain NBRC 100910 (MG971399.1). The similar 16S rDNA gene sequences from GenBank were used to carry out phylogenetic analysis, which also confirmed that the isolate K1 belongs to *Staphylococcus aureus*; therefore, it was named *Staphylococcus aureus* strain K1 (KX685332). This was done in order to remain confident that the bacterial strain used in this study was *Staphylococcus aureus* strain K1, as culture media can sometimes be contaminated with other bacteria.

Table 2. Biochemical and morphological characteristics of *S. aureus* strain K1.

Sr. No.	Characteristic	*Staphylococcus aureus* K1
1	Morphology	Convex, round
2	Color	Yellowish, golden
3	Gram-reaction	+ve
4	Catalase	+ve
5	Coagulase plasma reaction	−ve

3.2. Effect of S. aureus K1 Contact Time on Chromium (Cr^{6+}) Reduction

Staphylococcus aureus K1 exhibited optimum growth at pH 8 and 35 °C. Under optimum growth conditions, the effect of contact time on bacterial ability to reduce the hexavalent Cr in the medium was observed. It was observed that the Cr reduction of *S. aureus* K1 increased with increasing contact time (Figure 1). It was found that 26%, 45%, 71%, 80% and 99% Cr^{6+} (initial metal concentration = 1 mM) was removed from the medium by *Staphylococcus aureus* K1 after 2, 4, 8, 16 and 24 h of incubation, respectively (Figure 1).

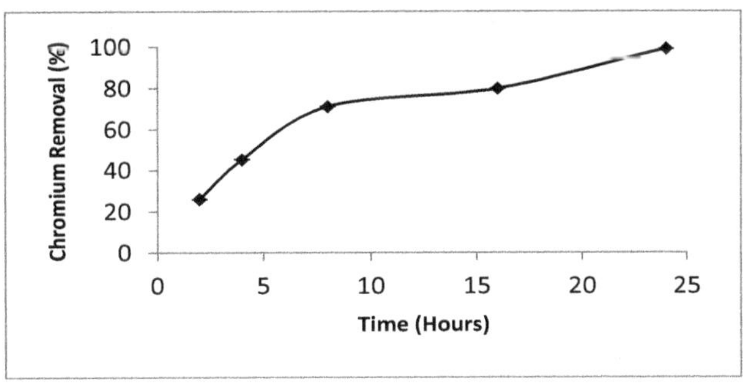

Figure 1. The impact of contact time (hours) impact on the Cr removal ability of *Staphylococcus aureus* K1.

3.3. Effect of S. aureus K1 on Plant Growth Promotion

Chromium stress substantially decreased the growth of wheat plants. A significant reduction in the length of shoots (31.18%), roots (32.02%) and spikes (40.70%) and the dry weight of shoots (34.29), roots (44.17) and grains (31.06%) of the plant was observed at 100 mg·kg^{-1} Cr concentration alone as compared to *S. aureus* K1 inoculated seeds + 100 mg·kg^{-1} Cr concentration (Figure 2). A significant change in shoot and root length was observed in inoculated plants as compared to uninoculated plants at all levels of Cr. Wheat plants stressed with 50 mg·kg^{-1} of Cr showed an observable reduction in growth attributes; however, this decrease was minimized in inoculated plants compared to uninoculated plants, as shown in Figure 2. The growth was gradually decreased when the Cr concentration in the growth medium increased from 25 to 100 mg·kg^{-1} (Figure 2A–D). Moreover, the maximum growth reduction was noticed with 100 mg·kg^{-1} of Cr stress. The data regarding plant growth attributes indicated that inoculation with *S. aureus* K1 significantly improved the wheat growth and dry biomass under Cr stress conditions.

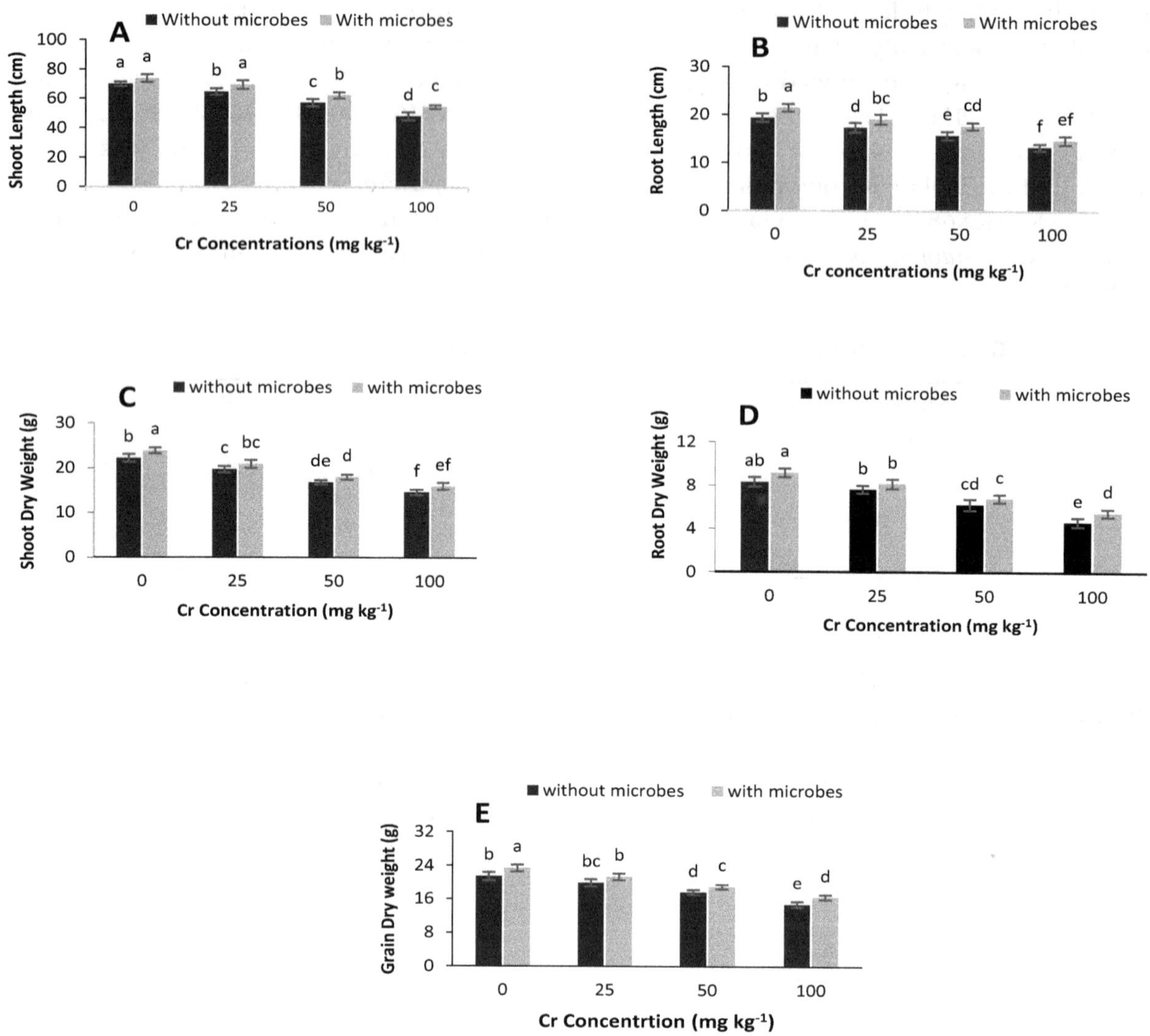

Figure 2. Influence of the different Cr levels (0, 25, 50, 100 mg·kg^{-1}), with and without peat-moss-based microbial inoculation, on length of shoot (**A**) and root (**B**), dry weight of shoot (**C**) and root (**D**) and grain dry weight (**E**) of wheat. Bars indicate the mean of three replicates with standard deviation (SD). Different bars with lowercase letters show noteworthy changes among various treatments at $p < 0.05$.

3.4. IRGA Parameters and Chlorophyll Contents

IRGA parameters such as transpiration rate, stomatal conductance and photosynthetic rate gradually reduced under increased Cr concentrations alone. The transpiration rate was greater at 25 mg·kg^{-1} of Cr stress and decreased with increasing Cr stress levels at concentrations from 50 to 100 mg·kg^{-1}. Without microbial inoculation, transpiration rate decreased by 12%, 21% and 32% under 25, 50 and 100 mg·kg^{-1} Cr stress, respectively, as compared to control (Figure 3A). Similarly, stomatal conductance and photosynthetic rate in uninoculated plants also reduced with increasing Cr concentrations. Stomal conductance decreased by 9%, 25%, 45% and photosynthetic rate decreased by 12%, 25% and 46% under 25, 50 and 100 mg·kg^{-1} Cr stress, respectively, as shown in Figure 3B,C. These results explain the effective role of bacterial inoculation in improving gas exchange attributes in Cr-stressed wheat plants by comparing uninoculated plants. Similarly, chlorophyll a contents decreased by 9.40%, 26.21% and 40.08% in inoculated plants and by 10.66%, 28.02% and 41.87% in uninoculated plants under 25, 50 and 100 mg·kg^{-1} Cr stress, respectively, as shown in Figure 3D.

On the other hand, as compared to untreated control, chlorophyll b was reduced by 15.36%, 27.27% and 40.80% in uninoculated wheat plants and by 14.44%, 27.24% and 40.63% in inoculated wheat plants under 25, 50 and 100 mg·kg^{-1} Cr stress, respectively, as shown in Figure 3E. A gradual decrease in carotenoid contents was also observed in inoculated and uninoculated plants with increasing level of Cr stress, where inoculated plants showed 6%, 19% and 27% reduction in carotenoid contents while uninoculated plants showed 9%, 19% and 28% reduction under 25, 50 and 100 mg·kg^{-1} Cr stress, respectively (Figure 3F)

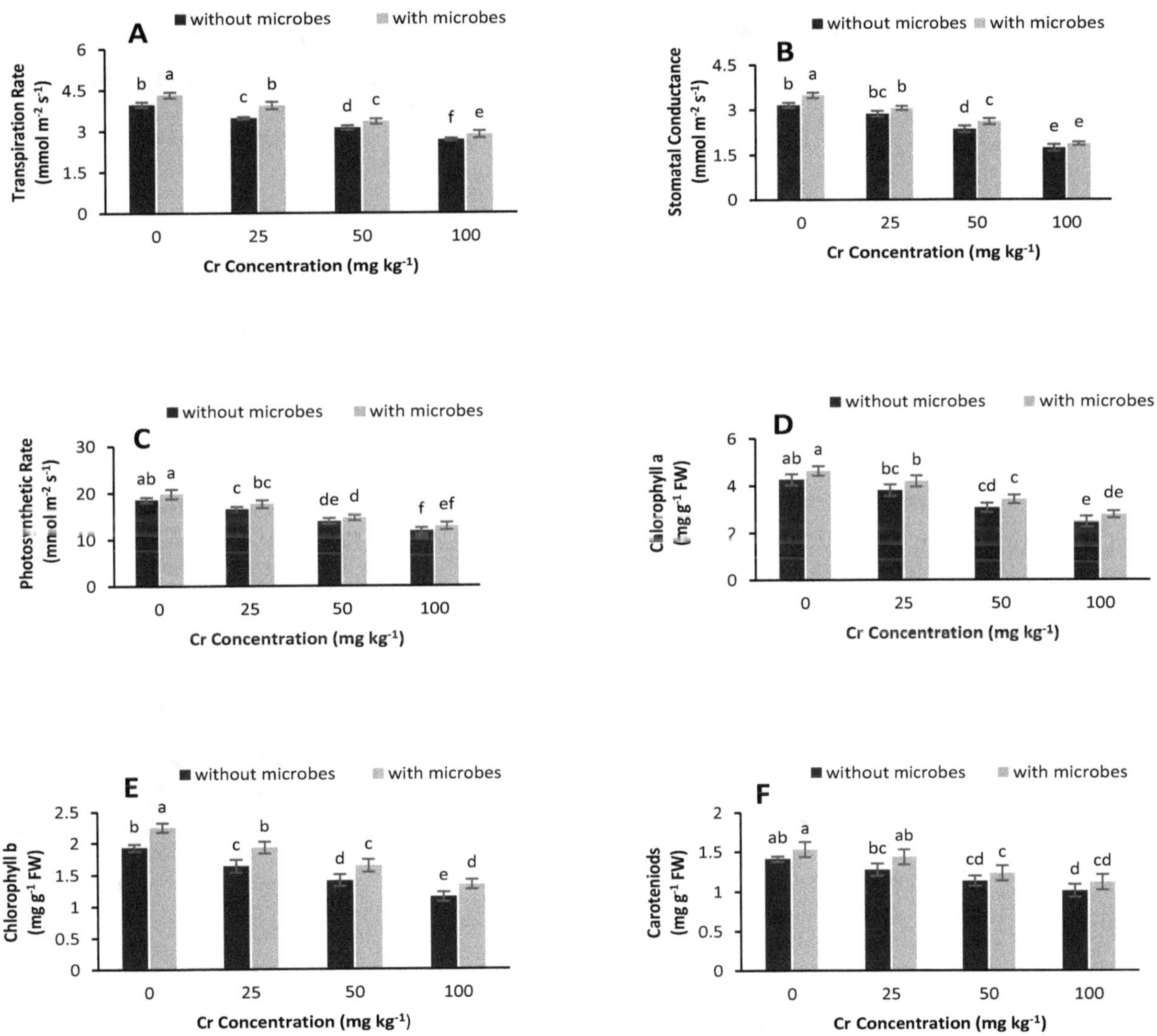

Figure 3. Influence of the different Cr levels (0, 25, 50, 100 mg·kg^{-1}), with and without peat-moss-based microbial inoculation, on transpiration rate (**A**), stomatal conductance (**B**), photosynthetic rate (**C**), chlorophyll a (**D**), chlorophyll b (**E**) and carotenoids (**F**) of wheat plants. Bars indicate the mean values and standard deviation of three replicates. Different bar letters show significant changes among various treatments at $p < 0.05$.

3.5. Estimation of EL, MDA and H_2O_2

A substantial increase in EL was noted in both roots and shoots of wheat plants under Cr stress, as shown in Figure 4A,B. Uninoculated wheat plants showed more EL in leaves and roots under all Cr levels (0, 25, 50 and 100 mg·kg^{-1}) as compared to inoculated plants. EL in uninoculated leaves was increased by 17.98%, 36.40% and 56.52% and EL in uninoculated roots increased by 9%, 32% and 53% under 25, 50 and 100 mg·kg^{-1} Cr, respectively. On the other hand, inoculation with *S. aureus* K1 increased EL in leaves by 15.83%, 33.26% and 55.90% and in roots by 13%, 33% and 56%, under 25, 50 and 100 mg·kg^{-1} Cr, respectively (Figure 4A,B).

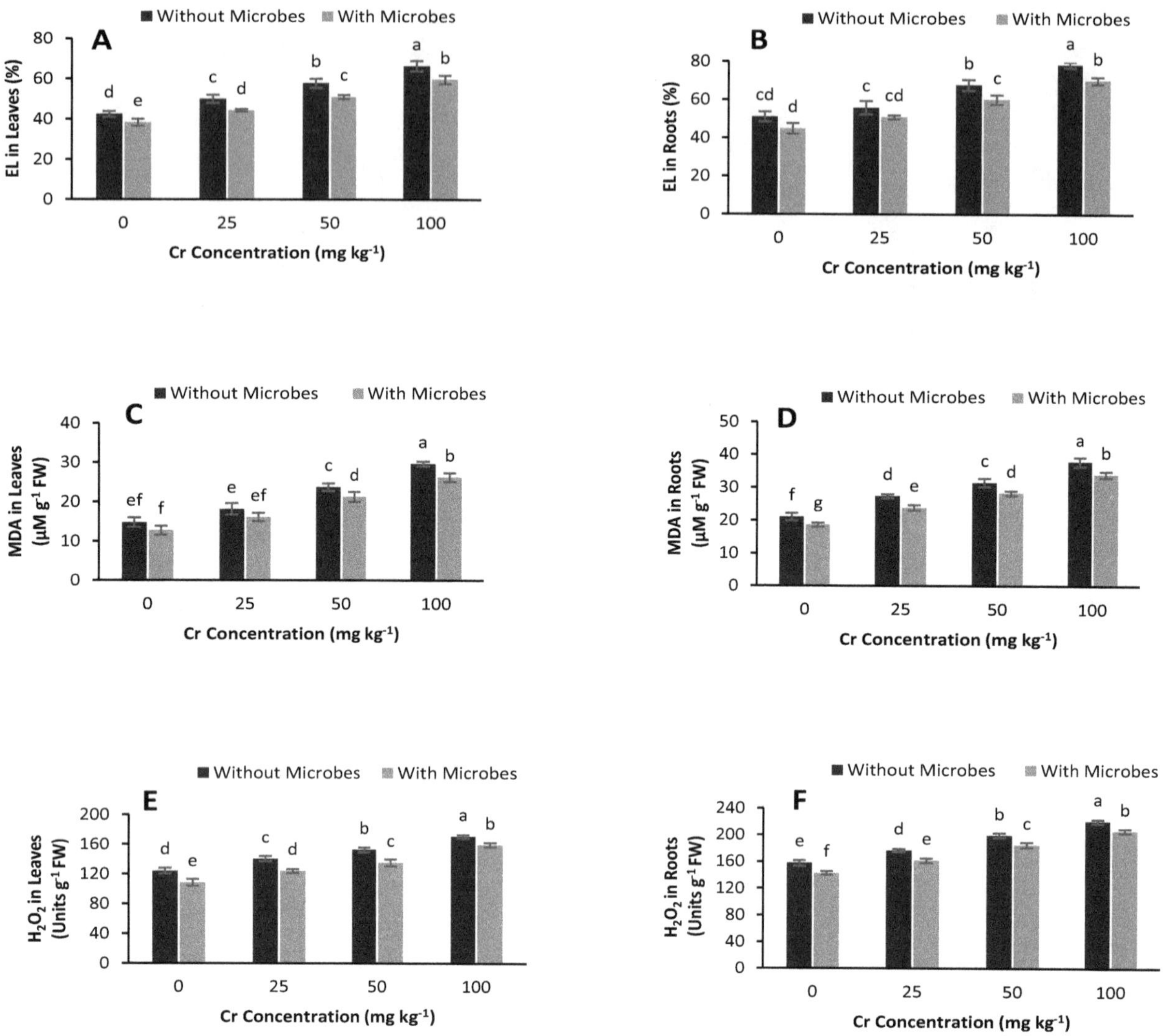

Figure 4. Influence of the different Cr levels (0, 25, 50, 100 mg.kg−1), with and without peat-moss-based microbial inoculation, on EL in leaves (**A**), EL in roots (**B**), MDA in leaves (**C**), MDA in roots (**D**), H_2O_2 in leaves (**E**) and H_2O_2 in roots (**F**) of wheat plants. Bars indicate the mean values and standard deviation of three replicates. Different bar letters show significant changes among various treatments at $p < 0.05$.

There was a noticeable increase in MDA content of leaves, showing lipid peroxidation due to high level of Cr stress, as shown in Figure 4C,D. Maximum MDA contents were observed in leaves and roots of uninoculated plants under 100 mg·kg^{-1} Cr stress as compared to their respective controls. However, inoculation with *S. aureus* K1 reduced MDA content in all the plants of varying level of Cr stress compared to uninoculated plants. Likewise, a gradual rise in H_2O_2 of wheat leaves was observed with increasing levels of Cr (Figure 4E,F). Furthermore, a noteworthy decrease in H_2O_2 content was observed in *S. aureus* K1 inoculated plants, both Cr-stressed and control.

3.6. *Effect of S. aureus on Antioxidant Enzyme Activities*

The findings revealed that SOD activity in leaves and roots was significantly higher at the 25 mg·kg^{-1} Cr level but gradually decreased with increasing Cr levels, both in uninoculated and inoculated plants. SOD activity increased by 19.59%, 5.22% and 6.98% in uninoculated plant leaves and by 17.58%, 5.22% and 3.08% in uninoculated plant roots under 25, 50 and 100 mg·kg^{-1} Cr treatments, respectively. However, inoculation with *S. aureus* K1 enhanced the SOD activity by 24.71%, 9.64% and 3.51% in leaves and 20.83%, 9.49%, and 4.34% in roots under 25, 50 and 100 mg·kg^{-1} Cr, respectively (Figure 5A,B). As compared to noncontaminated treatments (control), a decline in the CAT activity was observed under Cr contamination (Figure 5C,D). Inoculation with *S. aureus* K1 provoked a substantial increase in the activity of the CAT enzyme in wheat leaves (Figure 5C). CAT activity in roots also improved (114.31 Units·g^{-1} FW) under bacterial inoculation as compared to uninoculated plants (102.66 Units g^{-1} FW) at 25 mg·kg^{-1} Cr (Figure 5D). Moreover, abridged CAT activity was noticed at the highest level of Cr stress (100 mg·kg^{-1}); activity at this level was increased by 5.52% in leaves and 3.63% in roots for uninoculated plants, while inoculated plants showed increase of 5.06% in leaves and 1.37% in roots, as shown in Figure 5C,D. The POD activity substantially ($p < 0.05$) increased due to addition of Cr as compared to control (Figure 5E,F). There was a noticeable reduction in POD activity in leaves under bacterial inoculation with *S. aureus* strain K1 (22.27%, 11.99% and 0.21%) as compared to uninoculated treatments (21.63%, 10.12% and 2.92%) (Figure 5E). There was a substantial increase in the activity of the APX enzyme observed under Cr stress in wheat plants, as shown in Figure 5G,H. There was increase in APX activity in plant shoots and roots, with the maximum production occurring at the Cr concentration of 25 mg·kg^{-1}, and the APX activity decreased at the highest Cr level in the growth medium (Figure 5G,H). Furthermore, the maximum APX activity was observed in roots without inoculation at Cr concentration of 25 mg·kg^{-1}, as shown in Figure 5H.

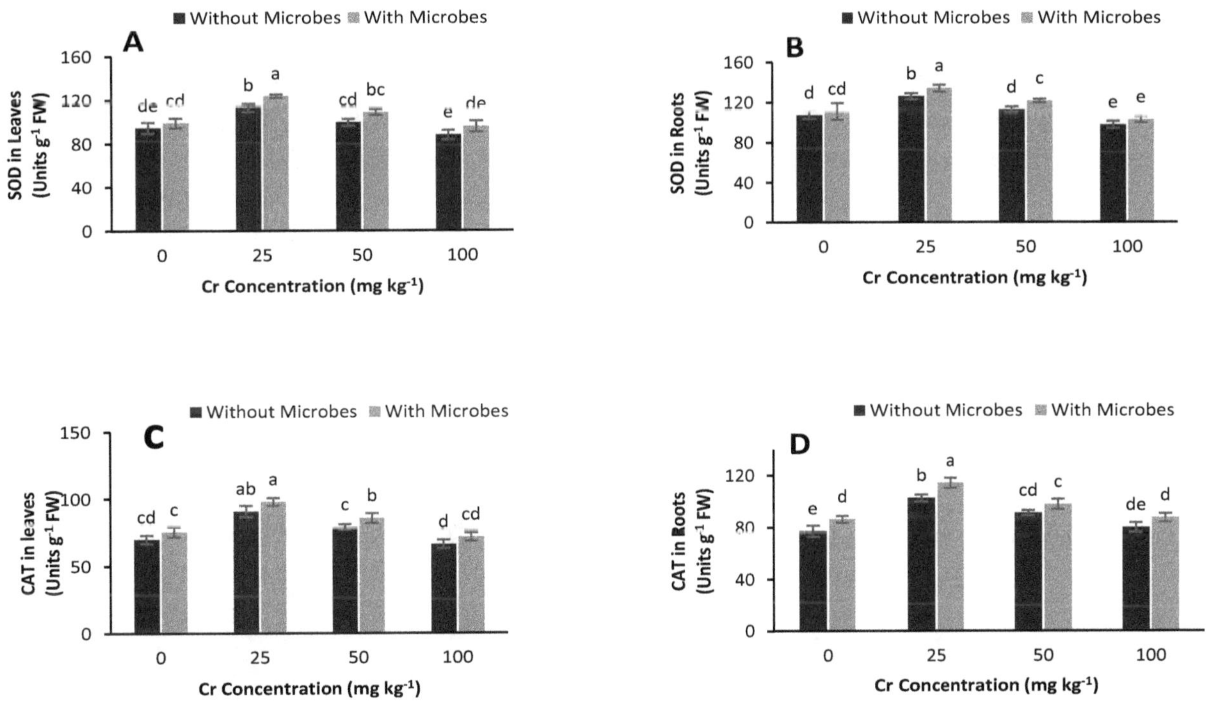

Figure 5. *Cont.*

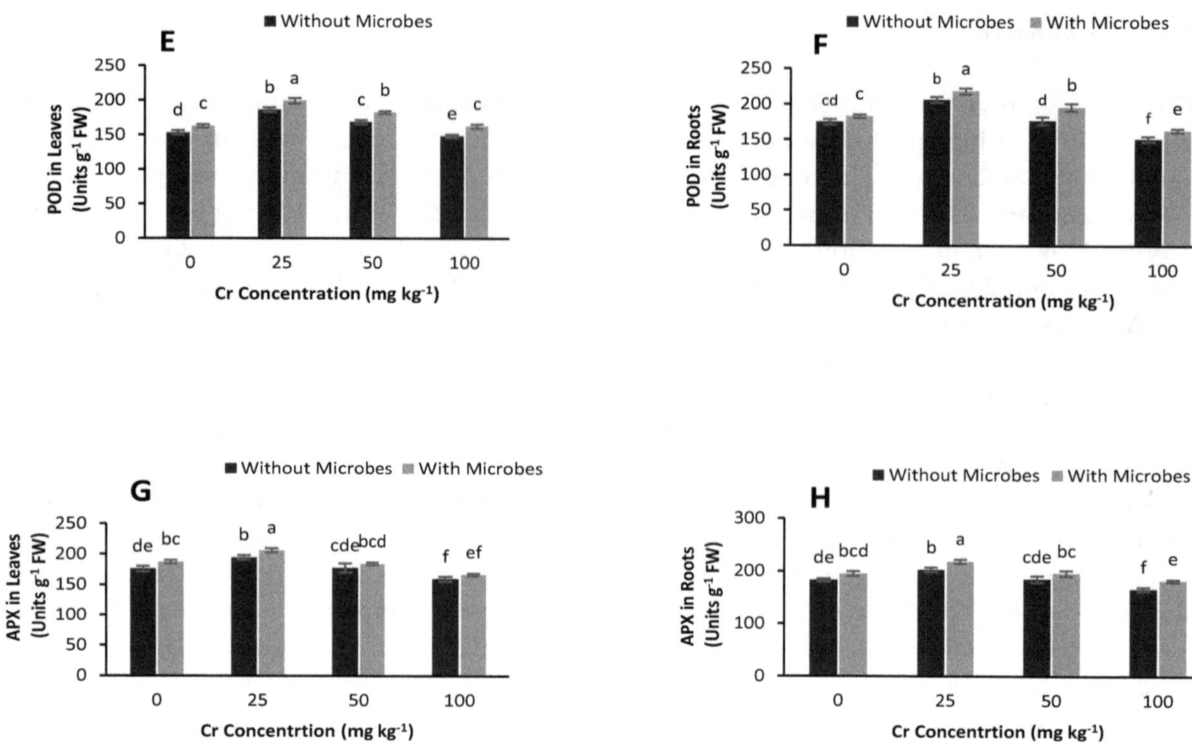

Figure 5. Influence of the different Cr levels (0, 25, 50, 100 mg·kg^{-1}), with and without peat-moss-based microbial inoculation, on SOD in leaves (**A**), SOD in roots (**B**), CAT in leaves (**C**), CAT in roots (**D**), POD in leaves (**E**), POD in roots (**F**), APX in leaves (**G**) and APX in roots (**H**) of wheat plants. Bars indicate the mean values with standard SD of three replicates. Different bar letters show significant changes among various treatments at $p < 0.05$.

3.7. Cr Accumulation in Plants

The data regarding Cr accumulation in shoots and roots of the wheat plants are shown in Figure 6A,B. With increasing concentration of applied Cr, a gradual increase in Cr concentrations was observed in roots and shoots in a dose-additive manner. In addition, inoculation of *S. aureus* K1 significantly decreased the Cr concentrations both in shoots and roots as compared to uninoculated plants.

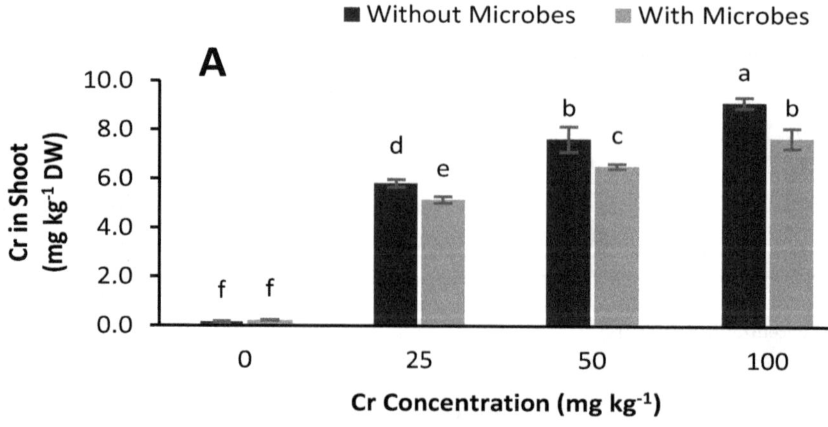

Figure 6. *Cont.*

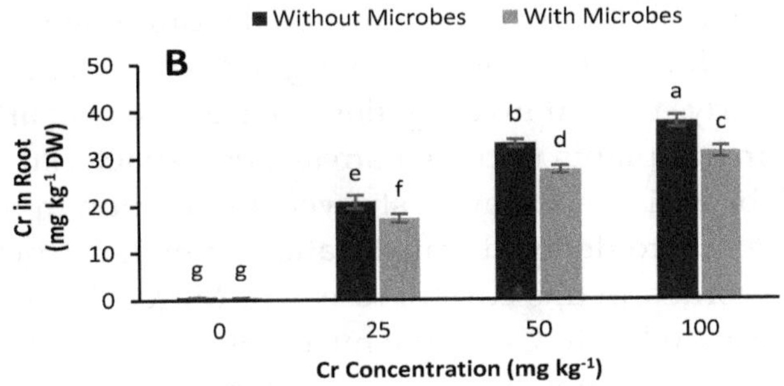

Figure 6. Influence of the different Cr levels (0, 25, 50, 100 mg·kg^{-1}), with and without peat-moss-based microbial inoculation, on Cr concentrations in shoots (**A**) and roots (**B**) of wheat plants. Bars indicate the mean values with standard SD of three replicates. Different bar letters show significant changes among various treatments at $p < 0.05$.

4. Discussion

The major objective of our research was to appraise the effectiveness of *Staphylococcus aureus* K1 treatment in reducing the toxic effects of Cr stress in wheat plants. An indigenous bacterial strain, *Staphylococcus aureus* K1 (GenBank accession no. KX685332), capable of tolerating up to 22 mM of Cr^{6+} was isolated from a metal-polluted environment. Numerous research studies with similar metal-tolerant bacterial isolations from metal-contaminated sites have been reported [35,50,51]. Our results also supported the findings of Mustapha and Halimoon [52], who isolated a total of 21 isolates from electroplating industries and reported that merely 5 of them were Cr-tolerant (up to 50 mg·L^{-1}). The results of the current study show that *S. aureus* K1 increased plant growth parameters under Cr metal stress (Figure 2).

4.1. *Detoxification of Metals by S. aureus K1*

Microbes have a number of metal resistance mechanisms involving chromosomes, transposon-encoded genes or plasmids. These mechanisms are mostly plasmid-facilitated and show resistance to some particular anion or cation [53]. Metals can have different impacts inside cells depending upon their concentration [53]; once a certain level is exceeded, bacteria respond with the initiation of a number of resistance mechanisms, including metallothioneins, P-type ATPases, CDF transporters and RND efflux pumps [54]. The genes located on plasmids, chromosomes or transposons that are responsible for resistance can easily be transferred to new community members from their point of location [53,55].

The genotype of bacteria, the nature and type of the metal and the pH of the culturing media are among the factors responsible for showing the degree of tolerance of microbes to various metals (Hg, Co, Pb, Ag, Zn, Mn, Cu, Cr) [56]. This kind of resistance against toxic heavy metals might be recognized by employing a number of potential methods like bioaccumulation of heavy metals by microbes, ion exclusion and low-molecular-weight binding protein production [57,58]. Elevated levels of metal resistance systems in bacterial cells are an indication of environmental heavy metal bioavailability [59]. The results of Chudobova et al. [60] showed a maximum resistance and capability of *S. aureus* strains under Cd^{2+} and Zn^{2+} ions. This resistance observed in *S. aureus* might be due to the efflux system containing a P-type ATPase transport system acting against Cd^{2+} ions [53,61].

4.2. *Effect of S. aureus on Plant Growth Promotion under Cr Metal Stress*

Different wheat varieties may differ in their response to different concentration of Cr in the soil. This could be attributed to various biological aspects of wheat varieties, as different wheat varieties show differences in growth parameters (e.g., leaf size). A heavy metal like Cr can easily make its way to aerial portions of plants, where it will affect their shoot metabolism at the cellular level and cause severe damage to minerals, water and nutrients, consequently retarding plant growth [10,62].

However, bacterial inoculation may improve the nutritional requirements of both micro- (Mn, Zn, Cu and Fe) and macronutrients (N, P and K) by modifying host physiology, which results in changed uptake pattern of roots. Similarly, a recent investigation done by Islam et al. [63] showed an increase in Fe and K concentrations in maize plants under Cr stress due to bacterial inoculations. According to an observation, plants with bacterial inoculation showed a reduction in metal accumulation in their aerial parts, which might be due to delayed translocation of metals from roots to upper parts [64]. Similar observations were recorded in this current research. Moreover, we isolated *S. aureus* K1 from wastewater that was contaminated with Cr, so the microbes may have the capability of performing metal detoxification as a part of their metabolic system. There was substantial improvement in plant growth and leaf pigments due to inoculation of specific microbes [63].

4.3. Chlorophyll Contents

Higher chlorophyll contents were observed in plants with bacterial inoculation compared to uninoculated plants (Figure 3). However, with further increasing metal concentrations, a reduction in chlorophyll contents was noted. This is in agreement with the findings of another research study, where chlorophyll a and chlorophyll b in wheat plants decreased with increasing concentrations of Pb in the growth medium [65].

4.4. ROS Species and Antioxidant Enzyme Production

Reactive oxygen species can be produced in plants when exposed to Cr^{6+}, which may damage the photosynthetic apparatus and protein complex of thylakoid membranes and result in inhibition of chlorophyll production [66]. In adverse conditions, plants release MDA contents; this reveals the level of lipid peroxidation, as MDA is the last decomposition product of membrane lipid peroxidation [67]. The increase in MDA contents found in the present study is indicative of imbalance between the generation and removal of free radicals in the cells [68]. The decreased lipid peroxidation with *S. aureus* K1 inoculation under Cr stress could be due to the increase in ROS-scavenging enzyme production in plants. This may be supported by a previously published study which revealed that the gene profile of metal detoxifying enzymes was activated by bacterial inoculation to deal with metal stress [69]. Reactive oxygen species are generated in response to stress caused by heavy metals like hexavalent Cr, and plants have a detoxifying antioxidant enzyme system for their maintenance. These enzymes are POD, SOD, APX and CAT, and they work alongside other non-enzymatic antioxidants. The activities performed by antioxidant enzymes in plants under metal stress are extremely variable and dependent on plant species, metal concentration, metal ions and exposure time period [70]. At low metal concentration, SOD activity may increase, but it becomes constant with increased metal concentration [71]. The enhancement in CAT activity was also noted in a number of plants under metal stress [72]. An increase in CAT activity was also observed as an adaptive trait of isolate CPSB21 [73]. Increased antioxidant enzyme activities in plants with inoculation of CPSB21 may be due to increases in mRNA/gene expression of antioxidant enzymes as compared to uninoculated plants [74].

4.5. Reduction of Cr Concentration in Plants by Bacterial Inoculation

A significant difference was found between uninoculated and *S. aureus* K1 inoculated plants in terms of Cr concentration. In contaminated soil, the results showed that the level of Cr was higher in the roots of wheat plants than it was in the shoots, which may be due to decreased translocation of Cr from roots to shoots of plants [75,76]. Immobilization of Cr in root cell vacuoles may lead to higher Cr accumulation in roots, which can cause toxicity in plants [77]. In the present study, inoculation of wheat plants with Cr-resistant microbes decreased the Cr concentration and its translocation from soil to roots and upper parts of wheat plants. The reduction of hexavalent Cr (Cr^{6+}) to trivalent Cr (Cr^{3+}) by bacterial isolates may be the reason for the improved growth of wheat plants [78] and hence the decreased level of the Cr contents in soil. Hasnain and Sabri [79] also reported a pattern of decreased Cr uptake and accumulation in roots and shoots of wheat plants inoculated with *Pseudomonas* sp.

A decrease in Cr concentration in soil was observed after wheat plant harvesting. This decrease was recorded in uninoculated Cr-contaminated wheat plants as a result of increased accumulation and uptake of Cr in roots and shoots [80]. Such decrease may also be due to Cr^{6+} reduction into Cr^{3+} under the influence of bacterial inoculation [78,81]. Scientists are also considering the use genetically engineered microorganisms (GEM), which may be well adjusted to their local environment (both climatic and soil) for effective elimination of heavy metals from contaminated soils [58,82,83].

5. Conclusions

The outcomes of this study indicate that the application of the peat-moss-based microbial inoculum improved plant growth and yield parameters and comparatively decreased metal accumulation by the plants. Overall, gas exchange attributes and chlorophyll contents increased with *S. aureus* K1 inoculation. This research study concluded that *S. aureus* K1 reduced the toxicity of Cr in wheat plants. The Cr-resistant *S. aureus* K1 supported the plant growth, decreased and detoxified Cr in plants and allowed better production of wheat in a Cr-contaminated environment. However, in-depth exploration (i.e., at the molecular level) of the alleviative mechanisms in plants should be conducted in future studies.

Author Contributions: F.Z., M.R., T.Y., S.A., M.A.E.-S., M.N.A. and L.W.; Data curation, M.Z., M.W., A.A. (Alia Anayat) and T.Y.; Formal analysis, M.Z., M.W., A.A. (Alia Anayat), M.R. and T.Y.; Funding acquisition, S.A., M.N.A. and L.W.; Investigation, M.R., A.A. (Awais Ahmad), T.Y. and M.A.E.-S.; Methodology, M.A., M.W., M.R., A.A. (Awais Ahmad), T.Y. and S.A.; Project administration, F.Z., T.Y. and S.A.; Resources, F.Z., M.W., S.A., M.A.E.-S., M.N.A. and L.W.; Software, F.Z., M.Z., M.A.E.-S. and L.W.; Supervision, M.R., T.Y. and S.A.; Validation, M.Z., M.W., A.A. (Alia Anayat) and A.A. (Awais Ahmad); Visualization, M.Z., M.W., A.A. (Alia Anayat) and A.A. (Awais Ahmad); Writing—original draft, F.Z., M.Z., S.A. and L.W.; Writing—review & editing, F.Z., T.Y. and M.N.A. All authors have read and agreed to the published version of the manuscript.

Acknowledgments: We are grateful to Government College University, Faisalabad, Pakistan, for its support. The authors would like to extend their sincere appreciation to the Researchers Supporting Project Number (RSP-2020/182), King Saud University, Riyadh, Saudi Arabia.

References

1. Singh, H.P.; Mahajan, P.; Kaur, S.; Batish, D.R.; Kohli, R.K. Chromium toxicity and tolerance in plants. *Environ. Chem. Lett.* **2013**, *11*, 229–254. [CrossRef]
2. Kunjam, M. Studies on selected heavy metals on seed germination and plant growth in pea plant (*Pisum sativum*) grown in solid medium. *J. Pharmacogn. Phytochem.* **2015**, *3*, 85–87.
3. Habiba, U.; Ali, S.; Farid, M.; Shakoor, M.B.; Rizwan, M.; Ibrahim, M.; Abbasi, G.H.; Hayat, T.; Ali, B. EDTA enhanced plant growth, antioxidant defense system, and phytoextraction of copper by *Brassica napus* L. *Environ. Sci. Pollut. Res.* **2014**, *22*, 1534–1544. [CrossRef] [PubMed]
4. Adrees, M.; Ali, S.; Iqbal, M.; Bharwana, S.A.; Siddiqi, Z.; Farid, M.; Ali, Q.; Saeed, R.; Rizwan, M. Mannitol alleviates chromium toxicity in wheat plants in relation to growth, yield, stimulation of anti-oxidative enzymes, oxidative stress and Cr uptake in sand and soil media. *Ecotoxicol. Environ. Saf.* **2015**, *122*, 1–8. [CrossRef] [PubMed]
5. Mathur, S.; Kalaji, H.M.; Kalaji, H.M. Investigation of deleterious effects of chromium phytotoxicity and photosynthesis in wheat plant. *Photosynthetica* **2016**, *54*, 185–192. [CrossRef]
6. Kamran, M.A.; Mufti, R.; Mubariz, N.; Syed, J.H.; Bano, A.; Javed, M.T.; Munis, M.F.H.; Tan, Z.; Chaudhary, H.J. The potential of the flora from different regions of Pakistan in phytoremediation: A review. *Environ. Sci. Pollut. Res.* **2013**, *21*, 801–812. [CrossRef]
7. Daud, M.K.; Mei, L.; Variath, M.T.; Ali, S.; Li, C.; Rafiq, M.T.; Zhu, S.J. Chromium (VI) Uptake and Tolerance Potential in Cotton Cultivars: Effect on Their Root Physiology, Ultramorphology, and Oxidative Metabolism. *BioMed Res. Int.* **2014**, *2014*, 1–12. [CrossRef]

8. Afshan, S.; Ali, S.; Bharwana, S.A.; Rizwan, M.; Farid, M.; Abbas, F.; Ibrahim, M.; Mehmood, M.A.; Abbasi, G.H. Citric acid enhances the phytoextraction of chromium, plant growth, and photosynthesis by alleviating the oxidative damages in *Brassica napus* L. *Environ. Sci. Pollut. Res.* **2015**, *22*, 11679–11689. [CrossRef]
9. Nagajyoti, P.C.; Lee, K.D.; Sreekanth, T.V.M. Heavy metals, occurrence and toxicity for plants: A review. *Environ. Chem. Lett.* **2010**, *8*, 199–216. [CrossRef]
10. Shanker, A. Chromium toxicity in plants. *Environ. Int.* **2005**, *31*, 739–753. [CrossRef]
11. Kamran, M.A.; Eqani, S.A.M.A.S.; Bibi, S.; Xu, R.-K.; Monis, M.F.H.; Katsoyiannis, A.; Bokhari, H.; Chaudhary, H.J. Bioaccumulation of nickel by E. sativa and role of plant growth promoting rhizobacteria (PGPRs) under nickel stress. *Ecotoxicol. Environ. Saf.* **2016**, *126*, 256–263. [CrossRef] [PubMed]
12. Dotaniya, M.L.; Das, H.; Meena, V.D. Assessment of chromium efficacy on germination, root elongation, and coleoptile growth of wheat (*Triticum aestivum* L.) at different growth periods. *Environ. Monit. Assess.* **2014**, *186*, 2957–2963. [CrossRef] [PubMed]
13. Lenka, S. Impacts of fertilizers use on environmental quality. In Proceedings of the National Seminar on Environmental Concern for Fertilizer Use in Future, Bidhan Chandra Krishi Viswavidyalaya, Kalyani, India, 26 February 2016.
14. Saha, J.; Selladurai, R.; Coumar, M.V.; Dotaniya, M.L.; Kundu, S.; Patra, A.K. *Industrial Activities in India and Their Impact on Agroecosystem*; Springer Science and Business Media LLC: Berlin/Heidelberg, Germany, 2017; Volume 10, pp. 229–249.
15. Khan, M.Y. Effect of microbial inoculation on wheat growth and phytostabilization of chromium contaminated soil. *Pak. J. Bot.* **2013**, *45*, 27–34.
16. Rizwan, M.; Ali, S.; Abbas, T.; Zia-Ur-Rehman, M.; Hannan, F.; Keller, C.; Al-Wabel, M.I.; Ok, Y.S. Cadmium minimization in wheat: A critical review. *Ecotoxicol. Environ. Saf.* **2016**, *130*, 43–53. [CrossRef] [PubMed]
17. Shewry, P.R.; Hey, S.J. The contribution of wheat to human diet and health. *Food Energy Secur.* **2015**, *4*, 178–202. [CrossRef]
18. Awan, N.; Fatima, A.; Farhan, M.; Awan, N. Comparative Analysis of Chromium and Cadmium in Various Parts of Wheat and Maize. *Pol. J. Environ. Stud.* **2019**, *28*, 1561–1566. [CrossRef]
19. Carlin, D.J.; Naujokas, M.F.; Bradham, K.D.; Cowden, J.; Heacock, M.; Henry, H.F.; Lee, J.S.; Thomas, D.J.; Thompson, C.; Tokar, E.J.; et al. Arsenic and Environmental Health: State of the Science and Future Research Opportunities. *Environ. Health Perspect.* **2016**, *124*, 890–899. [CrossRef]
20. Abbas, Q.; Yousaf, B.; Liu, G.; Zia-Ur-Rehman, M.; Ali, M.U.; Munir, M.A.M.; Hussain, S.A. Evaluating the health risks of potentially toxic elements through wheat consumption in multi-industrial metropolis of Faisalabad, Pakistan. *Environ. Sci. Pollut. Res.* **2017**, *24*, 26646–26657. [CrossRef]
21. Lytle, C.M.; Lytle, F.W.; Yang, N.; Qian, J.-H.; Hansen, D.; Zayed, A.; Terry, N. Reduction of Cr(VI) to Cr(III) by Wetland Plants: Potential for In Situ Heavy Metal Detoxification. *Environ. Sci. Technol.* **1998**, *32*, 3087–3093. [CrossRef]
22. Mishra, J.; Singh, R.; Arora, N.K. Alleviation of Heavy Metal Stress in Plants and Remediation of Soil by Rhizosphere Microorganisms. *Front. Microbiol.* **2017**, *8*, 1706. [CrossRef]
23. Ashraf, S.; Ali, Q.; Zahir, Z.A.; Ashraf, S.; Asghar, H.N. Phytoremediation: Environmentally sustainable way for reclamation of heavy metal polluted soils. *Ecotoxicol. Environ. Saf.* **2019**, *174*, 714–727. [CrossRef] [PubMed]
24. Kong, Z.; Glick, B.R. *The Role of Plant Growth-Promoting Bacteria in Metal Phytoremediation*; Elsevier: Amsterdam, The Netherlands, 2017; pp. 97–132.
25. Ryan, R.P.; Monchy, S.; Cardinale, M.; Taghavi, S.; Crossman, L.; Avison, M.B.; Berg, G.; Van Der Lelie, D.; Dow, J.M. The versatility and adaptation of bacteria from the genus Stenotrophomonas. *Nat. Rev. Genet.* **2009**, *7*, 514–525. [CrossRef] [PubMed]
26. Taj, Z.Z.; Rajkumar, M. *Perspectives of Plant Growth-Promoting Actinomycetes in Heavy Metal Phytoremediation*; Springer Science and Business Media LLC: Berlin/Heidelberg, Germany, 2016; pp. 213–231.
27. Mishra, S.; Bharagava, R.N. Toxic and genotoxic effects of hexavalent chromium in environment and its bioremediation strategies. *J. Environ. Sci. Health Part C* **2015**, *34*, 1–32. [CrossRef] [PubMed]
28. Wirtanen, G.; Salo, S. Biofilm risks. In *Handbook of Hygiene Control in the Food Industry*; Elsevier: Amsterdam, The Netherlands, 2016; pp. 55–79.

29. Ahemad, M. Enhancing phytoremediation of chromium-stressed soils through plant-growth-promoting bacteria. *J. Genet. Eng. Biotechnol.* **2015**, *13*, 51–58. [CrossRef]
30. Jing, Y.X.; Yan, J.L.; He, H.D.; Yang, D.J.; Xiao, L.; Zhong, T.; Yuan, M.; De Cai, X.; Bin Li, S. Characterization of Bacteria in the Rhizosphere Soils ofPolygonum Pubescensand Their Potential in Promoting Growth and Cd, Pb, Zn Uptake byBrassica napus. *Int. J. Phytoremediation* **2013**, *16*, 321–333. [CrossRef]
31. Mukhtar, S. Potential of sunflower (*Helianthus annuus* L.) for phytoremediation of nickel (Ni) and lead (Pb) contaminated water. *Pak. J. Bot.* **2010**, *42*, 4017–4026.
32. Soltanpour, P.N. Use of ammonium bicarbonate DTPA soil test to evaluate elemental availability and toxicity 1. *Commun. Soil Sci. Plant Anal.* **1985**, *16*, 323–338. [CrossRef]
33. Moodie, C.D.; Smith, H.W.; McCreery, R.A. Laboratory Manual for Soil Fertility. *Soil Sci.* **1951**, *71*, 400. [CrossRef]
34. Lucious, S. Heavy metal tolerance and antibiotic sensitivity of bacterial strains isolated from tannery effluent. *Asian J. Exp. Biol. Sci.* **2013**, *4*, 597–606.
35. Zhang, X.; Krumholz, L.R.; Yu, Z.; Chen, Y.; Liu, P.; Li, X. A Novel Subspecies of *Staphylococcus aureus* from Sediments of Lanzhou Reach of the Yellow River Aerobically Reduces Hexavalent Chromium. *J. Bioremediation Biodegrad.* **2013**, *4*. [CrossRef]
36. Turner, S.; Pryer, K.M.; Miao, V.P.W.; Palmer, J.D. Investigating deep phylogenetic relationships among cyanobacteria and plastids by small subunit rRNA sequence analysis. *J. Eukaryot. Microbiol.* **1999**, *46*, 327–338. [CrossRef] [PubMed]
37. Moyersoen, B.; E Beever, R.; Martin, F. Genetic diversity of Pisolithus in New Zealand indicates multiple long-distance dispersal from Australia. *New Phytol.* **2003**, *160*, 569–579. [CrossRef]
38. Tamura, K.; Peterson, D.; Stecher, G.; Nei, M.; Kumar, S.; Peterson, N. MEGA5: Molecular Evolutionary Genetics Analysis Using Maximum Likelihood, Evolutionary Distance, and Maximum Parsimony Methods. *Mol. Boil. Evol.* **2011**, *28*, 2731–2739. [CrossRef] [PubMed]
39. Hadi, F.; Bano, A. Effect of diazotrophs (Rhizobium and Azatebactor) on growth of maize (*Zea mays* L.) and accumulation of lead (Pb) in different plant parts. *Pak. J. Bot.* **2010**, *42*, 4363–4370.
40. Wu, S.; Cheung, K.; Luo, Y.; Wong, M.H. Effects of inoculation of plant growth-promoting rhizobacteria on metal uptake by Brassica juncea. *Environ. Pollut.* **2006**, *140*, 124–135. [CrossRef] [PubMed]
41. Lichtenthaler, H.K. Chlorophylls and carotenoids: Pigments of photosynthetic biomembranes. In *Methods in Enzymology*; Elsevier: Amsterdam, The Netherlands, 1987; pp. 350–382.
42. Dionisio-Sese, M.L.; Tobita, S. Antioxidant responses of rice seedlings to salinity stress. *Plant Sci.* **1998**, *135*, 1–9. [CrossRef]
43. Plumb-Dhindsa, P.; Dhindsa, R.S.; Thorpe, T.A. Leaf Senescence: Correlated with Increased Levels of Membrane Permeability and Lipid Peroxidation, and Decreased Levels of Superoxide Dismutase and Catalase. *J. Exp. Bot.* **1981**, *32*, 93–101. [CrossRef]
44. Zhang, J.; Kirkham, M. Drought-Stress-Induced Changes in Activities of Superoxide Dismutase, Catalase, and Peroxidase in Wheat Species. *Plant Cell Physiol.* **1994**, *35*, 785–791. [CrossRef]
45. Jana, S.; Choudhuri, M.A. Glycolate metabolism of three submersed aquatic angiosperms: Effect of heavy metals. *Aquat. Bot.* **1981**, *11*, 67–77. [CrossRef]
46. Zhang, X. The measurement and mechanism of lipid peroxidation and SOD, POD and CAT activities in biological system. In *Research Methodology of Crop Physiology*; Agriculture Press: Beijing, China, 1992; pp. 208–211.
47. Aebi, H. Catalase in vitro. *Methods Enzym.* **1984**, *105*, 121–126.
48. Nakano, Y.; Asada, K. Hydrogen Peroxide is Scavenged by Ascorbate-specific Peroxidase in Spinach Chloroplasts. *Plant Cell Physiol.* **1981**, *22*, 867–880. [CrossRef]
49. Rehman, M.Z.-U.; Rizwan, M.; Ghafoor, A.; Naeem, A.; Ali, S.; Sabir, M.; Qayyum, M.F. Effect of inorganic amendments for in situ stabilization of cadmium in contaminated soils and its phyto-availability to wheat and rice under rotation. *Environ. Sci. Pollut. Res.* **2015**, *22*, 16897–16906. [CrossRef] [PubMed]
50. Khan, Z.; Nisar, M.A.; Hussain, S.Z.; Arshad, M.N.; Rehman, A. Cadmium resistance mechanism in Escherichia coli P4 and its potential use to bioremediate environmental cadmium. *Appl. Microbiol. Biotechnol.* **2015**, *99*, 10745–10757. [CrossRef] [PubMed]

51. Oaikhena, E.E.; Makaije, D.B.; Denwe, S.D.; Namadi, M.M.; Haroun, A.A. Bioremediation Potentials of Heavy Metal Tolerant Bacteria Isolated from Petroleum Refinery Effluent. *Am. J. Environ. Prot.* **2016**, *5*, 29. [CrossRef]
52. Mustapha, M.U.; Halimoon, N. Screening and Isolation of Heavy Metal Tolerant Bacteria in Industrial Effluent. *Procedia Environ. Sci.* **2015**, *30*, 33–37. [CrossRef]
53. Bruins, M.R.; Kapil, S.; Oehme, F.W. Microbial Resistance to Metals in the Environment. *Ecotoxicol. Environ. Saf.* **2000**, *45*, 198–207. [CrossRef]
54. Nies, D.H. Efflux-mediated heavy metal resistance in prokaryotes. *FEMS Microbiol. Rev.* **2003**, *27*, 313–339. [CrossRef]
55. Sørensen, S.J.; Bailey, M.; Hansen, L.H.; Kroer, N.; Wuertz, S. Studying plasmid horizontal transfer in situ: A critical review. *Nat. Rev. Genet.* **2005**, *3*, 700–710. [CrossRef]
56. Kashif, M.; Ngaini, Z.; Harry, A.V.; Vekariya, R.L.; Ahmad, A.; Zuo, Z.; Alarifi, A. An experimental and DFT study on novel dyes incorporated with natural dyes on titanium dioxide (TiO_2) towards solar cell application. *Appl. Phys. A* **2020**, *126*, 1–13. [CrossRef]
57. Nies, D.H. Microbial heavy-metal resistance. *Appl. Microbiol. Biotechnol.* **1999**, *51*, 730–750. [CrossRef]
58. Das, S.; Dash, H.R.; Chakraborty, J. Genetic basis and importance of metal resistant genes in bacteria for bioremediation of contaminated environments with toxic metal pollutants. *Appl. Microbiol. Biotechnol.* **2016**, *100*, 2967–2984. [CrossRef] [PubMed]
59. Roosa, S.; Wattiez, R.; Prygiel, E.; Lesven, L.; Billon, G.; Gillan, D.C. Bacterial metal resistance genes and metal bioavailability in contaminated sediments. *Environ. Pollut.* **2014**, *189*, 143–151. [CrossRef] [PubMed]
60. Chudobova, D.; Dostalova, S.; Ruttkay-Nedecký, B.; Guran, R.; Rodrigo, M.A.M.; Tmejová, K.; Krizkova, S.; Zitka, O.; Adam, V.; Kizek, R. The effect of metal ions on *Staphylococcus aureus* revealed by biochemical and mass spectrometric analyses. *Microbiol. Res.* **2015**, *170*, 147–156. [CrossRef]
61. Nies, D.H.; Silver, S.D. Ion efflux systems involved in bacterial metal resistances. *J. Ind. Microbiol. Biotechnol.* **1995**, *14*, 186–199. [CrossRef] [PubMed]
62. Chatterjee, J.; Kumar, P.; Sharma, P.N.; Tewari, R.K. Chromium toxicity induces oxidative stress in turnip. *Indian J. Plant Physiol.* **2015**, *20*, 220–226. [CrossRef]
63. Islam, F.; Yasmeen, T.; Arif, M.S.; Riaz, M.; Shahzad, S.M.; Imran, Q.; Ali, I. Combined ability of chromium (Cr) tolerant plant growth promoting bacteria (PGPB) and salicylic acid (SA) in attenuation of chromium stress in maize plants. *Plant Physiol. Biochem.* **2016**, *108*, 456–467. [CrossRef]
64. Islam, F.; Yasmeen, T.; Ali, Q.; Ali, S.; Arif, M.S.; Hussain, S.; Rizvi, H. Influence of Pseudomonas aeruginosa as PGPR on oxidative stress tolerance in wheat under Zn stress. *Ecotoxicol. Environ. Saf.* **2014**, *104*, 285–293. [CrossRef]
65. Lamhamdi, M.; El Galiou, O.; Bakrim, A.; Nóvoa-Muñoz, J.C.; Arias-Estévez, M.; Aarab, A.; Lafont, R. Effect of lead stress on mineral content and growth of wheat (*Triticum aestivum*) and spinach (*Spinacia oleracea*) seedlings. *Saudi J. Boil. Sci.* **2013**, *20*, 29–36. [CrossRef]
66. Oliveira, H. Chromium as an Environmental Pollutant: Insights on Induced Plant Toxicity. *J. Bot.* **2012**, *2012*, 1–8. [CrossRef]
67. Arif, M.S.; Yasmeen, T.; Shahzad, S.M.; Riaz, M.; Rizwan, M.; Iqbal, S.; Asif, M.; Soliman, M.H.; Ali, S. Lead toxicity induced phytotoxic effects on mung bean can be relegated by lead tolerant Bacillus subtilis (PbRB3). *Chemosphere* **2019**, *234*, 70–80. [CrossRef]
68. Sytar, O.; Kumar, A.; Latowski, D.; Kuczyńska, P.; Strzałka, K.; Prasad, M.N.V. Heavy metal-induced oxidative damage, defense reactions, and detoxification mechanisms in plants. *Acta Physiol. Plant.* **2012**, *35*, 985–999. [CrossRef]
69. Islam, F.; Yasmeen, T.; Ali, Q.; Mubin, M.; Ali, S.; Arif, M.S.; Hussain, S.; Riaz, M.; Abbas, F. Copper-resistant bacteria reduces oxidative stress and uptake of copper in lentil plants: Potential for bacterial bioremediation. *Environ. Sci. Pollut. Res.* **2015**, *23*, 220–233. [CrossRef]
70. Sharma, S.S.; Dietz, K.-J. The relationship between metal toxicity and cellular redox imbalance. *Trends Plant Sci.* **2009**, *14*, 43–50. [CrossRef] [PubMed]
71. Štolfa, I.; Pfeiffer, T.Z.; Špoljarić, D.; Teklić, T.; Lončarić, Z. *Heavy Metal-Induced Oxidative Stress in Plants: Response of the Antioxidative System*; Springer Science and Business Media LLC: Berlin/Heidelberg, Germany, 2015; pp. 127–163.

72. Fayiga, A.O.; Ma, L.Q.; Cao, X.; Rathinasabapathi, B. Effects of heavy metals on growth and arsenic accumulation in the arsenic hyperaccumulator *Pteris vittata* L. *Environ. Pollut.* **2004**, *132*, 289–296. [CrossRef] [PubMed]
73. Vital, S.A.; Fowler, R.W.; Virgen, A.; Gossett, D.R.; Banks, S.W.; Rodriguez, J. Opposing roles for superoxide and nitric oxide in the NaCl stress-induced upregulation of antioxidant enzyme activity in cotton callus tissue. *Environ. Exp. Bot.* **2008**, *62*, 60–68. [CrossRef]
74. Gururani, M.A.; Upadhyaya, C.P.; Baskar, V.; Venkatesh, J.; Nookaraju, A.; Park, S.W. Plant Growth-Promoting Rhizobacteria Enhance Abiotic Stress Tolerance in Solanum tuberosum Through Inducing Changes in the Expression of ROS-Scavenging Enzymes and Improved Photosynthetic Performance. *J. Plant Growth Regul.* **2012**, *32*, 245–258. [CrossRef]
75. Huffman, E.W.D.; Allaway, W.H. Chromium in plants. Distribution in tissues, organelles, and extracts and availability of bean leaf chromium to animals. *J. Agric. Food Chem.* **1973**, *21*, 982–986. [CrossRef]
76. Zayed, A.; Lytle, C.M.; Qian, J.-H.; Terry, N. Chromium accumulation, translocation and chemical speciation in vegetable crops. *Planta* **1998**, *206*, 293–299. [CrossRef]
77. Shanker, A.; Djanaguiraman, M.; Sudhagar, R.; Chandrashekar, C.; Pathmanabhan, G. Differential antioxidative response of ascorbate glutathione pathway enzymes and metabolites to chromium speciation stress in green gram ((L.) R.Wilczek. cv CO 4) roots. *Plant Sci.* **2004**, *166*, 1035–1043. [CrossRef]
78. Salunkhe, P.; Dhakephalkar, P.; Paknikar, K.M. Bioremediation of hexavalent chromium in soil microcosms. *Biotechnol. Lett.* **1998**, *20*, 749–751. [CrossRef]
79. Hasnain, S.; Sabri, A.N. Growth stimulation of *Triticum aestivum* seedlings under Cr-stresses by non-rhizospheric pseudomonad strains. *Environ. Pollut.* **1997**, *97*, 265–273. [CrossRef]
80. Mishra, S.; Shanker, K.; Srivastava, M.; Srivastava, S.; Shrivastav, R.; Dass, S.; Prakash, S. A study on the uptake of trivalent and hexavalent chromium by paddy (*Oryza sativa*): Possible chemical modifications in rhizosphere. *Agric. Ecosyst. Environ.* **1997**, *62*, 53–58. [CrossRef]
81. Cheung, K.H.; Gu, J.-D. Reduction of chromate (CrO42−) by an enrichment consortium and an isolate of marine sulfate-reducing bacteria. *Chemosphere* **2003**, *52*, 1523–1529. [CrossRef]
82. Gupta, S.; Singh, D. Role of Genetically Modified Microorganisms in Heavy Metal Bioremediation. In *Advances in Environmental Biotechnology*; Springer Science and Business Media LLC: Berlin/Heidelberg, Germany, 2017; Volume 3, pp. 197–214.
83. Ali, B.; Wang, B.; Ali, S.; Ghani, M.A.; Hayat, M.T.; Yang, C.; Xu, L.; Zhou, W. 5-Aminolevulinic Acid Ameliorates the Growth, Photosynthetic Gas Exchange Capacity, and Ultrastructural Changes Under Cadmium Stress in *Brassica napus* L. *J. Plant Growth Regul.* **2013**, *32*, 604–614. [CrossRef]

10

Isolation and Characterization of Plant Growth Promoting Endophytic Bacteria from Desert Plants and their Application as Bioinoculants for Sustainable Agriculture

Muneera D. F. ALKahtani [1], Amr Fouda [2], Kotb A. Attia [3], Fahad Al-Otaibi [4,5], Ahmed M. Eid [2], Emad El-Din Ewais [2], Mohamed Hijri [4,6], Marc St-Arnaud [4], Saad El-Din Hassan [2], Naeem Khan [7], Yaser M. Hafez [8] and Khaled A. A. Abdelaal [8,*]

1 Biology Department, College of Science, Princess Nourah Bint Abdulrahman University, Riyadh POX 102275-11675, Saudi Arabia; mdfkahtani@gmail.com

2 Botany and Microbiology Department, Faculty of Science, AL-Azhar University, Nasr City, Cairo 11884, Egypt; amr_fh83@azhar.edu.eg (A.F.); aeidmicrobiology@azhar.edu.eg (A.M.E.); ewais_e@yahoo.com (E.E.-D.E.); saad.el-din.hassan@umontreal.ca (S.E.-D.H.)

3 Center of Excellence in Biotechnology Research, King Saud University, Riyadh POX 2455-11451, Saudi Arabia; kattia1.c@ksu.edu.sa

4 Biodiversity Centre, Institut de Recherche en Biologie Végétale, Université de Montréal and Jardin botanique de Montréal, Montréal, QC 22001, Canada; fahad.alotaibi@umontreal.ca (F.A.-O.); mohamed.hijri@umontreal.ca (M.H.); marc.st-arnaud@umontreal.ca (M.S.-A.)

5 Department of Soil Science, King Saud University, Riyadh 11564, Saudi Arabia

6 AgroBioSciences, Mohammed VI Polytechnic University (UM6P), 43150 Ben Guerir, Morocco

7 Department of Agronomy, Institute of Food and Agricultural Sciences, University of Florida, Gainesville, FL 32611, USA; naeemkhan@ufl.edu

8 Excellence Center (EPCRS), Plant Pathology and Biotechnology Laboratory, Faculty of Agriculture, Kafrelsheikh University, Kafr El Sheikh 33516, Egypt; hafezyasser@gmail.com

* Correspondence: khaled.elhaies@gmail.com

Abstract: Desert plants are able to survive under harsh environmental stresses inherent to arid and semiarid regions due to their association with bacterial endophytes. However, the identity, functions, and the factors that influence the association of bacterial endophytes with desert plants are poorly known. These bacterial endophytes can be used as an untapped resource to favor plant growth and development in agro-ecosystems of arid regions. The present study is therefore focused on the isolation and identification of bacterial endophytes from two native medicinal plants (*Fagonia mollis* Delile and *Achillea fragrantissima* (Forssk) Sch. Bip.) growing spontaneously in the arid region of the South Sinai (Egypt), and characterization of their plant growth promoting (PGP) traits. Thirteen putative bacterial endophytes were isolated from the leaves of both plant species and characterized for their plant growth promoting abilities using molecular and biochemical approaches, as well as greenhouse trials. Selected endophytic bacterial strains were applied to maize plants (*Zea mays* L. var. Single cross Pioneer 30K08) to further evaluate their PGP abilities under greenhouse conditions. Isolated bacterial strains have variable plant growth promoting activities. Among these activities, isolated bacterial endophytes have the efficacy of phosphate solubilizing with clear zones ranging from 7.6 ± 0.3 to 9.6 ± 0.3 mm. Additionally, the obtained bacterial endophytes increased the productivity of indole acetic acid (IAA) in broth media from 10 to 60 $\mu g{\cdot}mL^{-1}$ with increasing tryptophan concentration from 1 to 5 $mg{\cdot}mL^{-1}$. *Bacillus* and *Brevibacillus* strains were frequently isolated from the leaves of both plant species, and had significant positive effects on plant growth and shoot phosphorus (P) and nitrogen (N) contents. Results suggest that these endophytes are good candidates as plant growth promoting inoculants to help reduce chemical input in conventional agricultural practices and increase nutrient uptake and stress resilience in plant species.

Keywords: *Zea mays* L.; environmental stresses; endophytic bacteria; plant growth promoting ability

1. Introduction

The enhancement of crop productivity is required for feeding the increasing population in developing countries and often relies on the use of chemical fertilizers. However, long-term use of these fertilizers was shown to decrease bacterial diversity in soil [1,2] and can also have harmful effects on the environment, such as leaching of phosphorus and nitrogen into groundwater, and increasing soil and groundwater pollution [3]. One way to increase the sustainability of agricultural practices is the use of efficient, nutrient mobilizing microorganisms to reduce the need and dependency on chemical fertilizers [4,5]. Plant growth promoting bacteria (PGPB) that form symbiotic interactions with their host plants are crucial to improve plant productivity and health under various environmental conditions [4,6–8]. Bacterial endophytes colonize plant tissues without any apparent pathogenic symptoms and establish beneficial associations with their plant host through phytohormone synthesis, enzyme production, and nutrient mobilization and translocation, such as phosphate (PO_4^{-3}) solubilization, nitrogen fixation, and ammonia (NH_3) production [9–11]. Moreover, many endophytes display various applications such as antimicrobial mechanisms, which reduce crop losses caused by pathogens [12–16], and its metabolites integrated into different biotechnological applications [17–20].

The Sinai Peninsula is located in the Sahara-Arabian deserts and represents approximately 6% of the total land area of Egypt. The semi-arid to arid climate and winter precipitations are the main characteristics of the Sinai Peninsula desert. Plants growing in desert conditions were found to harbor a microbiome that increased the biomass during drought stress periods [21]. Medicinal plants from desert farming in Sekem (Egypt) were shown that their roots are strongly associated with bacteria [22–24]. Although the Sinai desert has diverse medicinal plants, very few studies have focused on the associated bacterial endophytes and their PGP activities. Hanna et al. [25] collect 43 different plant species from the North Sinai desert, and reported that *Fagonia mollis* was the highest plant species harboring culturable bacteria. Among these bacteria, *Gluconacetobacter diazotrophicus* was the lowest endophytic species exhibit N_2-fixing activity. In the same regards, 132 endophytic strains were isolated from 18 Egyptian medicinal plants, including nine fungal strains isolated from *Achillea fragrantissima* and exhibiting inhibitory activities against different pathogenic bacteria and yeasts [26]. Application of the bacterial endophytes (*Bacillus thuringiensis*) led to improved plant growth and increased relative water content, chlorophyll content, chlorophyll fluorescence parameter (Fv/Fm ratio), and fruit yield of sweet pepper plants [27].

Maize has become a staple food in many parts of the world, with the total production of maize surpassing that of wheat or rice. Maize crop has several uses, such as food stuff for human or as animal feed because of its high nutritional value. Maize has also been used for corn ethanol and other maize products, such as fructose, corn starch, corn oil, and corn syrup [28]. The Arab Republic of Egypt is the largest country consuming maize at the level of the African continent. However, the production of the Arab Republic of Egypt reached to about 1% of the total global production during (2005–2013), while Egypt represents the third place at the level of the African continent [29]. Therefore, the current study aimed to improve the performance maize growth under the optimal conditions and at normal habitat.

Fagonia mollis and *Achillea fragrantissima* are medicinal plants frequently found in the Sinai Peninsula, and their bacterial endophytes could partially be responsible for the production of various bioactive compounds [30], and the ability of these plants to withstand the harsh, drought condition of the Sinai Peninsula. Therefore, this study focused on the isolation and characterization of putative bacterial endophytes from *F. mollis* and *A. fragrantissima*, which are native inhabitants of the arid and extremely harsh conditions of the Sinai desert. Plant growth promoting (PGP) properties of the bacterial endophytes involving extracellular enzymes (amylase, cellulase, protease, pectinase, and xylanase) production, antimicrobial activity against selected pathogenic bacteria and fungi, indole-3-acetic acid (IAA) and NH_3 production, and P-solubilization ability were evaluated. In addition, their effect on maize growth, plant biomass production, and nutrients content in plant shoots were also investigated in order to evaluate their potentials as bioinoculants for sustainable agriculture practices.

2. Materials and Methods

2.1. Plant Sampling and Study Area

Fagonia mollis Delile (family *Zygophyllaceae*) and *Achillea fragrantissima* (Forssk.) Sch.Bip. (family *Asteraceae*) were collected from two sites, Wadi al-Zwatin (latitude 28.539290° to 28.53919° N, longitude 33.930784° to 33.92044° E) and Wadi Selebat (latitude 28.545493 to 28.543339 N, longitude 33.933707 to 33.932984 E), Saint Katherine Protectorate, South Sinai, Egypt (Figure 1). Four individual plants from each species were collected per site. The plant samples were carefully placed in sterile polyethylene bags and brought back to the laboratory in a portable cooler maintained at 4 °C using ice packs. The formal identification of the plant specimens was carried out at the herbarium of Botany and Microbiology Department of Al-Azhar University, where plant herbarium specimens were also deposited.

Figure 1. (**A**) *Fagonia mollis* Delile. and (**B**) *Achillea fragrantissima* Forssk.

2.2. Isolation of Bacterial Endophytes

On each plant, the first five leaves from a shoot tip were excised and washed under running tap water. Sterilization of leaf surfaces was done by soaking the tissues in a series of baths: sterile distilled water for 1 min, 70% ethanol for 1 min, 2.5% sodium hypochlorite for 4 min, 70% ethanol for 30 s, and a final series of rinsing thrice in sterile distilled water in three different containers. A 0.1 mL aliquot of the final rinse water was plated onto nutrient agar plates to confirm the success of surface sterilization.

The sterilized plant leaves were then cut into 5 mm segments, and twenty leaf segments per individual plant were placed in four petri dishes (9 cm; five segments/plate) containing luria broth (LB) media (tryptone 10 $g{\cdot}L^{-1}$; yeast extract 5 $g{\cdot}L^{-1}$; NaCl 10 $g{\cdot}L^{-1}$; agar 15 $g{\cdot}L^{-1}$; and 1 L dis. H_2O, adjusted to pH 7) supplemented with nystatin (25 $\mu g{\cdot}mL^{-1}$) to suppress fungal growth, and incubated in the dark at 35 ± 2 °C. Another twenty segments of sterilized leaves per individual plant were together crushed in 10 mL sterile saline solution using a sterile digital homogenizer (PRO25D, Pro Scentific, 120 V, Willenbrock, Oxford, CT, USA), and 1 mL of the suspension was serially diluted until 10^{-3} from which a 0.1 mL aliquot was spread onto each of the three Petri plates containing LB medium and incubated in the dark at 35 ± 2 °C [31]. The cultures were regularly observed for bacterial growth, for a period of 96 hours. Bacteria growing from the previous steps were streaked on fresh LB plates to obtain single colonies, which picked up and inoculated on LB slants and stored at 4 °C until further study.

2.3. Molecular Identification of Bacterial Endophytes

Bacterial identification was based on 16S rRNA gene sequence analysis. Genomic DNA of each isolate was extracted following the method of Miller et al. [32], with some modifications. Briefly, individual colonies from an agar plate were picked up either using a sterile toothpick or an inoculating loop and resuspended in 50 μL of sterile deionized water. The cell suspension was placed in a water bath at 97 °C and heated for 10 min, the cell lysate was centrifuged (15,000× *g*, 10 min), and the supernatant containing the DNA was recovered. DNA concentration was determined by measuring its absorbance at UV spectrum of 260 nm using a spectrophotometer (JENWAY 6350, 230 V/50 Hz, Staffordshire, UK). A partial 16S rDNA fragment was PCR amplified using the bacterial universal primers 27f (5′-AGAGTTTGATCCTGGCTCAG-3′) and 1492r (5′-GGTTACCTTGTTACGACTT-3′) [33]. The PCR reaction contained: 1 × PCR buffer, 0.5 mM $MgCl_2$, 2.5 U Taq DNA polymerase (QIAGEN), 0.25 mM dNTP, 0.5 μM of each primer, and approximately 5 ng of bacterial genomic DNA. The PCR cycling conditions were 94 °C for 3 min, followed by 30 cycles of 94 °C for 0.5 min, 55 °C for 0.5 min, and 72 °C for 1 min, followed by a final extension performed at 72 °C for 10 min. The PCR products were forward and reverse sequenced using the Applied Biosystems 3730xl DNA Analyzer technology at the Genome Quebec Innovation Center (Montreal, QC, Canada). The sequences generated in this study were deposited in GenBank under accession numbers KY555785 to KY555797. The 16S rRNA sequences were then compared against the GenBank database using the NCBI BLAST nucleotide search. A multiple sequence alignment was constructed on approximately 1200 bp of 16S rRNA gene fragments using the ClustalX 1.8 software package (http://www.clustal.org/clustal2) and a phylogenetic tree was constructed using the neighbor-joining method in the MEGA v6.1 software (www.megasoftware.net), with confidence tested by bootstrap analysis (1000 repeats).

2.4. Screening the Extracellular Enzymatic Activities of Bacterial Endophytes

The production and activity of extracellular enzymes (amylase, cellulase, protease, pectinase, and xylanase) of isolated bacterial endophytes were assessed by growing the isolates in a mineral salt (MS) media ($NaNO_3$ 5 g·L^{-1}; KH_2PO_4 1 g·L^{-1}; K_2HPO_4 2 g·L^{-1}; $MgSO_4.7H_2O$ 0.5 g·L^{-1}; KCl 0.1 g·L^{-1}; $CaCl_2$ 0,01 g·L^{-1}; $FeSO_4.7H_2O$ 0.02 g·L^{-1}; agar 15 g·L^{-1}; and 1 L dis. H_2O) complemented with various additives, depending on the enzyme being tested, as detailed below. Control treatments consisted of the same media without bacterial inoculation. After incubation for 24–48 h depending on the growth rates of the bacterial endophytes at 35 ± 2 °C, specific reagents were added (see paragraph below), and the size of the clear zone surrounding the bacterial colony was measured, indicating extracellular enzymatic activities. All assays were performed in triplicates.

Amylolytic and cellulase activity were assessed by growing the endophytic bacterial isolates on MS agar medium supplemented with 1% soluble starch and 1% cellulose or carboxy-methylcellulose (CMC) respectively. After incubation, the plates were flooded with 1% iodine. MS agar medium containing 1% gelatine was used to determine the bacterial proteolytic activity. After incubation, the degradation of gelatine was highlighted using acidic mercuric chloride as an indicator. Pectinolytic activity was determined by growing bacteria in MS medium containing 1% pectin. After the incubation period, the plates were flooded with 1% aqueous solution of hexadecyl trimethyl ammonium bromide. MS agar medium supplemented with 1% xylan from corncobs was used to measure bacterial xylanolytic activity. After the incubation period, the xylanase activity was assessed after flooding with absolute ethyl alcohol to indicate biodegradation [34].

2.5. Antimicrobial Activity of Bacterial Endophytes

To test the antimicrobial activity of the bacterial endophytes, the isolated strains were cultured in nutrient broth medium for 6 days at 35 ± 2 °C on a shaker (LABOAO, LH-2102C, Zhengzhou, China) at 180 rpm. Crude fermentation broth was blended thoroughly and centrifuged at 4000 rpm for 5 min. Liquid supernatant was extracted twice with an equal volume of ethyl acetate. The organic solvent extract was then evaporated under reduced pressure using a rotary evaporator (RE-801, BM-500 water bath (4 L), glassware C set, Yamato scientific, Tokyo, Japan). The crude extracts were dissolved in dimethyl sulfoxide (DMSO) and used for antimicrobial screening using a well diffusion method [35]. Nutrient broth media without bacterial inoculation were extracted and dissolved in DMSO and were used as controls.

Microbial strains used for antimicrobial assays were: *Staphylococcus aureus* ATCC 6538 and *Bacillus subtilis* ATCC 6633 (Gram-positive bacteria), *Escherichia coli* ATCC 8739, *Pseudomonas aeruginosa* ATCC 9027 and *Salmonella typhimurium* ATCC 14028 (Gram-negative bacteria), and *Candida albicans* ATCC 10231 (yeast). Test organisms were inoculated in Petri dishes containing Muller–Hinton agar medium (Sigma-Aldrich) for bacteria or Sabouraud agar medium (Sigma-Aldrich) for yeast [19,36,37]. Three wells of 1 cm diameter were cut in the tested organism colony using a sterile cork borer and filled with 40 μL of endophytic bacterial extract. Negative control wells were filled with 40 μL of control extract. The plates were kept at 4 °C for 4 h to allow diffusion of antimicrobial compounds, and then incubated at 35 ± 2 °C for bacteria and 28 ± 2 °C for *C. albicans* for 24 h [38,39]. The inhibition zones around the wells were measured to assess the antimicrobial activity of bacterial extracts. All antimicrobial activity assays were performed in triplicates.

2.6. *Screening for In Vitro Plant Growth Promoting (PGP) Traits*

2.6.1. Phosphate Solubilization

The bacterial endophytic isolates were screened for P-solubilization as follows. Pikovskaya medium (glucose 10 $g{\cdot}L^{-1}$; $Ca_3(PO_4)$ 2·5 $g{\cdot}L^{-1}$; $(NH_4)2SO_4$ 0.5 $g{\cdot}L^{-1}$; NaCl 0.2 $g{\cdot}L^{-1}$; $MgSO_4{\cdot}7H_2O$ 0.1 $g{\cdot}L^{-1}$; KCl 0.2 $g{\cdot}L^{-1}$; $FeSO_4{\cdot}7H_2O$ 0.002 $g{\cdot}L^{-1}$; yeast extract 0.5 $g{\cdot}L^{-1}$; $MnSO_4{\cdot}2H_2O$ 0.002 $g{\cdot}L^{-1}$; agar 15 $g{\cdot}L^{-1}$; and 1 L dis. H_2O) was prepared and bromophenol blue was added as an indicator. The medium was inoculated with endophytic isolates and incubated for 48 h. The Pikovskaya medium without bacterial growth was used as a control. The formation of clear zones around the colony, due to the utilization of tricalcium phosphate, was measured to assess the ability of endophytes to solubilize phosphate [40].

2.6.2. Ammonia Production

The ability of the isolated endophytic bacterial strains to produce NH_3 was assessed after growing the bacterial strains in peptone water (peptone 10 $g{\cdot}L^{-1}$; NaCl 5 $g{\cdot}L^{-1}$; and 1 L dis. H_2O) for 72 h at 35 ± 2 °C. Peptone water without bacterial inoculation was used as a control. The addition of 1 mL of Nessler's reagent in the peptone liquid medium was used to assess the ammonia production. A color change to faint yellow indicated the minimum ammonia production while deep yellow to brownish color indicated the maximum ammonia production [41].

2.7. *Quantitative Screening for Indole-3-acetic acid (IAA) Production*

The ability of bacterial endophytes to produce IAA was determined in nutrient broth at 35 ± 2 °C for 24 h. One milliliter of each bacterial suspension was added to 20 mL of nutrient broth medium containing 0, 1, 2, or 5 $mg{\cdot}mL^{-1}$ tryptophan, and incubated for 14 days. Controls consisted of nutrient broth media containing 0, 1, 2, or 5 $mg{\cdot}mL^{-1}$ tryptophan but without bacterial inoculation. Five milliliters of each culture were collected from the incubating broth after 14 days and centrifuged at 6000 rpm for 30 min. One milliliter of the supernatant was mixed with 1 drop of orthophosphoric acid and 2 mL of Salkowski's reagent (300 mL concentrated sulfuric acid, 500 mL distilled water, and 15 mL 0.5 M $FeCl_3$). Development of a pink color indicated IAA production. The optical density at 530 nm was measured using a spectrophotometer (Jenway 6305 UV spectrophotometer, 230 V/50 Hz, Staffordshire, UK), and the amount of IAA produced was estimated using a standard curve for authentic IAA [42].

Five isolates were chosen based on their ability to produce IAA for further analysis. Equal amount of each inoculum was added to nutrient broth medium containing 5 mg mL^{-1} tryptophan and incubated for 14 days at 35 ± 2 °C. IAA concentration was determined at 2 days intervals up to the 14th day after inoculation. Samples were centrifuged at 6000 rpm for 30 min and IAA production was determined as mentioned above. All the IAA production assays were performed in triplicates.

2.8. *Effect of Bacterial Isolates on Zea mays L. Growth*

2.8.1. Experimental Design

A pot experiment was conducted in a completely randomized design with five replicates of each treatment. Plants were inoculated with one of five individual bacterial isolates (*Brevibacillus* spp. Af.13, and Af.14, *Bacillus* spp. Fm.3, Fm.4, and Fm.6) or with a bacterial consortium formed of an equal amount of the five bacteria isolates. A control treatment consisted of uninoculated plants.

2.8.2. Culture Conditions

A loamy soil was collected from an agricultural field in the El-Menoufia governorate. Physical and chemical characteristics of the soil are shown in Table 1. The soil was air-dried, sieved with a 2 mm sieve, mixed with quartz sand at a soil: sand ratio of 3:1 and autoclaved twice for one hour at 121 °C. The five most potent IAA producing bacterial isolates (as listed above) were inoculated in nutrient broth and incubated at 35 ± 2 °C for 24 h on a shaker (LABOAO, LH-2102C, Zhingzhou, China) at 180 rpm. Seeds of maize (*Zea mays*, Cultivar Giza 9) were surface sterilized by soaking in 2.5% sodium hypochlorite for 3 minutes and then washed 5 times in sterile distilled water. Six groups of pregerminated seeds were separately incubated in 50 mL aliquots of the culture medium inoculated each with one of the five bacterial strains or with the bacterial consortium, and incubated at room temperature for 4 h on a shaker at 180 rpm. After incubation, the soaked seeds were sown in 1 L plastic pots filled with 900 g of sterilized soil-sand mixture. Each pot received three germinated seeds. Plants were grown in a greenhouse with a temperature of 25–30 °C and were irrigated with tap water as required without fertilization.

Table 1. Physical and chemical characteristics of the soil used in the greenhouse experiment.

Parameters	Soil Analysis
Soil Texture	Loamy Sand
Physical characters (%)	
Sand	76.8
Silt	10.9
Clay	12.2
Chemical characters (mg kg^{-1})	
P	24
K	14.075
Na	186.44
Ca	27.25
Cl	134.35

2.9. *Plant Tissue Analysis*

After 30 days, plants were harvested, shoot and root systems were separated, and roots were washed carefully with tap water to remove the attached soil particles. The dry weight of shoots and roots were measured after drying for 48 h at 60 °C. Phosphorus, nitrogen, and potassium contents were determined according to the methods described by AOAC international [43] and Rice et al. [44].

2.10. *Statistical Analysis*

Data were statistically analyzed using SPSS v17 (SPSS Inc., Chicago, IL, USA). When the normality and homogeneity of variance hypotheses were satisfied, one-way analysis of variance (ANOVA) was used to compare the bacterial isolates for extracellular enzymes production, antimicrobial activity,

IAA and ammonia production, P-solubilization ability, and the effect of these endophytes on maize growth performance. A posteriori multiple comparisons were done using Tukey's range tests at $p < 0.05$. All results are the means of three to five independent replicates, as specified above.

3. Results

Thirteen endophytic bacterial strains were isolated from the leaves of the two medicinal plants (Table 2). Nine strains were isolated from *F. mollis* plants and identified as *Bacillus* spp. (eight strains), and *Paenibacillus* sp. (one strain), while four bacterial strains were isolated from *A. fragrantissima* plants and identified as *Paenibacillus* sp. (one strain) and *Brevibacillus* sp. (three strains). The 16S rRNA gene sequences of strains Fm.2 to Fm.9 showed 96–99% sequence similarity with the sequences of *Bacillus amyloliquefaciens*, *Bacillus thuringiensis*, and *Bacillus cereus* (Figure 2 and Table 2). The 16S rRNA gene sequences of strains Fm.1 and Af.12 showed 99% of sequence similarity with *Paenibacillus barengoltzii*. Isolates Af.13 to Af.15 showed between 93 and 99% of 16S rRNA sequence similarity with *Brevibacillus agri*.

All bacterial endophytes isolated from *F. mollis* were positive for amylase, pectinase, carboxymethyl cellulase, cellulose, xylanase, and gelatinase, while those isolated from *A. fragrantissima* showed activities for only one to four enzymes (Table 3). The highest activities of cellulase and carboxymethyl cellulase were observed with *Bacillus* sp. Fm.5 (22.0 ± 1.1 and 21.3 ± 1.2 mm, respectively), while *Bacillus* sp. Fm.2 showed the highest activities of pectinase, xylanase, and gelatinase enzymes with clear zone 17.6 ± 0.6, 19.6 ± 0.3, and 22.0 ± 0.5 mm respectively. The highest gelatinase activity (22.3 ± 1.4 mm) was measured for *Bacillus* sp. Fm.3.

The antimicrobial activity of the bacterial endophytes against selected pathogenic bacterial and yeast strains are given in Table 4. The crude extract of *Brevibacillus* sp. Af.13 suppressed the growth of five tested pathogenic microorganisms, while *Bacillus* sp. Fm.8 inhibited the growth of *P. aeruginosa*, *S. typhi*, and *E. coli*. Endophytic strains *Bacillus* sp. Fm.2 and *Brevibacillus* sp. Af.13 were the only endophytes whose crude extracts showed an inhibitory effect against the pathogenic yeast *C. albicans* ATCC 10231 with clear zone 15 and 18 mm respectively. While the filtrates extracted from all strains showed some inhibition of *P. aeruginosa* ATCC 9027, the highest growth inhibition was noted from strains Fm.6, Fm.7, Fm.8, Fm.9, and Af.14 with inhibition zones ranging between 15 to 30 mm.

Table 2. The 16S rRNA sequence identification of endophytic bacterial strains from two different medicinal plants.

Plant Species	Bacterial Strain Code	Homologue Sequences (Sequence Identity %)	NCBI Accession Numbers
Fagonia mollis	Fm.1	*Paenibacillus barengoltzii* (99%)	NR_042756
	Fm.2	*Bacillus amyloliquefaciens* (98%)	NR_117946
	Fm.3	*Bacillus thuringiensis* (97%)	NR_043403
	Fm.4	*Bacillus cereus* (98%)	NR_115526
	Fm.5	*Bacillus cereus* (99%)	NR_115526
	Fm.6	*Bacillus thuringiensis* (98%)	NR_114581
	Fm.7	*Bacillus amyloliquefaciens* (97%)	NR_117946
	Fm.8	*Bacillus cereus* (97%)	NR_115526
	Fm.9	*Bacillus cereus* (96%)	NR_115526
Achillea fragrantissima	Af.12	*Paenibacillus barengoltzii* (99%)	NR_113988
	Af.13	*Brevibacillus agri* (95%)	NR_113767
	Af.14	*Brevibacillus agri* (93%)	NR_113767
	Af.15	*Brevibacillus agri* (99%)	NR_113767

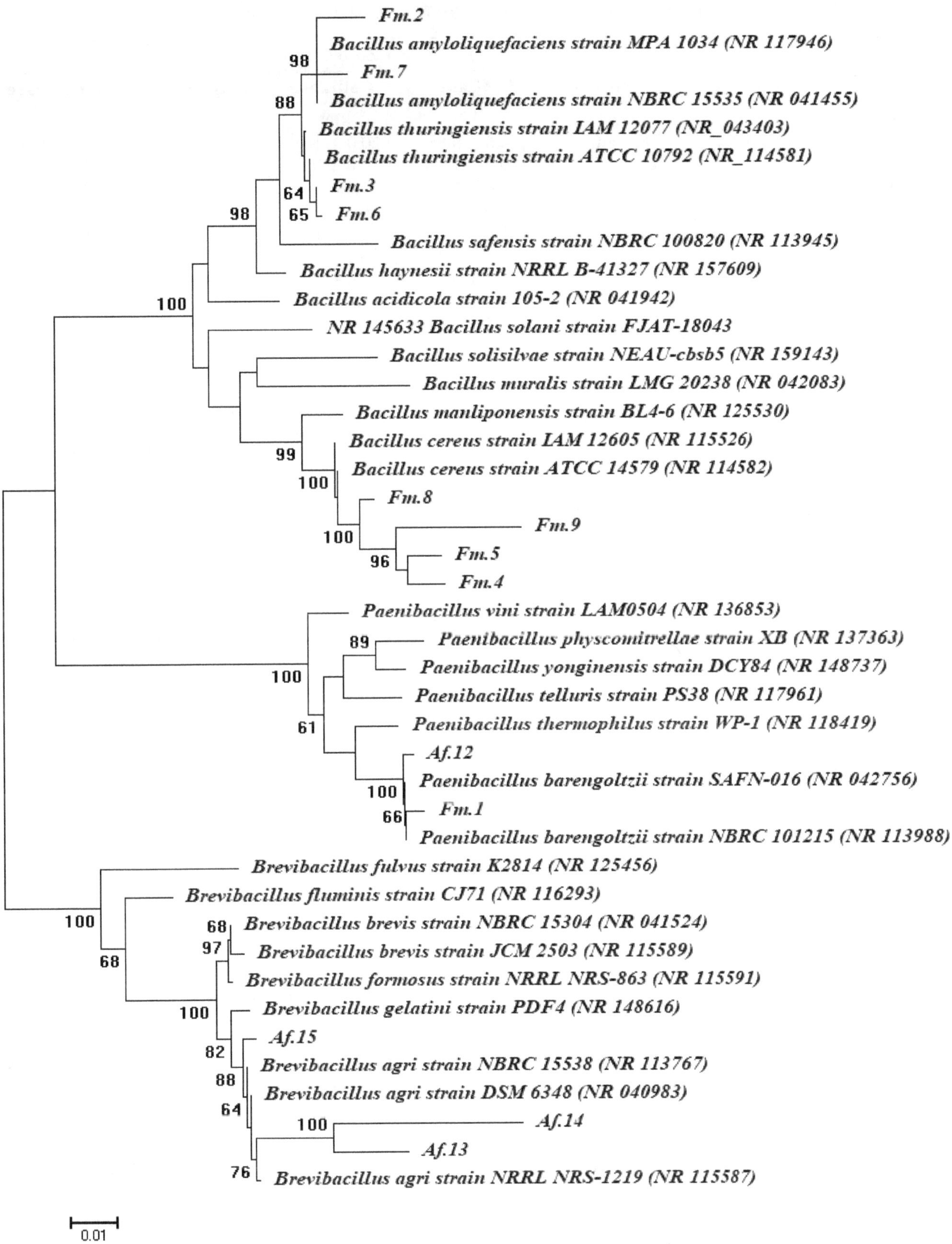

Figure 2. Phylogenetic analysis of 16S rRNA sequences of bacterial strains with reference sequences from NCBI. Fm.1–Fm.9 refers to 16S rRNA sequences of bacteria isolated from *Fagonia mollis* plants, whereas Af.13–Af.15 are the sequences from isolates from *Achillea fragrantissima*. Identity of the bacterial isolates is available in Table 2. The analysis was performed in MEGA 6 using the neighbor-joining method.

Table 3. Extracellular enzymatic activities of bacterial endophytes.

Bacterial Strains [1]	Diameter of Clear Zones (mm) [2]					
	Amylase	Pectinase	CMCase [3]	Cellulase	Xylanase	Gelatinase
C	0 [d]	0 [d]	0 [e]	0 [d]	0 [f]	0 [f]
Fm.1	17.6 ± 1.2 [a]	16.3 ± 0.3 [b]	18.6 ± 0.6 [b]	20.0 ± 1.7 [b]	17.0 ± 1.0 [b]	21.0 ± 0.5 [a]
Fm.2	17.3 ± 0.8 [a]	17.6 ± 0.6 [a]	20.6 ± 0.6 [a]	18.6 ± 0.8 [c]	19.6 ± 0.3 [a]	22.0 ± 0.5 [a]
Fm.3	17.0 ± 1.5 [a]	17.0 ± 0.5 [a]	18.0 ± 1.0 [b]	18.6 ± 1.2 [c]	16.0 ± 00.0 [b]	22.3 ± 1.4 [a]
Fm.4	17.6 ± 0.8 [a]	17.0 ± 0.5 [a]	19.0 ± 0.5 [b]	19.6 ± 1.7 [b]	14.6 ± 0.3 [c]	20.6 ± 0.6 [b]
Fm.5	17.6 ± 0.8 [a]	16.3 ± 0.8 [b]	21.3 ± 1.2 [a]	22.0 ± 1.1 [a]	18.6 ± 0.3 [a]	21.6 ± 0.8 [a]
Fm.6	17.0 ± 1.1 [a]	15.6 ± 0.3 [c]	18.3 ± 0.8 [b]	16.3 ± 0.3 [d]	16.3 ± 0.8 [b]	17.0 ± 1.5 [d]
Fm.7	17.3 ± 0.6 [a]	17.0 ± 0.5 [a]	15.6 ± 0.8 [c]	16.6 ± 1.2 [d]	14.3 ± 0.6 [c]	15.0 ± 1.5 [e]
Fm.8	17.6 ± 0.6 [a]	14.6 ± 0.3 [c]	18.0 ± 1.15 [b]	18.3 ± 0.3 [c]	14.3 ± 0.6 [c]	18.3 ± 1.6 [c]
Fm.9	15.3 ± 1.7 [b]	15.6 ± 0.3 [c]	18.6 ± 0.3 [b]	17.6 ± 0.8 [c]	9.0 ± 1.1 [e]	18.0 ± 1.5 [c]
Af.12	9.3 ± 0.3 [c]	0 [d]	19.3 ± 0.3 [b]	20.6 ± 0.6 [b]	13.6 ± 0.3 [d]	0 [f]
Af.13	0 [d]	0 [d]	0 [e]	0 [d]	0 [f]	19.0 ± 1.5 [c]
Af.14	0 [d]	14.0 ± 0.2 [d]	0 [e]	0 [d]	0 [f]	0 [f]
Af.15	0 [d]	17.6 ± 0.3 [a]	11.0 ± 1.0 [d]	17.33 ± 1.76 [c]	0 [f]	20.3 ± 2.3 [b]

[1] C: controls without bacterial inoculation. Identity of the bacterial isolates is available in Table 2. [2] Different letters between lines denote that mean values are significantly different ($p \leq 0.05$) by Tukey's test, means ± Standard Error (SE) ($n = 3$). [3] Carboxymethyl cellulase.

Table 4. Antimicrobial activities of bacterial endophytes.

Bacterial Strains [1]	Diameter of Clear Zone (mm)					
	P. aeruginosa	*S. typhi*	*E. coli*	*S. aureus*	*B. subtilis*	*C. albicans*
C	–	–	–	–	–	–
Fm.1	17	–	–	–	–	–
Fm.2	15	–	–	–	–	15
Fm.3	15	–	–	–	–	–
Fm.4	15	–	–	–	–	–
Fm.5	17	–	–	–	–	–
Fm.6	20	–	–	–	–	–
Fm.7	22	–	18	–	–	–
Fm.8	30	15	18	–	–	–
Fm.9	25	–	20	–	–	–
Af.12	15	–	–	–	–	–
Af.13	15	11	11	–	17	18
Af.14	20	–	–	–	–	–
Af.15	15	–	–	–	–	–

[1] C: controls without bacterial inoculation. Identity of the bacterial isolates is available in Table 2.

All endophytes identified as *Brevibacillus* sp. produced the highest amount of ammonia compared to *Bacillus* spp. strains (Table 5). Moreover, nine endophytes (Fm.2 to Fm.9 and Af.12) displayed significant ability to solubilize inorganic phosphate with clear zone on the Pikovskaya medium ranging from 7.6 ± 0.3 to 9.6 ± 0.3 mm. Results showed that all the isolated strains were IAA producers, with or without tryptophan (Figure 3). However, increasing tryptophan concentration from 1 to 5 mg·mL^{-1} resulted in increased bacterial ability to produce IAA from 10 to 60 µg·mL^{-1}. Strains of *Brevibacillus* spp. Af.14, Af.13, *Bacillus* sp. Fm.6, *Bacillus* sp. Fm.4, and *Bacillus* sp. Fm.3 produced the highest amount of IAA, and were selected for further analysis to measure the production of IAA at 2 day intervals in a time course over 14 days. The results indicated that the maximum IAA production with tryptophan was 5 mg mL^{-1} after 10 days. The results revealed that *Brevibacillus* sp. Af.14 produced the highest amount of IAA 59.7 µg·mL^{-1} ($p \leq 0.05$; Figure 4).

Table 5. Ammonia production and phosphate solubilization of endophytic bacterial strains.

Bacterial Strains [1]	Ammonia Production [2]	*P solubilization* Diameter of Clear Zone (mm) [3]
C	-	0 [d]
Fm.1	–	0 [d]
Fm.2	+	8.6 ± 0.3 [b]
Fm.3	++	9.0 ± 0 [a]
Fm.4	-	7.6 ± 0.3 [c]
Fm.5	++	9.6 ± 0.3 [a]
Fm.6	+	8.6 ± 0.3 [b]
Fm.7	+	8.3 ± 0.3 [b]
Fm.8	-	9.3 ± 0.3 [a]
Fm.9	++	9.3 ± 0.3 [a]
Af.12	++	9.3 ± 0.3 [a]
Af.13	++	0 [d]
Af.14	++	0 [d]
Af.15	++	0 [d]

[1] C: controls without bacterial inoculation. Identity of the bacterial isolates is available in Table 2. [2] -, +, and ++ denote no, low, and high ammonia production, respectively. [3] Different letters between columns denote that mean values are significantly different ($p \leq 0.05$) by Tukey's test, means ± SE ($n = 3$).

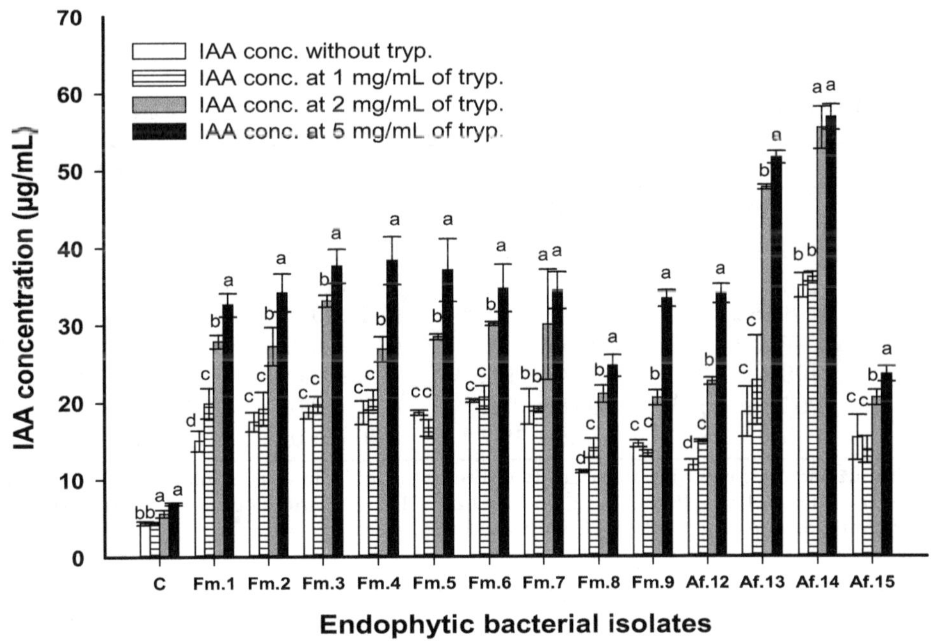

Figure 3. Quantitative production of IAA by endophytic bacterial strains with and without tryptophan. C, controls without bacterial inoculation. Identity of the bacterial isolates are available in Table 2. Data are statistically different at $p \leq 0.05$ by Tukey's test, ($n = 3$); error bars are means ± SE. Bars with the same letter for each endophytic isolate did not differ significantly, different letters on bars denote that mean values are significantly different at significant level of ($P \leq 0.05$), error bars are means ± SE.

In the greenhouse experiment, all maize plants inoculated with bacterial endophytes yielded significantly higher dry shoot weights ($F_{6,28} = 11.09$ and 10.33 respectively; $p \leq 0.001$) compared to the uninoculated control plants (Table 6). Plants inoculated with bacterial endophytes produced dry root weight higher than those recorded in control plants, but the differences were not significant ($F_{6,28} = 1.51$; $p = 0.21$).

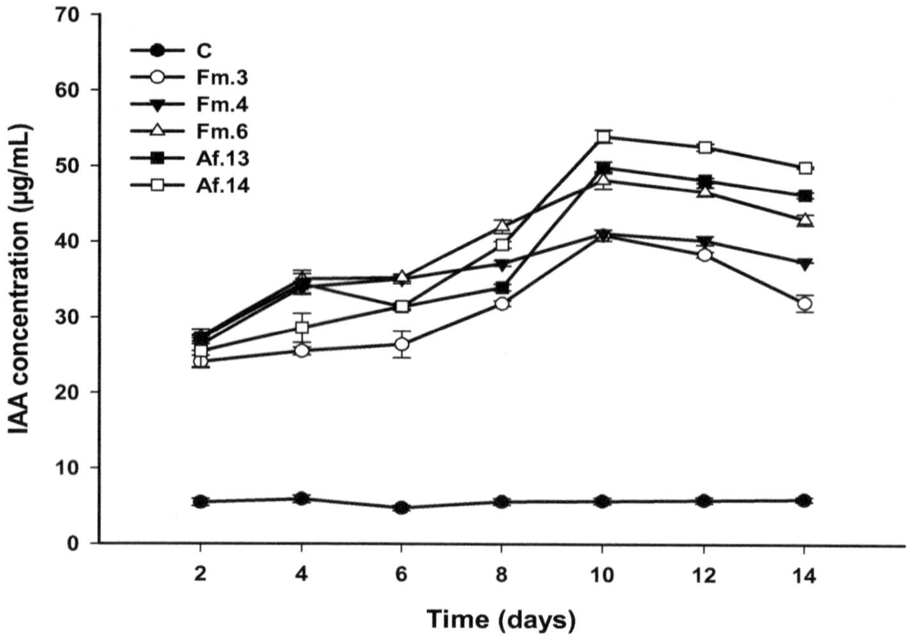

Figure 4. IAA production by the most potent bacterial strains in the presence of 5 mg mL^{-1} tryptophan and over 14 days. C, controls without any bacterial inoculation. Identity of the bacterial isolates is available in Table 2. At each time point, bars with the same letter did not differ significantly at a significant level of ($p \leq 0.05$) by Tukey's test, ($n = 3$).

Table 6. Effect of bacterial inoculations on the growth properties of maize plants.

Bacterial Strains [1]	Dry Weight (mg) [2]		Shoot Nutrients Content (mg)		
	Shoot	Root	P	N	K
C	82 ± 3.56 [c]	252.8 ± 19.5 [a]	0.42 ± 0.01 [c]	2.2 ± 0.17 [c]	8.70 ± 0.03 [b]
Fm.3	110.3 ± 5.5 [a]	322.64 ± 16.0 [a]	1.05 ± 0.07 [a]	3.9 ± 0.19 [b]	11.35 ± 0.92 [a]
Fm.4	81.8 ± 4 [c]	315.8 ± 27.9 [a]	0.72 ± 0.03 [b]	3.1 ± 0.60 [bc]	8.80 ± 0.62 [b]
Fm.6	103.2 ± 1.8 [a]	295.66 ± 30.1 [a]	0.76 ± 0.02 [b]	4.4 ± 0.18 [b]	11.11 ± 0.31 [a]
Af.13	108.2 ± 1.05 [a]	286.86 ± 19.3 [a]	0.44 ± 0.01 [c]	7.1 ± 0.18 [a]	10.57 ± 0.60 [ab]
Af.14	90.78 ± 4.55 [b]	303.8 ± 27.3 [a]	0.40 ± 0.02 [c]	7.2 ± 0.43 [a]	9.97 ± 0.13 [ab]
Mix	95.2 ± 2.8 [b]	249.6 ± 19.7 [a]	0.40 ± 0.01 [c]	3.4 ± 0.27 [bc]	10.45 ± 0.36 [ab]

[1] C: controls without bacterial inoculation. Identity of the bacterial isolates is available in Table 2. Mix, bacterial consortium consists of Fm.3, Fm.4, Fm.6, Af.13, and Af.14. [2] Different letters between columns denote that mean values are significantly different ($p \leq 0.05$) by Tukey's test, means ± SE ($n = 5$).

Inoculation of maize plants with *Bacillus* spp. Fm.3, Fm.4, and Fm.6 significantly ($F6,14 = 49.07$; $p \leq 0.001$) increased P shoot contents (1.05 ± 0.07, 0.72 ± 0.03, and 0.76 ± 0.02 mg respectively) as compared to the un-inoculated control plants (0.42 ± 0.01 mg), while *Brevibacillus* spp. Af.13, Af.14, and the bacterial consortium formed by a mixture of the five isolates did not affect shoot P content compared to control plants (Table 6). Analysis showed that bacterial inoculation significantly increased N shoot contents compared to the control plants ($F6,14 = 35.76$, $p \leq 0.001$; Table 6). Plants inoculated with *Bacillus* spp. Fm.3, and Fm.6, *Brevibacillus* spp. Af.13, and Af.14 had significantly higher ($p \leq 0.05$) N contents (range of 3.9 ± 0.19 to 7.2 ± 0.43 mg) than those treated with other strains or uninoculated. The bacterial strains *Bacillus* spp. Fm.3 and Fm.6 significantly ($F6,14 = 4.15$; $p = 0.013$) increased K shoot contents (11.35 ± 0.92 and 11.11 ± 0.31 mg) in comparison with uninoculated control plants (8.70 ± 0.03 mg).

4. Discussion

In this study, 13 putative bacterial endophytic strains were isolated from two medicinal plants growing under the adverse conditions of the Sinai desert. Nine bacterial endophytes were isolated from *F. mollis* and identified as different species of *Bacillus*, and *Paenibacillus*, and four bacterial endophytes were isolated from *A. fragrantissima* and identified as *Paenibacillus* spp. and *Brevibacillus* spp. (Table 2). The plant growth promoting (PGP) activities of these bacterial strains were characterized, including extracellular enzyme production, antimicrobial action, IAA and ammonia production, and P-solubilization. In the same regard, Eida et al. [45] reported isolation of endosphere and rhizosphere bacterial groups associated with four native Saudi desert plants and proved their plant growth promotion potential including phosphate solubilization and IAA production. Based on PGP characteristics, five endophytic bacterial strains were selected to evaluate their effects on plant growth and development. Results showed that the selected endophytes have key PGP properties, and significantly increased dry weight of tissues, and P concentrations in shoots of maize plants compared with uninoculated controls. Corresponding with our results, Marag and Suman [42] isolated six bacterial endophytes including *Bacillus cereus* from two cultivars of maize, and the pot experiment indicates the efficacy of the isolates in improving biomass parameters of inoculated maize plants, in addition to compensating for approximately 25% of the NPK fertilizer input.

The bacterial endophytes exhibited different enzymatic activities involving cellulase, pectinase, xylanase, amylase, and gelatinase production [46,47]. Cellulolytic and pectinolytic activities are known to enable microorganisms to penetrate plant tissues and establish a symbiotic relationship with their host plants. The *Bacillus* spp. strain isolated in this study showed high hydrolytic activity for cellulose and pectin, as well as proteolytic activity. Similarly, different endophytic strains of *Bacillus* were shown to be strong producers of cellulase and pectinase [48]. The extracellular hydrolytic enzymes produced by endophytes contribute indirectly to plant growth promotion and protection against pathogens [49,50]. The endophytes can be described as bioproducers for amylases and xylanases based on their amylolytic and xylanolytic activities. Similarly, bacterial endophytes isolated from mangrove plants had activities associated with amylases [50]. The diverse enzymatic activities of the isolated endophytes showed their capability to catalyze different biochemical reactions and their potential for agricultural and industrial applications. Likewise, Castro et al (2014) isolated endophytic *Bacillus* from two Brazilian mangrove species, the isolates displayed extracellular amylase, esterase, lipase, protease, and endoglucanase activities and thus can be used in industrial applications [50]. Moreover, theses enzymes could enable endophytes to penetrate plant tissues and build a symbiotic relationship with their host plant, besides protecting the host from pathogens by hydrolysis of the pathogen cell wall [10].

Antimicrobial activities of the isolated endophytes were evaluated based on the suppression of microbial growth caused by the crude extracts. The estimation of antimicrobial activity of crude extracts is the initial step required for the discovery of new antimicrobial compounds. Selection of bacterial isolates as inoculants based on their PGP traits, and on their inhibitory effect against different pathogens, has received attention and has been suggested as an approach to enhance plant growth and protect plants against diseases [51]. In the current study, the isolated endophytes showed a significant antagonistic effect against different pathogenic microorganisms. Endophytic bacteria can indirectly assist plant growth through the production of substances, which inhibit plant pathogens [52,53]. Endophytes isolated from other medicinal plants have also produced novel bioactive compounds [49,54]. Hassan [10] isolated six bacterial endophytes including *Bacillus cereus* and *Bacillus subtilis* from the native desert medicinal plant *Teucrium polium* L., the isolates manifested variable broad-spectrum activity against *Pseudomonas aeruginosa, Salmonella typhimurium, Escherichia coli, Staphylococcus aureus, Bacillus subtilis*, and *Candida albicans*. Accordingly, suggesting their application as biocontrol agents [10].

Sun et al. [55] showed that 10 endophytic bacterial strains of *Bacillus* and *Streptomyces* isolated from *Polygonum cuspidatum* exhibited antagonistic effects against different plant pathogens. *Bacillus licheniformis* and *Bacillus pumilus* endophytic isolates from *Platycodon grandiflorum* roots also exhibited a significant antifungal action against *Phytophthora capsici, Fusarium oxysporum, Rhizoctonia solanic*, and *Pythium ultimum*. The endophytic bacterial strain *Paenibacillus* sp. IIRAC-30 was isolated from cassava and suppressed the growth of *Rhizoctonia solani* [56]. *Bacillus amyloliquefaciens* was isolated from the Chinese medicinal plant *Scutellaria baicalensis* Georgi. A crude extract of this strain exhibited antagonistic effects against some plant pathogens, food-borne pathogenic and spoilage microorganisms [53].

The PGP properties of bacteria have been investigated to select bacteria with high potential to be used as biofertilizers. These tests are critical in light of the fact that they identify bacteria with higher benefits for plants before testing them in field traits [57]. Ammonia and IAA production, as well P-solubilization, are among various mechanisms exhibited by bacteria that enhance plant growth [58]. Here, most endophytic bacterial isolates were able to produce different amounts of ammonia. It is often found that ammonia-producing bacteria can supply ammonia as a nitrogen source for plant growth [59]. Bacterial endophytes can enhance plant growth through the production of ammonia through the hydrolysis of urea into ammonia and carbon dioxide [60]. With regard to P-solubilization, most of the isolated endophytes showed variable capacity to solubilize phosphate. Rodrigues et al. [61] found that about 47% of bacterial endophytes isolated from sugarcane have low P-solubilizing indices. At soil with low phosphate supply, inoculation of P-solubilizing endophytic bacteria leading to increase of pant growth performance.

Indole-3-acetic acid (IAA) is a phytohormone that can be produced by plants and various microorganisms. This hormone not only enhances plant growth but also contributes in the interaction between plants and microorganisms [62]. In this study, all endophytic bacterial strains had the ability to produce IAA in the absence and presence of tryptophan, the precursor for IAA production. Although most microorganisms utilize tryptophan in IAA synthesis [63,64], the advantage of bacterial endophytes is that they can produce IAA without tryptophan supplementation. Rodrigues et al. [61] showed that 57% of bacterial endophytes secreted high IAA concentration of 21.05–139.21 $\mu g\ mL^{-1}$ in 72 h in the presence of 5 mM tryptophan. Endophytic bacterial strains were shown to produce higher IAA concentrations than rhizospheric strains, suggesting a closer link, and potential symbiosis, between endophytes and their hosts [12]. Thus, in the current study, a higher capacity to produce IAA was used to select five bacterial strains to determine their effect on maize growth performance. Bokhari et al. [65] isolated *Bacillus circulans* PK3-138 from plants grown in Pakistan desert, reported the potency of this isolate for IAA production. Similarly, four bacterial endophytes (*Sphingomonas* sp., *Bacillus* sp., *Pantoea* sp., and *Enterobacterc* sp.) isolated from the roots of elephant grass showed valuable PGP traits including IAA production at a range of 10.50–759.19 mg/L, and ammonia production capacity. So, these inoculants could be used for increasing crop yield in a sustainable mode [58].

We found that inoculated plants produced more biomass than uninoculated plants. Plant–microbe interactions are well known to influence nutrient transfer between microorganisms and plants [66]. Therefore, it is possible that plant biomass production varied with different microbial taxa assemblages in the roots due to their various abilities to supply nutrients to their host. The results showed that the shoot P concentration was significantly increased in plants inoculated with Fm.3, Fm.4, and Fm.6 compared to the uninoculated plants. P-solubilizing bacteria help plants to access insoluble forms of phosphate, such as apatite, through excretion of protons and organic acids, mainly gluconic acid, rendering phosphate available to plants for uptake [11,67]. These bacteria can also produce enzymes that mineralize organic phosphorus, which also render it available for plants [67]. The capacity of microorganisms to absorb immobile nutrients such as P from soils and transfer it to their host plants is one of the main effects of microbial symbiosis; however, microbial capacity for nutrient transfer varies with different microorganisms [68]. Basically, plant roots can be colonized simultaneously by multiple microorganisms, which can positively or negatively benefit the host plant [69,70].

Importantly, not only did the isolated endophytic *Bacillus* and *Brevibacillus* species display the highest level of IAA and ammonia production, but they also had various plant growth-promoting traits. Bacteria that were isolated and characterized in the present study are potential candidates for plant bioinoculation in agricultural practices, in particular those that inhibited pathogens and harbored the highest levels of IAA production and of nutrient uptake.

5. Conclusions

Very few isolates were obtained herein to claim that, bacterial endophytes inhabiting the two studied medicinal plants, *F mollis* and *A. fragrantissima,* mainly belong to *Bacillus* and *Brevibacillus* spp. Bacterial endophytes characterization including extracellular enzymatic activity, antimicrobial actions, P-solubilization activity, ammonia, and IAA production were performed in terms of their plant growth-promoting abilities in-vitro and in plants. Five bacterial strains identified as *Brevibacillus agri* Af.13, and Af.14, *Bacillus* sp. Fm3, *Bacillus* sp. Fm.4, and *Bacillus* sp. Fm.6 were selected and inoculated into maize plant to increase their growth performance under normal conditions. These endophytic bacterial isolates significantly promote plant growth and increase P and N shoot contents of maize plant. However, in order to demonstrate the beneficial role of these bacterial endophytes in plant growth promotion of their host plants, particularly under real field conditions, further investigation of their mechanisms of colonization and competition against other soil microbial communities will be required.

Author Contributions: Conceptualization, A.F., A.M.E., E.E.-D.E., S.E.-D.H., K.A.A.A., Y.M.H., K.A.A. and M.D.F.A.; methodology, A.F., A.M.E., E.E.-D.E., S.E.-D.H., K.A.A.A., Y.M.H., K.A.A. and M.D.F.A.; software, A.F., A.M.E., F.A.-O., M.H., M.S.-A., E.E.-D.E., S.E.-D.H., K.A.A.A., Y.M.H., K.A.A. and M.D.F.A.; formal analysis, A.F., A.M.E., E.E.-D.E., S.E.-D.H., F.A.-O., M.H., M.S.-A., K.A.A.A., Y.M.H., K.A.A., A.M.E. and N.K.; investigation, A.F., A.M.E., E.E.-D.E., S.E.-D.H., F.A.-O., M.H., M.S.-A., K.A.A.A., and N.K.; resources, A.F., A.M.E., E.E.-D.E., S.E.-D.H., K.A.A.A., Y.M.H.; data curation, A.F., A.M.E., E.E.-D.E., S.E.-D.H., F.A.-O., M.H., M.S.-A., K.A.A.A., Y.M.H. and K.A.A.; writing—original draft preparation, A.F., A.M.E., E.E.-D.E., F.A.-O., M.H., M.S.-A., S.E.-D.H and K.A.A.A.; writing—review and editing, A.F., A.M.E., E.E.-D.E., S.E.-D.H., K.A.A.A., Y.M.H., K.A.A., A.M.E., M.A. and N.K.; visualization, A.F., A.M.E., E.E.-D.E., S.E.-D.H., K.A.A.A., Y.M.H. and K.A.A.; funding acquisition, M.D.F.A., K.A.A., Y.M. and K.A.A.A. All authors have equal contribution. All authors have read and agreed to the published version of the manuscript.

Acknowledgments: The authors extend their appreciation to the Deanship of Scientific Research at Princess Nourah bint Abdulrahman University for funding this research through the Fast-track Research Funding Program.

References

1. Sun, R.; Zhang, X.; Guo, X.; Wang, D.; Chu, H. Bacterial diversity in soils subjected to long-term chemical fertilization can be more stably maintained with the addition of livestock manure than wheat straw. *Soil Biol. Biochem.* **2015**, *88*, 9–18. [CrossRef]
2. Zhou, J.; Jiang, X.; Zhou, B.; Zhao, B.; Ma, M.; Guan, D.; Li, J.; Chen, S.; Cao, F.; Shen, D.; et al. Thirty four years of nitrogen fertilization decreases fungal diversity and alters fungal community composition in black soil in northeast China. *Soil Biol. Biochem.* **2016**, *95*, 135–143. [CrossRef]
3. Rafi, M.M.; Krishnaveni, M.S.; Charyulu, P.B.B.N. Phosphate-Solubilizing Microorganisms and Their Emerging Role in Sustainable Agriculture. In *Recent Developments in Applied Microbiology and Biochemistry*; Buddolla, V., Ed.; Academic Press: Cambridge, MA, USA; Elsevier: Amsterdam, The Netherlands, 2019; Volume 17, pp. 223–233. [CrossRef]
4. De La, T.; Neyser, V.R.; Clara, I.R.; Martha, R.; Carlos, A.; Federico, A.G.; Héctor, P.; Reiner, R. Effect of plant growth-promoting bacteria on the growth and fructan production of *Agave americana* L. *Brazilian J. Microbiol.* **2016**, *47*, 587–596. [CrossRef] [PubMed]
5. Gamez, R.; Cardinale, M.; Montes, M.; Ramirez, S.; Schnell, S.; Rodriguez, F. Screening, plant growth promotion and root colonization pattern of two rhizobacteria (*Pseudomonas fluorescens* Ps006 and *Bacillus amyloliquefaciens* Bs006) on banana cv. Williams (*Musa acuminata* Colla). *Microbiol. Res.* **2019**, *220*, 12–20. [CrossRef] [PubMed]
6. Abdelaal, K.A.A.; Tawfik, S.F. Response of Sugar Beet Plant (*Beta vulgaris* L.) to Mineral Nitrogen Fertilization and Bio-Fertilizers. *Int. J. Curr. Microbiol. Appl. Sci.* **2015**, *4*, 677–688.
7. Abdelaal, K.A.A. Pivotal Role of Bio and Mineral Fertilizer Combinations on Morphological, Anatomical and Yield Characters of Sugar Beet Plant (*Beta vulgaris* L.). *Middle East J. Agric. Res.* **2015**, *4*, 717–734.

8. Abdelaal, K.A.A.; Badawy, S.A.; Abdel Aziz, R.M.; Neana, S.M.M. Effect of mineral nitrogen levels and PGPR on morphophysiological characters of three sweet sorghum varieties (*Sorghum bicolor* L. Moench). *J. Plant Prod.* **2015**, *6*, 189–203. [CrossRef]
9. Hassan, S.E.; Salem, S.S.; Fouda, A.; Awad, M.A.; El-Gamal, M.S.; Abdo, A.M. New approach for antimicrobial activity and bio-control of various pathogens by biosynthesized copper nanoparticles using endophytic actinomycetes. *J. Radiation Res. Appl. Sci.* **2018**, *11*, 262–270. [CrossRef]
10. Hassan, S.E. Plant growth-promoting activities for bacterial and fungal endophytes isolated from medicinal plant of *Teucrium polium* L. *J. Adv. Res.* **2017**, *8*, 687–695. [CrossRef]
11. Naseem, H.; Ahsan, M.; Shahid, M.A.; Khan, N. Exopolysaccharides producing rhizobacteria and their role in plant growth and drought tolerance. *J. Basic Microbiol.* **2018**, *58*, 1009–1022. [CrossRef]
12. Taktek, S.; St-Arnaud, M.; Yves Piché, J.; Fortin, A.; Antoun, H. Igneous phosphate rock solubilization by biofilm-forming mycorrhizobacteria and hyphobacteria associated with *Rhizoglomus irregulare* DAOM 197198. *Mycorrhiza* **2017**, *27*, 13–22. [CrossRef] [PubMed]
13. Abbamondi, G.R.; Tommonaro, G.; Weyens, N.; Thijs, S.; Sillen, W.; Gkorezis, P.; Iodice, C.; Rangel, M.; Nicolaus, B.; Vangronsveld, J. Plant growth-promoting effects of rhizospheric and endophytic bacteria associated with different tomato cultivars and new tomato hybrids. *Chem. Biol. Technol. Agric.* **2016**, *3*, 1. [CrossRef]
14. Hafez, Y.M.; Attia, K.A.; Kamel, S.; Alamery, S.; El-Gendy, S.; Al-Dosse, A.; Mehiar, F.; Ghazy, A.; Abdelaal, K.A.A. *Bacillus subtilis* as a bio-agent combined with nano molecules can control powdery mildew disease through histochemical and physiobiochemical changes in cucumber plants. *Physiol. Mol. Plant Path.* **2020**, *111*, 101489. [CrossRef]
15. Hafez, Y.; Emeran, A.; Esmail, S.; Mazrou, Y.; Abdrabbo, D.; Abdelaal, K.A.A. Alternative treatments improve physiological characters, yield and tolerance of wheat plants under leaf rust disease stress. *Fresenius Environ. Bull.* **2020**, *29*, 4738–4748.
16. St-Arnaud, M.; Vujanovic, V. Effect of the arbuscular mycorrhizal symbiosis on plant diseases and pests. In *Mycorrhizae in Crop Production*; Hamel, C., Plenchette, C., Eds.; Haworth Press: Binghampton, NY, USA, 2007; pp. 67–122.
17. Hafez, Y.M.; Abdelaal, K.A.A.; Badr, M.M.; Esmaeil, R.A. Control of Puccinia triticina the causal agent of wheat leaf rust disease using safety resistance inducers correlated with endogenously antioxidant enzymes up-regulation. *Egyptian J. Biol. Pest Control* **2017**, *27*, 1–10.
18. Fouda, A.; Abdel-Maksoud, G.; Abdel-Rahman, M.A.; Eid, A.M.; Barghoth, M.G.; El-Sadany, M.A. Monitoring the effect of biosynthesized nanoparticles against biodeterioration of cellulose-based materials by Aspergillus niger. *Cellulose* **2019**, *26*, 6583–6597. [CrossRef]
19. Fouda, A.; Abdel-Maksoud, G.; Abdel-Rahman, M.A.; Salem, S.S.; Hassan, S.E.; El-Sadany, M.A. Eco-friendly approach utilizing green synthesized nanoparticles for paper conservation against microbes involved in biodeterioration of archaeological manuscript. *Int. Biodet. Biodegr.* **2019**, *142*, 160–169. [CrossRef]
20. Fouda, A.; Hassan, S.E.; Salem, S.S.; Shaheen, T.I. In-Vitro cytotoxicity, antibacterial, and UV protection properties of the biosynthesized Zinc oxide nanoparticles for medical textile applications. *Microb. Pathog.* **2018**, *125*, 252–261. [CrossRef]
21. Cherif, H.; Marasco, R.; Rolli, E.; Ferjani, R.; Fusi, M.; Soussi, A.; Mapelli, F.; Blilou, I.; Borin, S.; Boudabous, A.; et al. Oasis desert farming selects environment-specific date palm root endophytic communities and cultivable bacteria that promote resistance to drought. *Environ. Microb. Rep.* **2015**, *7*, 668–678. [CrossRef]
22. Khan, N.; Bano, A.; Rahman, M.A.; Rathinasabapathi, B.; Babar, M.A. UPLC-HRMS-based untargeted metabolic profiling reveals changes in chickpea (*Cicer arietinum*) metabolome following long-term drought stress. *Plant Cell Environ.* **2019**, *42*, 115–132. [CrossRef]
23. Köberl, M.; Müller, H.; Ramadan, E.M.; Berg, G. Desert Farming Benefits from Microbial Potential in Arid Soils and Promotes Diversity and Plant Health. *PLoS ONE* **2011**, *6*, e24452. [CrossRef] [PubMed]
24. Khan, N.; Bano, A.; Babar, M.A. The stimulatory effects of plant growth promoting rhizobacteria and plant growth regulators on wheat physiology grown in sandy soil. *Arch. Microbiol.* **2019**, *201*, 769–785. [CrossRef] [PubMed]
25. Hanna, A.L.; Youssef, H.H.; Amer, W.M.; Monib, M.; Fayez, M.; Hegazi, N.A. Diversity of bacteria nesting the plant cover of north Sinai deserts, Egypt. *J. Adv. Res.* **2013**, *4*, 13–26. [CrossRef] [PubMed]

26. Selim, K.A.; El-Beih, A.A.; AbdEl-Rahman, T.M.; El-Diwany, A.I. Biodiversity and antimicrobial activity of endophytes associated with Egyptian medicinal plants. *Mycosphere* **2011**, *2*, 669–678. [CrossRef]
27. ALKahtani, M.D.F.; Attia, K.A.; Hafez, Y.M.; Khan, N.; Eid, A.M.; Ali, M.A.M.; Abdelaal, K.A.A. Chlorophyll Fluorescence Parameters and Antioxidant Defense System Can Display Salt Tolerance of Salt Acclimated Sweet Pepper Plants Treated with Chitosan and Plant Growth Promoting Rhizobacteria. *Agronomy* **2020**, *10*, 1180. [CrossRef]
28. Kiarie, S.; Nyasani, J.O.; Gohole, L.S.; Maniania, N.K.; Subramanian, S. Impact of Fungal Endophyte Colonization of Maize (*Zea mays* L.) on Induced Resistance to Thrips and Aphid-Transmitted Viruses. *Plants* **2020**, *9*, 416. [CrossRef]
29. Abd el Fatah, H.Y.; Mohamed, E.A.; Hassan, M.B.; Mohamed, K.A. An Economic Analysis for Maize Market in Egypt. *Middle East J. Agric. Res.* **2015**, *4*, 873–878.
30. Berg, G.; Grube, M.; Schloter, M.; Smalla, K. Unraveling the plant microbiome: Looking back and future perspectives. *Front. Microbiol.* **2014**, *5*, 148. [CrossRef]
31. Arora, S.; Patel, P.N.; Vanza, M.J.; Rao, G.G. Isolation and characterization of endophytic bacteria colonizing halophyte and other salt tolerance plant species from coastal Gujarta. *Afr. J. Microb. Res.* **2014**, *8*, 1779–1788.
32. Miller, D.N.; Bryant, J.E.; Madsen, E.L.; Ghiorse, W.C. Evaluation and optimization of DNA extraction and purification procedures for soil and sediment samples. *Appl. Environ. Microbiol.* **1999**, *65*. [CrossRef]
33. Lane, D.J. 16S/23S rRNA sequencing. In *Nucleic Acids Techniques in Bacterial Systematics*; Stackebrandt, E., Goodfellow, M., Eds.; John Wiley & Sons: Chichester, UK, 1991.
34. Lv, Y.; Zhang, F.; Chen, J.; Cui, J.; Xing, Y.; Li, X.; Guo, S. Diversity and antimicrobial activity of endophytic fungi associated with the alpine plant *Saussurea involucrata*. *Biol. Pharm. Bull.* **2010**, *33*, 1300–1306. [CrossRef]
35. Yadav, R.; Chauhan, N.; Chouhan, A.S.; Soni, V.K.; Omray, L. Antimicrobial screening of various extracts of *Aphanmixis polystachya* stems bark. *Int. J. Adv. Pharm. Sci.* **2010**, *1*, 147–150.
36. Mohamed, A.A.; Fouda, A.; Abdel-Rahman, M.A.; Hassan, S.E.; El-Gamal, M.S.; Salem, S.S.; Shaheen, T.I. Fungal strain impacts the shape, bioactivity and multifunctional properties of green synthesized zinc oxide nanoparticles. *Biocat. Agric. Biotech.* **2019**, *19*, 101103. [CrossRef]
37. Sharaf, O.M.; Al-Gamal, M.S.; Ibrahim, G.A.; Dabiza, N.M.; Salem, S.S.; El-ssayad, M.F.; Youssef, A.M. Evaluation and characterization of some protective culture metabolites in free and nano-chitosan-loaded forms against common contaminants of Egyptian cheese. *Carbohyd. Polym.* **2019**, *223*, 115094. [CrossRef] [PubMed]
38. Alsharif, S.M.; Salem, S.S.; Abdel-Rahman, M.A.; Fouda, A.; Eid, A.M.; Hassan, S.E.; Awad, M.A.; Mohamed, A.A. Multifunctional properties of spherical silver nanoparticles fabricated by different microbial taxa. *Heliyon* **2020**, *6*, 3943. [CrossRef]
39. Fouda, A.; Hassan, S.E.-D.; Abdo, A.M.; El-Gamal, M.S. Antimicrobial, Antioxidant and Larvicidal Activities of Spherical Silver Nanoparticles Synthesized by Endophytic *Streptomyces* spp. *Biol. Trace Element Res.* **2020**, *195*, 707–724. [CrossRef]
40. Jasim, B.; John, C.; Jyothis, M.; Radhakrishnan, E.K. Plant growth promoting potential of endophytic bacteria isolated from *Piper nigrum*. *Plant Growth Regul* **2013**, *71*, 1–11. [CrossRef]
41. Singh, P.; Kumar, V.; Agrawal, S. Evaluation of phytase producing bacteria for their plant growth promoting activities. *Intern. J. Microbiol.* **2014**, 426483. [CrossRef]
42. Marag, P.S.; Suman, A. Growth stage and tissue specific colonization of endophytic bacteria having plant growth promoting traits in hybrid and composite maize (*Zea mays* L.). *Microbiol. Rese* **2018**, *214*, 101–113. [CrossRef]
43. AOAC. International Official Methods of Analysis of AOAC International. In *Official Methods of Analysis of AOAC International*; George, W., Ed.; Association of Official Analytical Chemists: Rockville, MD, USA, 2016; p. 3172.
44. Rice, E.W.; Baird, R.B.; Eaton, A.D. *Standard Methods for the Examination of Water and Wastewater*, 23rd ed.; American Public Health Association, American Water Works Association: Washington, DC, USA, 2017.
45. Eida, A.A.; Ziegler, M.; Lafi, F.F.; Michell, C.T.; Voolstra, C.R.; Hirt, H.; Saad, M.M. Desert plant bacteria reveal host influence and beneficial plant growth properties. *PLoS ONE* **2018**, *13*, e0208223. [CrossRef] [PubMed]
46. Eid, A.M.; Salim, S.S.; Hassan, S.S.; Ismail, M.A.; Fouda, A. Role of Endophytes in Plant Health and Abiotic Stress Management. In *Microbiome in Plant Health and Disease: Challenges and Opportunities*; Kumar, V., Prasad, R., Kumar, M., Choudhary, D.K., Eds.; Springer: Singapore, 2019; pp. 119–144.

47. Fouda, A.; Hassan, S.E.; Eid, A.M.; Ewais, E.E. The Interaction Between Plants and Bacterial Endophytes Under Salinity Stress. In *Endophytes and Secondary Metabolites*; Jha, S., Ed.; Springer Nature: Cham, Switzerland, 2019; pp. 591–607.
48. Choi, Y.W.; Hodgkiss, I.; Hyde, K. Enzyme production by endophytes of *Brucea javanica*. *J. Agric. Tech.* **2005**, *1*, 55–66.
49. Glick, B.R. Plant growth-promoting bacteria: Mechanisms and applications. *Scientifica* **2012**, 1–15. [CrossRef] [PubMed]
50. Castro, R.A.; Quecine, M.C.; Lacava, P.T.; Batista, B.D.; Luvizotto, D.M.; Marcon, J.; Ferreira, A.; Melo, I.S.; Azevedo, J.L. Isolation and enzyme bioprospection of endophytic bacteria associated with plants of Brazilian mangrove ecosystem. *SpringerPlus* **2014**, *3*, 382. [CrossRef] [PubMed]
51. Vurukonda, S.S.P.; Giovanardi, D.; Stefani, E. Plant growth promoting and biocontrol activity of *Streptomyces* spp. as endophytes. *Intern. J. Mol. Sci.* **2018**, *19*, 952. [CrossRef] [PubMed]
52. Chow, Y.Y.; Rahman, S.; Ting, A.S.Y. Understanding colonization and proliferation potential of endophytes and pathogen in planta via plating, polymerase chain reaction, and ergosterol assay. *J. Adv. Res.* **2017**, *8*, 13–21. [CrossRef]
53. Khan, N.; Zandi, P.; Ali, S.; Mehmood, A.; Adnan Shahid, M.; Yang, J. Impact of salicylic acid and PGPR on the drought tolerance and phytoremediation potential of Helianthus annus. *Front. Microbiol.* **2018**, *9*, 2507. [CrossRef]
54. Asraful, I.; Md, S.; Math, R.K.; Kim, J.M.; Yun, M.G.; Cho, J.J.; Kim, E.J.; Lee, Y.H.; Yun, H.D. Effect of plant age on endophytic bacterial diversity of balloon flower (*Platycodon grandiflorum*) root and their antimicrobial activities. *Curr. Microbiol.* **2010**, *61*, 346–356. [CrossRef]
55. Sun, L.; Lu, Z.; Bie, X.; Lu, F.; Yang, S. Isolation and characterization of a co-producer of fengycins and surfactins, endophytic *Bacillus amyloliquefaciens* ES-2, from *Scutellaria baicalensis* Georgi. *World J. Microbiol. Biotechnol.* **2006**, *22*, 1259–1266. [CrossRef]
56. Canova, S.P.; Petta, T.; Reyes, L.F.; Zucchi, T.; Moraes, L.A.; Melo, I. Characterization of lipopeptides from *Paenibacillus* sp. (IIRAC30) suppressing *Rhizoctonia solani*. *World J. Microbiol. Biotechnol.* **2010**, *26*, 2241–2247. [CrossRef]
57. Biasolo, G.; Kucmanski, D.A.; Salamoni, S.P.; Gardin, J.P.; Minotto, E.; Baratto, C.M. Isolation, characterization and selection of bacteria that promote plant growth in grapevines (*Vitis* sp.). *J. Agric. Sci.* **2017**, *9*, 184–194. [CrossRef]
58. Li, X.; Geng, X.; Xie, R.; Fu, L.; Jiang, J.; Gao, L.; Sun, J. The endophytic bacteria isolated from elephant grass (*Pennisetum purpureum* Schumach) promote plant growth and enhance salt tolerance of hybrid Pennisetum. *Biotech. Biofuels* **2016**, *9*, 190. [CrossRef] [PubMed]
59. Banik, A.; Dash, G.K.; Swain, P.; Kumar, U.; Mukhopadhyay, S.K.; Dangar, T.K. Application of rice (*Oryza sativa* L.) root endophytic diazotrophic *Azotobacter* sp. strain Avi2 (MCC 3432) can increase rice yield under green house and field condition. *Microbiol. Res.* **2019**, *219*, 56–65. [CrossRef] [PubMed]
60. Khan, N.; Bano, A.; Curá, J.A. Role of Beneficial Microorganisms and Salicylic Acid in Improving Rainfed Agriculture and Future Food Safety. *Microorganisms* **2020**, *8*, 1018. [CrossRef] [PubMed]
61. Rodrigues, A.A.; Forzani, M.V.; Soares, R.; Sibov, S.T.; Vieira, J. Isolation and selection of plant growth-promoting bacteria associated with sugarcane. *Agric. Res. Tropics* **2016**, *46*, 149–158. [CrossRef]
62. Lin, L.; Zhengyi, L.; Chunjin, H.; Zhang, X.; Chang, S.; Yang, L.; Yangrui, L.; Qianli, A. Plant growth-promoting nitrogen-fixing *Enterobacteria* are in association with sugarcane plants growing in Guangxi, China. *Microbes Environ.* **2012**, *27*, 391–398. [CrossRef]
63. Fouda, A.; Hassan, S.E.; Eid, A.M.; Ewais, E.E. Biotechnological applications of fungal endophytes associated with medicinal plant *Asclepias sinaica* (Bioss). *Ann. Agric. Sci.* **2015**, *60*, 95–104. [CrossRef]
64. Hassan, S.E.; Fouda, A.; Radwan, A.A.; Salem, S.S.; Barghoth, M.G.; Awad, M.A.; Abdo, A.M.; El-Gamal, M.S. Endophytic actinomycetes Streptomyces spp mediated biosynthesis of copper oxide nanoparticles as a promising tool for biotechnological applications. *JBIC* **2019**, *24*, 377–393. [CrossRef]
65. Bokhari, A.; Essack, M.; Lafi, F.F.; Andres-Barrao, C.; Jalal, R.; Alamoudi, S.; Saad, M.M. Bioprospecting desert plant Bacillus endophytic strains for their potential to enhance plant stress tolerance. *Sci. Rep.* **2019**, *9*. [CrossRef]
66. Orozco, M.; Ma del, C.; Ma del, C.R.; Bernard, R.G.; Gustavo, S. Microbiome engineering to improve biocontrol and plant growth-promoting mechanisms. *Microbiol. Res.* **2018**, *208*, 25–31. [CrossRef]

67. Khan, N.; Bano, A.; Babar, M.A. The root growth of wheat plants, the water conservation and fertility status of sandy soils influenced by plant growth promoting rhizobacteria. *Symbiosis* **2017**, *72*, 195–205. [CrossRef]
68. Gyaneshwar, P.; Naresh, K.G.; Parekh, L.J.; Poole, P.S. Role of soil microorganisms in improving P nutrition of plants. *Plant Soil* **2002**, *245*, 83–93. [CrossRef]
69. Hijri, M. Analysis of a large dataset of mycorrhiza inoculation field trials on potato shows highly significant increases in yield. *Mycorrhiza* **2016**, *26*, 209–214. [CrossRef] [PubMed]
70. Read, A.; Louise, F.; Taylor, H. The Ecology of Genetically Diverse Infections. *Science* **2001**, *292*, 1099. [CrossRef] [PubMed]

11

Flavonoids in Agriculture: Chemistry and Roles in Biotic and Abiotic Stress Responses and Microbial Associations

Ateeq Shah and Donald L. Smith *

Department of Plant Science, Macdonald Campus, McGill University, Sainte-Anne-de-Bellevue, QC H9X 3V9, Canada; Ateeq.Shah@mail.mcgill.ca
* Correspondence: donald.smith@mcgill.ca

Abstract: The current world of climate change, global warming and a constantly changing environment have made life very stressful for living entities, which has driven the evolution of biochemical processes to cope with stressed environmental and ecological conditions. As climate change conditions continue to develop, we anticipate more frequent occurrences of abiotic stresses such as drought, high temperature and salinity. Living plants, which are sessile beings, are more exposed to environmental extremes. However, plants are equipped with biosynthetic machinery operating to supply thousands of bio-compounds required for maintaining internal homeostasis. In addition to chemical coordination within a plant, these compounds have the potential to assist plants in tolerating, resisting and escaping biotic and abiotic stresses generated by the external environment. Among certain biosynthates, flavonoids are an important example of these stress mitigators. Flavonoids are secondary metabolites and biostimulants; they play a key role in plant growth by inducing resistance against certain biotic and abiotic stresses. In addition, the function of flavonoids as signal compounds to communicate with rhizosphere microbes is indispensable. In this review, the significance of flavonoids as biostimulants, stress mitigators, mediators of allelopathy and signaling compounds is discussed. The chemical nature and biosynthetic pathway of flavonoid production are also highlighted.

Keywords: flavonoids; biotic and abiotic stress; symbiosis; signaling; rhizobium; AMF; salinity; allelopathy

---

1. Introduction

Climate change is the most serious threat to current human culture. Escalating global food demand and ever-increasing global warming put humanity in jeopardy. According to ongoing global temperature analysis carried out by NASA's Goddard Institute for Space Studies (GISS) scientists, the average global temperature has increased by about 1 °C since 1880 [1], and it is estimated that every 2 °C rise in global temperature will cause on hundred million human deaths and bring millions of species to the brink of extinction [2]. After fossil fuel burning for energy generation, agriculture is the second-largest contributor to climate change through the emission of greenhouse gases (GHGs) including carbon dioxide (CO_2), methane (CH_4) and nitrous oxide (N_2O) [3]. Commercial fertilizers are very convenient, easy to handle and a rapid source of soil nutrient recharge, however, the toxic and residual effects of synthetic chemicals have altered thinking around this. It is estimated that reductions in mineral fertilizer use could lead to a 20% reduction in GHG emissions [4]. As the world warms, there is an immediate need to adjust what have become inadequate and inappropriate policies. There is an urgency to develop ecofriendly land practices and more sustainable agriculture. The implementation of biobased products, for example, ushering in the use of organic farming, biofertilizers and biocontrol techniques, will be a progressive step towards sustainable global food security. In this review, we focused on a specific type of biostimulant, flavonoids, and their role in sustainable agriculture. Flavonoids are examples of a versatile set of low molecular weight secondary metabolites with a polyphenolic structure, involved in plant physiological functions, often demonstrating protective effects against biotic and abiotic stresses including UV-B radiation [5], salt stress [6] and drought [7], at least in part by detoxifying the Reactive Oxygen Species (ROS) produced under stress conditions in plants [8,9]. Flavonoids also play a crucial role in plant–microbe associations, predominantly plant–rhizobia and arbuscular mycorrhizal symbioses [10].

Certain flavonoids act as signaling compounds triggering nodule induction by inducing transcription of *nod* genes in rhizobia, the first step in legume–rhizobia symbiotic relationships [11]. In addition, some flavonoids act to combat certain pests and pathogens [12]. Some classes of flavonoids act as color pigments, producing specific hues in leaves and flower petals, helping plants attract pollinators [13]. Moreover, flavonoids have indirect effects on nutrient supply and availability by enhancing mycorrhizal symbioses and colonization of the rhizosphere by beneficial microorganisms [14].

2. Biosynthesis and Classification of Flavonoids

The biosynthesis of distinct flavonoid-based compounds is the result of condensation of one molecule of 4-coumaroyl-CoA (6-carbon) and three molecules of malonyl-CoA, carried out by the enzyme chalcone synthase (CHS). The two major precursors originate from two different pathways of cellular metabolism: the acetate pathway and shikimate pathway providing ring A and ring B, respectively, with chain linkages forming ring C. Ring A is generated from malonyl-CoA synthesized by carboxylation of acetyl-CoA via the acetate pathway, however, ring B along with the linking chain (ring C) is synthesized from coumaroyl-CoA via the shikimate pathway (Figure 1). Coumaroyl-CoA is generated directly from the amino acid phenylalanine by three enzymatic reactions of the phenylpropanoid pathway [15].

The condensation of these aromatic rings by these pathways results in the synthesis of chalcone which will then form flavanone after isomerase-catalyzed cyclization. The later compounds undergo further modifications such as hydroxylation, glycosylation or methylation resulting in the enormous range of flavonoid colors we see today.

Flavonoids are the largest family of natural products; more than nine thousands of these phenolic substances have been found in various plants [16]. Flavonoids have a basic structure containing three phenolic rings, namely A (6 carbon) and B (6-carbon) linked with the central C (3-carbon) ring; C_6-C_3-C_6 which can produce several derivatives and sub-class compounds with distinct substitutions in the basic structure [17]. The major subgroups of flavonoids are; flavonols, flavones, flavanones, flavanonols, flavanols, anthocyanins, isoflavonoids and chalcones [18]. However, based on the attachment of the B ring to the C ring, flavonoids have been classified into three major subgroups: Flavonoids (2-phenylbenzopyrans): The B ring is attached on 2-position of ring C), Isoflavonoids (3-benzopyrans): The B ring is attached on 3-position of ring C) and Neoflavonoids (4-benzopyrans: unlike isoflavonoids; the B ring is attached at 4-position of C ring) [19].

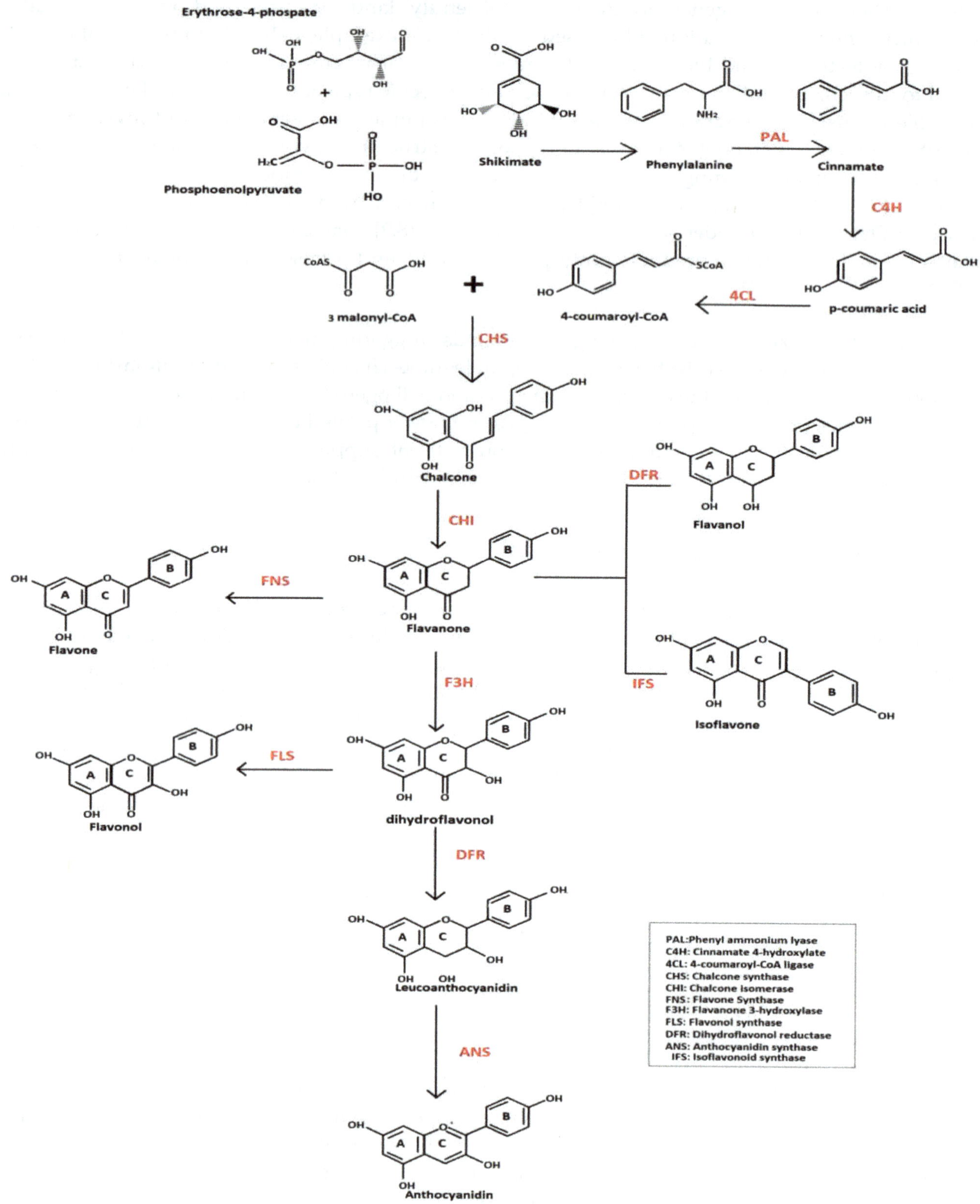

Figure 1. Biosynthesis of flavonoids.

2.1. Flavonols

Flavonols are the most abundant flavonoids in plants. The most studied subclasses of flavonols are the quercetins, kaempferols, myricetins and fisetins; distinctions in the structures of each subclass are shown in Figure 2. The substitution patterns in quercetins and kaempferols are 3,5,7,3′,4′-OH and 3,5,7,4′-OH, respectively. These are very often found in plants as glycosides. The major dietary sources of flavonols are fruits and vegetables, predominantly onions, but also including the apple, strawberry, lettuce and other leafy vegetables. In addition, black and green tea and red wine are also rich sources of flavonols. In general, soft fruits, leaves of medicinal plants and green leafy vegetables have greater levels of flavonoids than other vegetable and fruit plants [20]. However, cooking may lower the concentration of flavonols in vegetables such as tomato and onion [21].

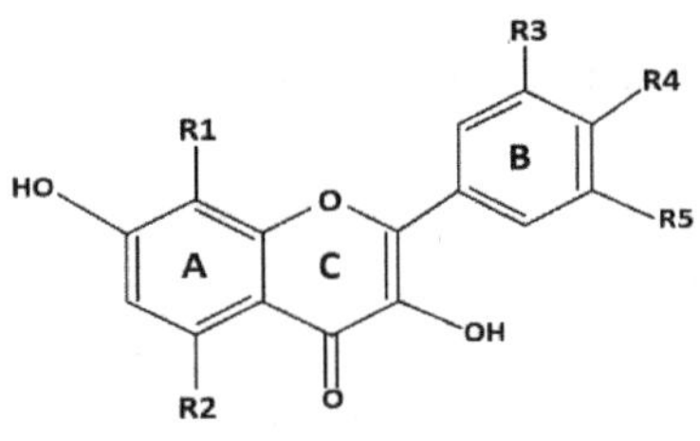

Flavonols

Subclass	R1	R2	R3	R4	R5
Kaempferol	H	OH	H	OH	H
Quercetin	H	OH	OH	OH	H
Myricetin	H	OH	OH	OH	OH

Figure 2. Flavonols: chemical structures, types and substitution positions in the basic skeleton.

2.2. *Flavones*

Flavones are one of the major subgroups of flavonoids. Fruits and vegetables including parsley, carrot, pepper, celery, olive oil, and peppermint are the main dietary sources of flavones [22]. Chrysin, apigenin, rutin (glycoside) and luteolin are the most studied subclasses of flavones. Substitution patterns in the basic structure of flavones are 5,7-OH (chrysin), 5,7,4′-OH (apigenin), and 5,7,3′,4′-OH (luteolin) [23] (Figure 3).

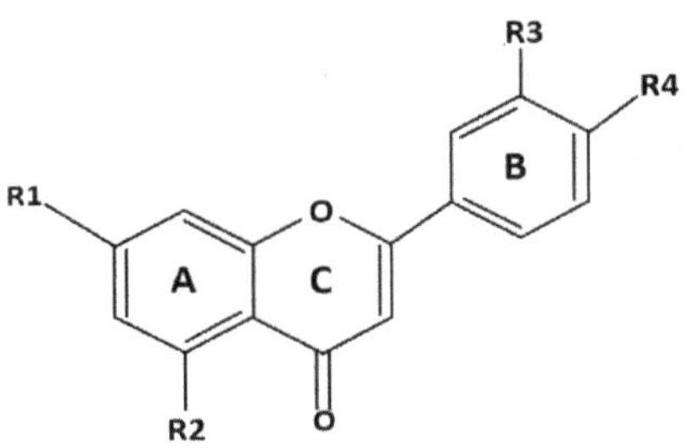

Flavones

Subclass	R1	R2	R3	R4
Chrysin	OH	OH	H	H
Luteolin	OH	OH	OH	OH
Apigenin	OH	OH	H	OH

Figure 3. Flavones: chemical structure, types and substitution positions in the basic skeleton.

2.3. *Flavanones*

Flavanones are different from flavones through their possession of a single bond between C2 and C3 of the C ring. Flavanones are most abundant in solid tissues of citrus fruits such as orange lemon and grape. The most studied types of flavanones are hesperidin and naringenin (Figure 4) The hydroxylation and substitution patterns in flavanone are 5,7,4′-OH (naringenin) and 5,3′-OH, 4′OMe, 7-rutinose (hesperidin) [23].

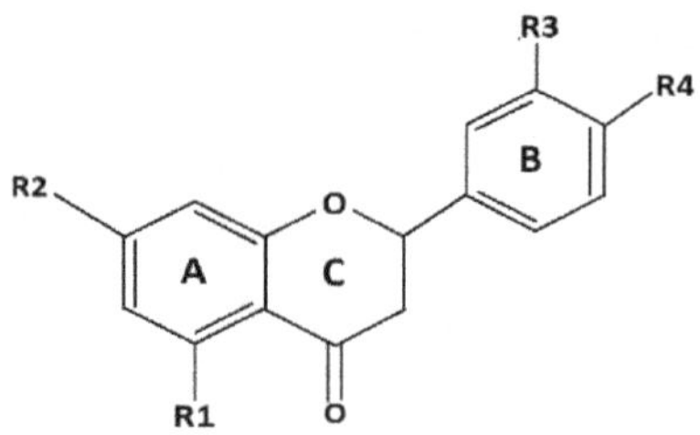

Flavonones

Subclass	R1	R2	R3	R4
Hesperitin	OH	OH	OH	O-Me
Naringenin	OH	OH	H	OH
Naringin	OH	O-Alk	H	OH

Figure 4. Flavonones: chemical structures, types and substitution positions in the basic skeleton.

2.4. Isoflavonoids

Isoflavonoids, with a B ring attached at 3-position of the C ring, are structurally different from other flavonoid classes. Isoflavonoids are found to be very helpful in microbial signaling and nodule induction in legume–rhizobia symbioses. Common examples of isoflavonoids are aglycone and glycosides of genistein and daidzein. The main natural sources of isoflavonoids are legumes such as soybean, as they are reported to exude these as signaling compounds to communicate with microbial symbionts [24]. The subclasses of isoflavonoids are shown in Figure 5.

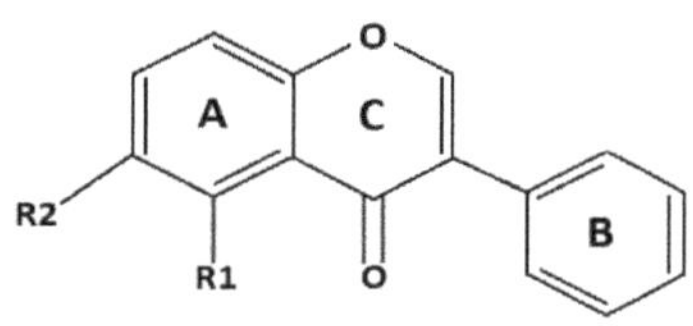

Isoflavonoids

Compound	R1	R2
Daidzein	H	H
Genistein	OH	H
Glycitein	H	OCH_3

Figure 5. Isoflavonoids: chemical structure, types and substitution positions in the basic skeleton.

2.5. Anthocyanidins

Anthocyanidins are responsible for the coloration of many fruits and vegetables. The red and blue colors in apple, grape and berries are due to anthocyanin or anthocyanin glycoside pigments [25]. The color depends on the structure of the compound which, usually changes due to hydroxylation and methylation at specific positions of the A and B rings [18]. Unlike other flavonoids, except flavanols, it carries no ketone group at the 4-position of C ring. Some of the anthocyanidin subclasses are shown in Figure 6.

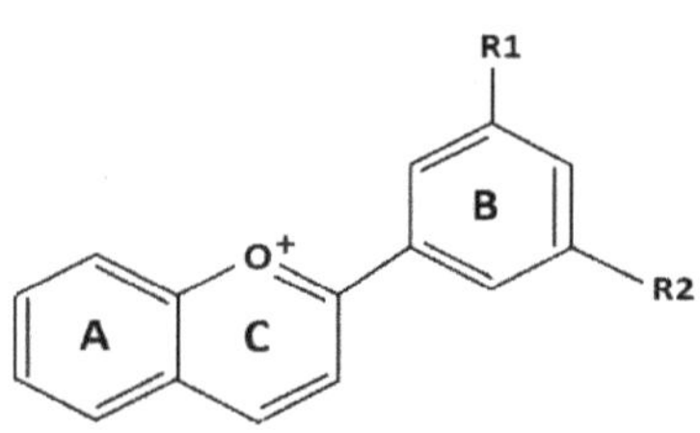

Anthocyanidins

Compound	R1	R2
Cyanidin	OH	H
Delphinidin	OH	OH
Pelargonidin	H	H

Figure 6. Anthocyanidins: chemical structures, types and substitution positions in the basic skeleton.

2.6. Flavanols

Flavanols have a missing ketone group at the 4-position of the heterocyclic ring C, like anthocyanidins. The major sources of flavanols are grape (seed, pulp, stem and skin), berries, tea, wine, apple, pear and peach [26,27]. The most common examples of flavanols are (+)-catechin and (–)-epicatechin (Figure 7).

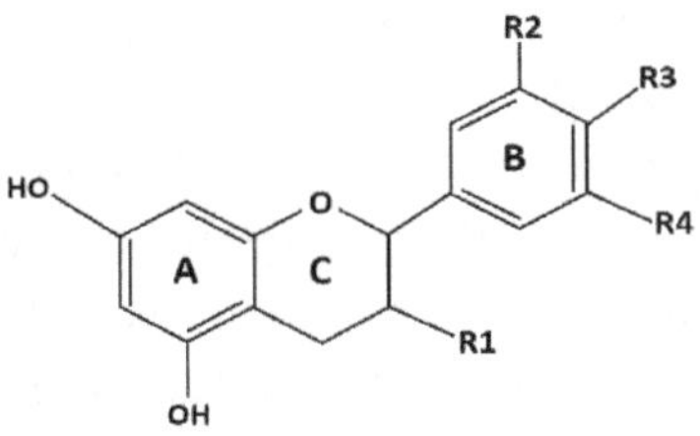

Flavanols

Compound	R1	R2	R3	R4
(+) Catechin	OH	OH	OH	H
(-) Epicatechin	OH	OH	OH	H
(-) Epigallocatechin	OH	OH	OH	OH

Figure 7. Flavanols: chemical structure, types and substitution positions in the basic skeleton.

2.7. Chalcones

Chalcone is the only class with an open ring; it serves as a precursor for various flavonoid classes. The missing C ring in the structure makes it quite different from other flavonoids. The major dietary sources of chalcones are apple, hops or beer [28], berries, tomato and certain wheats [18]. The most studied chalcone is chalco naringenin (Figure 8).

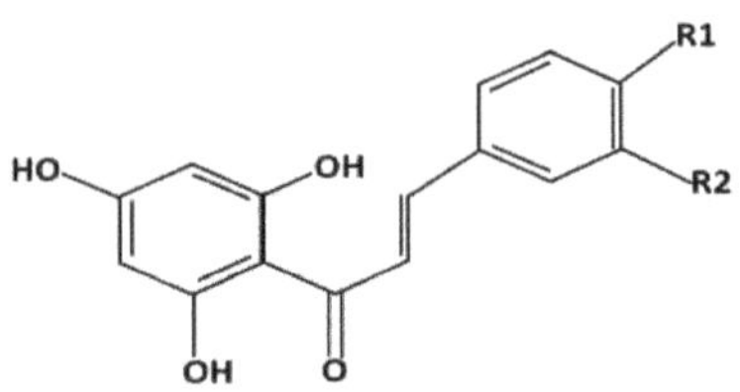

Chalcones

Compound	R1	R2
Naringenin chalcone	OH	H
Eriodictyol chalcone	OH	OH

Figure 8. Chalcones: chemical structure, types and substitution positions in the basic skeleton.

3. Role of Flavonoids in Plant Growth and Crop Yield

Plant growth and production are directly associated with soil productivity. In agronomic terms, it refers to the ability of soil to produce a certain yield of agricultural crops. However, the ability of soil is determined by numerous factors including soil physicochemical properties and management-related factors. Generally, it is the measure of inputs versus outputs, which in agronomic situations are related to water and nutrient supply (inputs) versus crop yield (output) [29]. In addition to these factors, the phytomicrobiome is becoming recognized as a key pillar and a major contributor to crop production. The soil is home to vast numbers of microbial species including free-living, symbionts, host-specific and non-host specific which could be either beneficial or harmful to plant growth and productivity. It is estimated that 1 g of soil contains 1 billion bacterial cells with 10,000 distinct genomes [30,31]. These microbial entities in phytomicrobial associations aid plants in their growth in both stressed and unstressed conditions, by a range of mechanisms. In the phytomicrobiome, associated entities trade for their benefits by aiding one another; microbes get reduced carbon-rich food materials exuded from plant roots and in turn, microbes aid plants in nutrient acquisition, disease resistance and stress mitigation by direct and indirect means. The contribution of the phytomicrobiome in relation to crop yield and production has been demonstrated by numerous studies [32–34]. However, this productive association is the result of crosstalk between trading partners, which is carried out by a complex signal cascade in the rhizosphere. In phytomicrobial associations, signaling is a key activity, and any interruption in signal transmission may disturb and/or reduce the interaction and efficacy of the symbionts.

Plants are equipped with biosynthetic machinery operating to supply thousands of bio-compounds that are required to perform vital functions. There is an array of compounds exuded from plant roots including sugars, amino acids, organic acids, phenolics, enzymes and growth regulators. These secretions serve as a source of carbon for many microorganisms associated with the plant root system. Besides providing a carbon-rich environment, plants generate chemical signals: a mode of communication with microbes in the rhizosphere to initiate colonization and other specific activities for mutual benefit [35,36]. The changing environment and anthropogenic global warming, and demand for sustainable agriculture for future security, have increased the significance of the phytomicrobiome in agricultural practices. Research on this area is becoming more prevalent in the agricultural scientific literature, and scientists are working to better understand plant–microbial associations and signaling to increase their efficacy in order to obtain substantial yield enhancements. In this review, we considered publications highlighting plant-microbial association in relation to signal compounds: flavonoids affecting plant growth and their production. Flavonoids have direct and indirect positive impacts on plant growth and development. The production of specific flavonoids and their accumulation in certain plant tissues following external or internal stimulation is largely unexplained. However, flavonoids are very effective in certain plant–microbe interactions in the rhizosphere and can have effects in combating biotic and abiotic stresses [37]. The roles of flavonoids, flavonoid subclasses and substitution patterns are extensively demonstrated in Table 1.

Table 1. Flavonoid subclasses, dietary sources, major functions in plants and structural substitutions in the basic skeleton.

	Compound	Flavonoid Class	Structure Substitution	Food Sources	Role in Plants	References
1	Quercetin	Flavonols	3,5,7,3′,4′-OH	Onions, apples, berries	Antioxidant, Allelopathic	[38,39]
2	Kaempferol	Flavonols	3,5,7,4′-OH	Tea, broccoli, cabbage, beans, tomato, strawberries and grapes	Antioxidant, antibacterial, insect repellent, abiotic stress mitigation	[40–43]
3	Chrysin	Flavones	5,7-OH	Honey, propolis	Antioxidant, UV-A/B Resistance, AMF symbiosis	[40,44,45]
4	Apigenin	Flavones	5,7,4′-OH	Parsley, Pepper	Antioxidant, AMF spore germination (symbiosis), phytoalexin	[46,47]
5	Naringenin	Flavanone	5,7,4′-OH	Grape, apple, orange	AntioxidantAMF Hyphal growth (Symbiosis)	[25,48]
6	Hesperidin	Flavanone	5,3′-OH, 4′OMe, 7-rutinose	Citrus, orange juice	Antioxidant	[25,49]
7	Genistein	Isoflavonoid	5,7,4′-OH	Currants, raisin, legumes	Nodule induction, signaling	[24,46]
8	Daidzein	Isoflavonoid	7,4′-OH	Currants, raisin, legumes	Nodule induction, signaling, chelation	[46,50]
9	Apigeninidin	Anthocyanin	7,4′-OH	Flowers, fruits	Color pigmentation, pollinator attractant, UV-B absorber	[19,25,51]
10	Fisetin	Flavonols	3,7,4′,5′-OH	Apple, strawberry, onion, cucumber	Antioxidant	[25,52]
11	Myricetin	Flavonols	3,5,7,3′,4′,5′-OH	Berries, tea, wine	Antioxidant	[53]
12	Luteolin	Flavones	5,7,3′,4′-OH	Broccoli, chilli, onion leaves bilimbi fruit and leaves, carrot, local celery	*Nod* gene inducer	[54]
13	Rutin	Flavones	5,7,3′,4′-OH, 3-rutinose	Parsley, Pepper, carrot	Mycorrhizae symbiosis, abiotic stress mitigation	[22,55]
14	(+)-catechin	Flavanol	3,5,7,3′,5′-OH	Grapes, pears, apples	Antioxidant, ROS scavengers	[26,56,57]
15	(−)-epicatechin.	Flavanol	3,5,7,3′,5′-OH	Strawberry, apple	Antioxidant	[56,57]

3.1. Flavonoids in the Rhizosphere

The rhizosphere is the most complex and intensive place for the interaction of plants with the external environment. It is the area of maximum biological community activity, nutrient acquisition (mobility, solubility and diffusion) and plant–microbial interaction, which may depend on the secretion of exudates containing large and small molecular weight organic and inorganic compounds including ions, phenolics, enzymes, secondary metabolites and carbohydrates [58].

Flavonoids (secondary metabolites), very often in both aglycone and glycoside forms, are likely to be exuded from root systems and have indirect effects on plant growth by mediating belowground interactions, including attracting compatible rhizosphere-dwelling rhizobia, stimulating mycorrhizal growth and hyphal branching, enhancing solubility of nutrients including phosphorus and iron and repulsion of pests and root pathogens [50,59]. Studies suggested that flavonoid secretions from roots are carried out through active transport (ATP dependent), catalyzed by ABC transporters [60]. However, flavonoid secretion can also be passive, through degrading of root cap and epidermal cells [61]. In the rhizosphere, flavonoid persistence and mobility may be influenced by solubility, structure, availability of microbes and binding sites, as these compounds can be adsorbed to cation binding sites of soil or cell walls. Flavonoid glycosides are sparingly soluble in water and expected to be less adsorbed to binding sites, improving mobility and availability [10]. In addition, flavonoid secretion may be influenced by environmental stresses, including nutrient supply (nitrogen and phosphorus) in the soil [62].

3.2. Flavonoids and Legume-Rhizobium Interaction

The evolution of intimate relationships that enable plants and microbes to coexist has been the subject of many studies; these have attempted to explain and simplify the interactions that occur between an individual or specific plants and their symbionts. However, in reality, these interactions are far more complex and involve a range of microbes associated with a single plant, exchanging chemical signals [63]. However, these interactions aid plants in many ways. Soil fertility and/or nutrient acquisition is one of the major services provided by soil microbes. Nitrogen deficiency, due to rapid nitrogen loss from soil by leaching, denitrification and immobilization, is a leading problem in crop production. Atmospheric nitrogen is fixed and becomes part of the soil nitrogen recharge, by biological and artificial means, of fixation from the atmosphere. However, the natural biological fixation of atmospheric nitrogen (N_2) contributes about 60% of the total atmospheric nitrogen fixation, which has gained the attention of researchers and growers. Below-ground interactions leading to the establishment of legume nitrogen-fixing symbioses are carried out by signal exchange, as a mode of communication between host and symbiont, for mutual benefit.

The plants release chemo-attractants, in the form of flavonoids, to initiate the symbiotic process [64]. Flavonoids are very often exuded in greatest concentrations from root tips, which is near the site of rhizobium attachment and infection [65]. These secondary metabolites serve as signaling compounds to attract rhizobia toward plant roots and to activate *nod* genes in the rhizobia, which then start the nodulation process in legumes [66]. The *nod* genes are responsible for making lipo-chitooligosaccharides (LCOs), referred to as nod factors, which are released in response to chemical stimulus (generally isoflavonoids) from plant roots. LCOs or other nod factors initiate root hair curling, the formation of infection threads and bacterial entrance into the host plant root cells [67]. In legume crops, almost half of the nitrogen required is fixed by nitrogen-fixing microbes, predominantly, *Rhizobium* and *Bradyrhizobium*; the rest is supplied by fertilizer supplements [68].

Isoflavones are considered to be very active and effective in plant–microbe interactions; they are very operative in signaling and enhancing nodulation by inducing *nod* gene systems. *Nod* gene inducing flavonoids (quercetin and luteolin) released from the seeds and roots of *Medicago sativa* L were investigated. Many, but not all, of the flavonoids, were found to induce *nod* genes in *Rhizobium melioti* [69], indicating that rhizobia is responsive to selective flavonoid signals. Nodulation related gene induction in *R. melioti* by flavonoids released from alfalfa was investigated. A chalcone (4,4′-dihydroxy-2′-methoxychalcone) released, was reported to be the primary *nod* gene inducer in the group. Moderate inducing activities for 4′,7-dihydroxyflavone and 4′-7-dihydroxyflavanone were also reported [70]. In addition, genistein (isoflavonoid), was tested under salt stress with inoculation of *Bradyrhizobium* to evaluate effects on nodulation, N_2-fixation and physiological changes. The results revealed that genistein increased photosynthesis levels, nodulation and nitrogen fixation under saline and non-saline conditions [71]. Isoflavonoids are hypothesized to induce *nod* gene expression and to control the concentration of auxin in soybean roots. Results from similar work provided genetic evidence of isoflavonoid involvement in soybean nodulation and assumed this to be essential for nodule induction, by inducing the *nod* genes in *Bradyrhizobium japonicum* [72].

These biological signals are crucial factors in plant–microbe associations and are very often disrupted by known and unknown causes. Studies on subtropical legume (soybean) nodulation indicated that root zone temperatures (RZTs) below 25 °C decrease nodule induction and N_2-fixation [73,74]. The appropriate range of RZT required for optimal nodulation is 25 to 30 °C. Below optimal RZT (25 °C), with each degree decrease in temperature there is a 2-day delay in the onset of N_2-fixation; below 17 °C each degree RZT decrease delays the onset of N_2-fixation by about one week [75]. However, suboptimal RZTs hinder root hair infection to a greater degree than nodule initiation and development [76]. In addition, N_2-fixation activity by the nitrogenase enzyme complex is delayed, as is nitrogen assimilation [77]. However, the interorganismal signaling disruption by suboptimal RZTs (17.5 and 15 °C) could be minimized by genistein application. Preincubation of *Bradyrhizobium japonicum* with genistein increased the number of nodules, N_2-fixation and plant total dry weight at suboptimal RZT. It is, at least in part, because rhizobial *nod* gene induction is temperature-dependent, making bradyrhizobia less sensitive to signal molecules. However, nodulation events began earlier at suboptimal RZT following genistein application and this stimulated the production of nod factors (LCO), nodule formation and nitrogen fixation [78].

The environmental growth conditions, in terms of nodulation and N_2-fixation, may affect the efficacy of applied signaling compounds and/or inoculated microbial strain, for example, under field conditions, the plant root system is surrounded by an array of phytomicrobiome members, which may compete for plant-supplied reduced carbon; in addition, other factors such as temperature, CO_2 limitation, water and nutrient supply may alter nodulation and the onset of nitrogen fixation [79,80]. However, similar results were observed by preincubation of *B. Japonicum* with genistein application under field conditions. Genistein application improved N_2-fixation (40%) and total nitrogen yield through increased nodule numbers and accelerated onset of N_2-fixation; however, these effects were greater in N-stressed plants. In addition, genistein preincubation of *B. japonicum* has meaningful impacts on yield components [81]. Genistein preincubated inoculum (*B. japonicum*) increased soybean grain yield and protein content by 16 and 70% respectively, as compared to those receiving only inoculum [82]. Similar studies indicated an increase in yield and protein content of soybean by 25.5 and 21.6%, respectively [78]. However, crop responses to genistein application vary with the genetic makeup of the crop cultivars. It was suggested that high yield potential cultivars respond more to genistein application [83]. In addition to crop responses, the cultivar differences also determine the concentration and accumulation of a range of secondary metabolites. The influences of soybean cultivars and selection for yield on the concentration of health beneficial compounds, including isoflavonoids (examples being genistein and daidzein) were determined. The findings revealed a positive correlation between yield and isoflavonoid concentration and suggested that breeders selecting for higher soybean yield may select for higher isoflavonoid concentration. However, isoflavonoid concentrations were negatively correlated with protein content, unlike oil content [84].

3.3. Flavonoids and Mycorrhizal Associations

Phosphorus (P) is the second most important plant nutrient, after nitrogen. However, P is one of the most deficient and least accessible primary nutrients and is often limiting to plant growth. Phosphorus solubilization and affinity for the soil matrix and organic complexes are crucial reasons for its unavailability for plant utilization. The rhizospheric microbial community plays a vital role in phosphorus availability and solubility by producing enzymes that can mobilize adsorbed phosphorus and enhance the process of mineralization. Moreover, beneficial soil microbial species have positive impacts on root growth and development and may increase root surface area, which ultimately enhances the phosphorus depletion zone and effective phosphorus mobility [85]. In plant–mycorrhizal symbioses, mycorrhizal fungi regulate plants to reduce root growth while increasing root extension through hyphal outgrowth (100 times longer than root hairs) and the depletion zone for phosphorus, increasing its availability by extending and proliferating in the soil far beyond the reach of root surfaces [86]. Arbuscular mycorrhizal (AM) associations are probably the most taxonomically extensive associations, formed by 70–90% of plant species. Arbuscular mycorrhizal associations are formed by a monophyletic group of fungi from the phylum Glomeromycota which is estimated to utilize about 20% of plant photosynthates, which is approximately equal to 5 billion t of carbon per year [87]. The mycorrhizal symbiosis is very effective at enhancing plant growth in drought conditions and nutrient-deficient soils, specifically those that are phosphorus-deficient. It was observed that AM inoculation increased potato yield by 9.5% in inoculated fields, as compared to uninoculated fields, with yields of 42.2 and 38.3 t ha^{-1}, respectively. Arbuscular mycorrhizal associations have substantial potential to increase crop productivity, but their efficacy and association in large-scale crop production systems are not yet fully investigated [34].

Root exudation, chemo-stimulation and presence of other microbial communities in the rhizosphere may affect mycorrhizal symbioses and their colonization. Root exuded flavonoids have been shown to enhance the mycorrhizal symbioses by stimulating fungal spore germination, hyphal growth and colonization of roots. Flavonoids are considered to be universal signaling compounds [88]. However, some scientists have a different perception regarding AM symbiotic signaling. The normal development of mycorrhizal relationships in the absence of flavonoid-based signaling compounds in carrot root extracts and deficient activity of chalcone synthase (necessary for flavonoid biosynthesis) brought them to a conclusion that "Flavonoids are not necessary plant signal compounds in Arbuscular Mycorrhizal symbiosis". However, if present, their influence greatly stimulates AM development [89] and plant growth depending on concentration, spore density and plant growth stage [90].

The effects of the AM stimulating flavonoid, formononetin, on potato yield mediated by native AMF were examined. The results revealed an increase in plant dry matter, tuber development and phosphorus use efficiency. Perhaps the effects were more prominent at low phosphorus (P) levels. In addition, formononetin increased soil sporulation more than 3 fold [91]. However, the variation in response of different cultivars to formononetin application indicated that signaling responses by participatory symbionts may depend on genetic characteristics [92]. Similar results suggested that formononetin seed application to soybean may reduce the need for phosphorus (P) fertilizer by 50% [93]. A similar study was conducted on the hyphal growth and root colonization of arbuscular mycorrhizal fungi (AMF) on tomato as affected by flavones and flavonols. A correlation was observed between the number of entry points and root colonization percentage for the specific tested AMF (*Gigaspora rosea*, *Gigaspora margarita*, *Glomus mosseae*, and *Glomus intraradices*). Flavones (chrysin and luteolin) and flavonols (morin) enhanced the colonization and number of entry points, whereas kaempferol and rutin have no effect on presymbiotic growth of AMF and subsequent root colonization [94]. The flavonoids (apigenin, hesperetin and naringenin) enhanced spore germination, hyphal growth and root colonization of *Gigaspora margarita* [95]. In addition, quercetin glycosides, exuded from alfalfa, were found to be effective in enhancing AMF symbioses by increasing hyphal growth and branching, and spore germination of *G. macrocarpum* and *G. etunicatum* [50].

Moreover, flavonoids have been found to solubilize phosphorus by enhancing mycorrhizal colonization of root systems and may help in nutrient availability and mineralization of nitrogen and other nutrients [96]. They may also act as metal chelating agents, making certain micronutrients more available for plants [97]. Flavonoids released from roots of white lupin cause significant increases in phosphorus acquisition [98]. An Isoflavonoid present in root exudates of alfalfa was shown to dissolve phosphates of iron, making both iron and phosphorus more available to plants. In addition, flavonoids, including genistein, quercetin and kaempferol, can make iron available by chelating and reducing iron oxides in the rhizosphere from Fe^{+3} to Fe^{+2} [99].

4. Flavonoids and Plant Abiotic Stresses

Plants, as they are sessile, are exposed to an array of unfavorable environmental conditions. Ecological variations and intense growth conditions affect plant hemostasis, physiology and growth, leading to diminished and stunted plants. A wide range of unfavorable biotic and abiotic stresses threaten sustainable agriculture and are often responsible for diminished crop production [100]. The mechanisms plants use to cope with abiotic stresses are coordinated among plant organs and tissues through chemical signals [101]. Most plant responses to stress conditions are not fully understood. However, the biosynthetic flavonoids and their ability to induce resistance against biotic and abiotic stresses have acquired some attention. Flavonoids are found to be supportive in abiotic stresses, including UV radiation, salt and drought stress.

4.1. Flavonoids as UV Scavengers

UV radiation is invisible, short wavelength and highly energetic radiation. UV radiation is divided into three segments based on the wavelength of the light: UV-A, B and C with a wavelength of 315–400 nm, 280–315 nm and less than 280 nm, respectively [19]. These wavelengths have enough energy to cause damage and abnormalities in plants by breaking chemical bonds through photochemical reactions [62]. The energy of the photon depends on the wavelength of the radiation; the shorter the wavelength, the more energetic the radiation will be. UV-C (less than 280 nm) is most energetic and can ionize certain molecules. UV-B can cause severe metabolic disruption in plants by negatively affecting photosynthesis, starch concentration and transpiration, and promoting cellular damage; it may also increase disease susceptibility by making defense mechanisms weak, affecting the process of cell division and inhibiting overall plant growth [65,102,103]. However, the absorbance of UV-B radiation by flavonoids permits little radiation in this wavelength range to pass through leaf epidermal cells [62]. Plant resistance to UV radiation is due to flavonoids (anthocyanins) filtering UV-B by absorbing such radiations and detoxifying the ROS produced by photochemical reactions [104].

The significance of UV-B scavenging flavonoids on two apple varieties (Granny Smith and Braeburn) exposed to sunlight and UV-B radiation was estimated. The results revealed different contents and compositions of UV-B absorbing compounds in the two varieties. The Granny Smith (anthocyanin-free) fruit showed significant decreases in chlorophyll and carotenoid contents. Conversely, Braeburn exposed to sunlight had higher contents of chlorophyll and carotenoids. However, quercetin glycosides were the principal compounds absorbing UV-A and UV-B radiations. [105]. Similar work has investigated tomato under controlled environment conditions; the findings demonstrated an increase of flower/fruit synchronization under high radiation, with minimal effect on vegetative plant parts. In addition, an increase of UV-B receptors and chlorophyll content was also observed, along with phenylpropanoid compounds responsible for UV absorption by-products of antioxidant pathways. [106]. UV-A/B, as an abiotic stress, can be used to enhance fruit quality by activating oxidation pathways in plants [107].

The effect of UV light on flavonoid content in barley leaves was tested, and results revealed significant increases in flavonoid (saponarin, lutonarin) content when leaves were exposed to UV-B radiation, as compared to control conditions (absence of UV-B) [108]. In addition, the increased flavonoid accumulation in response to UV-B radiation may reduce the damage in exposed leaves by absorbing specific radiation wavelengths. [109]. Therefore, plants grown in open-environment conditions, exposed to full sunlight, have greater flavonoid contents than plants grown in greenhouses.

4.2. Flavonoids in Managing Salt and Drought Stress

Salinity, one of the most concerning abiotic stressors, is a major constraint to global crop productivity. The worldwide extent of soil affected by salts is about 955 M ha, while 77 M ha are affected with secondary salinization, and of these 58% are irrigated lands. Almost 20% of irrigated land is affected by excess soluble salts [110]. Soil salinization and increased accumulation of soluble salts in the root zone, predominantly NaCl, is caused by natural and/or human activities which have resulted in degraded and abandoned formally fertile and productive agriculture lands [111]. Excess soluble salts in soil solution may limit plant growth, primarily through two mechanisms: osmotic stress and ion toxicity. First, low solute/osmotic potential due to increased ion concentration (NaCl) in soil water reduces the total soil-water potential (Ψ) which in turn reduces the ability of plant roots to uptake water, eventually leading to diminished plant growth. Second, ion toxicity in plant tissues, more frequently due to sodium accumulation, causes cellular damage by membrane disruption and disturbs plant physiological processes, including photosynthesis, respiration, transpiration and osmoregulation, resulting in necrosis or chlorosis, leading to reduced plant growth [112–114]. Depending on sensitivity and tolerance to salinity, plants are classified as either glycophytes or halophytes [115]. Most of the agricultural crops are glycophytes (low tolerance) and tend to exclude Na^+ and Cl^- from roots. Unlike glycophytes, halophytes are often native to saline growth conditions and tolerate salt concentrations that kill 99% of other vegetation. The salinity tolerance of halophytes relies on ionic homeostasis by controlled uptake and compartmentalization of ions (Na^+, K^+, and Cl^-) and accumulation of metabolically compatible solutes (organic) in the cytoplasm to balance the solute potential of ions accumulated in the vacuole [116,117]. Salt responses in plants follow a biphasic mechanism. The first, osmotic phase (rapid), begins immediately after root zone salinization increases to a threshold level, resulting in reduced shoot growth and leaf area, and causes stomatal closure. The second, ionic phase, begins with increased accumulation of ions (Na^+) to toxic levels in the cytoplasm, leading to chlorosis followed by leaf death. Osmotic stress not only exerts immediate effects on plant growth but is also more chronic than the ionic phase [118]. Increases in an array of compatible organic solutes is proposed to balance solute potential including sucrose, proline, glycine-betaine and sorbitol [119,120]. In addition to compartmentalization, some plants can prevent salt accumulation (whole plant or cellular level) and avoid toxic effects of ions on crop physiology including photosynthesis [113]. In response to salt stress, plants undergo several morphological, physiological and metabolic changes to cope with the stress conditions. These adaptations involve several biochemical pathways, sustained osmotic potential, ion compartmentalization and exclusion of toxin ions. Subsequent to ionic toxicity, specific toxic substances, ROS, including superoxide, singlet oxygen and hydrogen peroxide, cause oxidative damage in cells, which is considered a secondary effect of salinity [121,122]. However, plants are equipped with specific defense mechanisms to cope with such stress conditions, by initiating antioxidant pathways including enzymes and antioxidant agents; flavonoids, carotenoids and specific vitamins [121,123,124]. Reactive oxygen species or free radicals are molecular species that contain at least one unpaired electron in their atomic shells, making them highly reactive. Reactive species (RS) are quite unstable, most of them exist not more than 10^{-6} s in biological systems, and to be more stable they react with biomolecules by either donating or receiving an electron [125].

Oxygen is poisonous; aerobes are equipped with defense mechanisms mediated by antioxidants, which is how they survive such toxicity. The antioxidant defenses in biological systems are the result of several strategies [126]:

1. Suppressing RS formation either by uncoupling proteins triggered by superoxide, indicating that it may reduce mitochondrial ROS formation [127] or inhibition of enzymes involved in RS formation, for example, inhibition of cyclooxygenase, lipoxygenase and NADH oxidase by flavonoids [128]
2. Substitution of biomolecules vulnerable to oxidative damage with resistant ones
3. Antioxidants acting as "sacrificial agents" by reacting with reactive species to prevent them from reacting with important biomolecules [129].

Flavonoids have been found to play an important role as antioxidants by detoxifying and scavenging of ROS produced [130] as by-products of oxidative metabolism [121] during abiotic stresses including salt and drought. However, the accumulation of such metabolites advances when plants face any environmental uncertainty. It was observed that anthocyanin accumulation in response to salt stress increased by 40%, which could be a phytochemical strategy to combat salt stress and subsequent toxic reactions [131]. In addition, the protective nature of anthocyanin was also compared in two groups of rice genotypes: salt sensitive and salt resistant. The total anthocyanin content in salt-tolerant genotypes was higher than in salt-sensitive varieties, with antioxidant activities of 125–199% and 106–113%, respectively. It was therefore concluded that anthocyanins in rice contribute to cellular protection by detoxifying accumulated salts [132].

The effects of applying genistein (an isoflavonoid) to rhizobial culture on signal production and subsequent growth and yield of soybean have been investigated. Results demonstrated a significant increase in plant growth with increased leaf area and number of nodules. Genistein application enhanced crop yield by 21% under salt stress [133]. Salinity may inhibit signal exchange between host and symbiont, which is very important for initiating a functional symbiotic nitrogen-fixation relationship. The interaction of soybean and *B. japonicum* was evaluated under salt stress. The findings are consistent with similar studies. Genistein application enhanced the stimulation of growth and signaling between the symbiotic partners and, hence, increased nodulation and growth of the plant. The results may help in cultivating soybean in a more efficient and productive way under unfavorable environmental conditions [134]. Likewise, an increased concentration of flavonoids has been found in tomato plants when exposed to salt or drought stress, however, plant growth and chlorophyll content were significantly reduced, indicating no and/or insignificant effect of flavonoids on plant growth and physiology [121]. In contrast, findings from a similar study on flavonoid biosynthesis and accumulation in wheat leaves under drought stress suggested that drought resistance in wheat is closely related to increased flavonoid accumulation [135]. Similar results were observed in two native shrubs from Argentina. Flavonoid accumulation was observed throughout the year; however, a significant increase was noted during times of intensive drought [136]. Drought mitigation by flavonoids and flavonoid derivatives has been confirmed in *Arabidopsis thaliana*. The role of individual flavonoids was unclear, however, increased production of flavonoids in plants and associated drought resistance was confirmed [95].

5. Flavonoids against Plant Biotic Stress

Plants, as they are sessile, have no possible way to physically remove themselves from invading pests and pathogens. In nature, plants are exposed to an array of pathogenic fungi, bacteria and herbivore pests. However, plants have evolved strategies to combat such unwanted guests. Pathogens do not generally succeed in infecting plants that are not host species (non-host resistance) and/or resistant varieties (incompatible interaction), but intense damage after infection may be caused in susceptible plants (compatible). The damage caused by pathogens in most cases is inversely correlated with the hypersensitive response including reinforcement of cell wall, induction of lytic enzymes and production of phytoalexins [137]. One of the defense strategies adopted by plants in response to invading pests and pathogens is formation, accumulation and secretion of phytoalexins. Phytoalexins are chemicals released by plants in response to pests and pathogens, to ward off the disease and disease-causing agents. Flavonoids are the most-described secondary metabolites in plant defense systems [132]. Their role in plant physiology, morphology and communication and defense mechanisms is considerable. Certain flavonoids are found to be strong phytoalexins against pathogenic bacteria, fungi and nematodes, and may act as insect repellents. External morphological modifications in plants may also act as protective mechanisms against invading pests, predominantly feeding animals including insects, however the chemical tool of insect repellent is more effective. The synthesis of these antibiotic secondary metabolites in plants is due to infection caused by pathogens, bacteria, fungi and nematodes, and may also be induced by feeding insects [138,139].

5.1. *Phytoalexin Flavonoids as Nematicides*

Parasitic nematodes cause tremendous crop yield loss by forming cysts or galls on roots. Plants, in response to nematode invasion, produce several chemicals to increase resistance to or minimize the effect of, nematode presence. The synthesis and accumulation of flavonoids within plant root systems is often stimulated by nematode invasion. The induced synthesis of such phytoalexins assists plants in coping with nematode infections.

Coumestrol (an isoflavonoid) can act as a phytoalexin; it is synthesized in lima bean as nematicide when infected by *Pratylenchus penetrans*. Glyceollin is a protecting isoflavonoid synthesized in soybean roots when infected by the root-knot nematode; *Meloidogyne incognita*. The results demonstrated a decrease in nematode mobility and O_2 uptake [140]. The synthesis of glyceollin, resistance inducing flavonoids in soybean, in response to *Meloidogyne penetrans* infection minimized crop damage [141]. *Heterodera glycines* (soybean cyst nematode) is the most destructive parasitic nematode of soybean. Accumulation of the phytoalexin glyceollin in root systems of soybean after cyst nematode invasion was determined by HPLC (High Performance Liquid Chromatography): A form of liquid chromatography, used to separate, identify and quantify compounds in a solution. No glyceollin was found in control plants, however, the content of the phytoalexin glyceollin increased at the 2nd, 4th and 6th days after inoculation, by 12, 19 and 23 $\mu g\ g^{-1}$ root, respectively [142]. The major phytoalexin in oat, when infected with major nematodes of cereals, was identified as O-methyl-apigenin-C-deoxyhexoside-O-hexoside (a flavone-C-glycoside). The phytoalexin flavone, induced by nematode invasion, was extracted from oat roots and shoots, and was significantly reduced invasion by major cereal nematodes: *H. avenae* and *P. neglectus* [143]. The interaction of flavonoids with parasitic nematodes as defense mechanisms is unclear, however, flavonoids were found to be protecting agents as they inhibit nematode motility and chemotaxis [140].

5.2. *Flavonoids against Pathogenic Fungi*

Fungi, as the most dominant disease-causing agent in plants, adversely affect agricultural crop production. The wide range of diseases caused by fungi decreases crop production dramatically. However, plants have adapted themselves to produce resistance mechanisms against biotic stresses. The production of phytoalexins in response to pathogenic invasion, predominantly fungi, is an effective tool used by plants for combatting biotic stress. The effect of cucumber powdery mildew and subsequent biochemical changes in response to invading pathogen was investigated. Results revealed that silicon can contribute to powdery mildew resistance in cucumber by increasing the accumulation of a fungi-toxic phytoalexin, which was identified as the flavonol aglycone rhamnetin (a flavonoid) [144]. Brown rot lesion is a disease caused by the fungus *Phytophthora citrophthora* in citrus fruits. The correlation of infection caused by pathogen and level of phytoalexin flavones accumulation in host plant was examined. The increased accumulation of heptamethoxyflavone, nobiletin, sinensetin, and tangeretin was confirmed along with the antifungal effects of phytoalexin flavonoids. The most effective flavonoids against *P. citrophthora* were naringenin and hesperetin [145]. Similar results were found in tangelo fruit defense mechanisms against *P. citrophthora*. The accumulation of isoflavonoids was induced by 6-benzylaminopurine application, which enhanced fruit resistance to the pathogenic fungus by 60%. The most inhibiting of the accumulated phytoalexins were nobiletin, sinensetin, heptamethoxyflavone, followed by tangeretin [146]. Further research was carried out to evaluate phytoalexin accumulation in soybean cotyledons using four species of *Aspergillus*. All the pathogenic species induced accumulation of phytoalexins in soybean cotyledons, however the phytoalexins glyceollin at 955 $\mu g\ g^{-1}$ (fw) and coumestrol at 27.2 $\mu g\ g^{-1}$ (fw), following inoculation with *A. sojae* and *A. niger*, accumulated to the greatest degrees [147].

5.3. Antibacterial Effects of Flavonoids

The study of natural defense mechanisms of plants related to synthesizing antimicrobial phytoalexins in response to biotic stress demonstrated that phytoalexin level is increased as part of the resistance to phytopathogenic agents [148]. Information regarding phytoalexin accumulation in response to fungal invasion is considerable, however very little is known about phytoalexin synthesis in response to phytopathogenic bacterial invasion [149]. The production of antibacterial phytoalexins in bean leaves was studied by inoculating bean plants with *Pseudomonas spp.* Coumestrol, a phytoalexin isoflavonoid, was accumulated in infected bean leaves and inhibited the growth of two pathogenic bacterial species: *P. mars-prunorum and P. phaseolicola*. The coumestrol was obtained from hypersensitive and susceptible lesions at 1 and 5 days after inoculation. Their accumulation explains the inhibition of bacterial colonization in hypersensitive and susceptible lesions of bean leaves [150]. The accumulation of isoflavonoid in soybean leaves in response to *Pseudomonas glycinea* invasion was investigated. Coumestrol and daidzein were identified as the major phytoalexins accumulated in response to pathogenic (*P. glycinea*) and non-pathogenic (*P. lachrymans*) inoculations of soybean leaves. The data demonstrated inhibiting properties of coumestrol against pathogenic bacterial colonization and suggested that the resistance in soybean leaves against *P. glycinea* was due to induced accumulation of isoflavonoids [151]. In contrast, coumestrol was found ineffective against pathogenic bacterial strains when tested with another five isoflavonoids on twenty isolates of pathogenic and saprophytic bacteria, including species of *Pseudomonas, Xanthomonas* and *Achromobacter.* However, phaseollinisoflavan and kievitone showed antibacterial activity by strongly inhibiting the population of *Xanthomonas* and *Achromobacter* species [152]. Recent research toward finding natural solutions against biotic stress introduced new products against certain disease-causing agents. A new compound isolated from the roots of *Erythrina poeppigiana*, identified as an isoflavonoid was isolated against *Staphylococcus aureus,* and compared with five other root isolates. Results revealed strong inhibiting activities against inoculated pathogens. The minimum inhibitory concentration was 12.5 g mL^{-1} against thirteen (13) strains of S. *aureus.* It was also assumed that new compounds could act as potent antibiotics against infections caused by *S. aureus* [153].

5.4. Flavonoids as Insect/Herbivore Repellents

Plants, as sessile beings, act as an available food source for herbivores, including insect pests. Plants have evolved defense strategies to avoid and/or deal with such biotic stresses, by secreting and accumulating repellent molecules or signals, which are either plant constitutes, in some cases produced inductively in response to pest invasion [154]. The defense mechanism may initiate from undamaged tissues by secreting phytoalexins in response to chemical signals from wounded tissues, which may repel or intoxicate insects [155]. Most of the insect repellent or antifeedant molecules are alkaloids, flavonoids and other secondary metabolites [156]. Morphological modifications like thorns and waxes can make feeding difficult for insect pests. Plants usually use two different strategies: direct and indirect methods, as defense mechanisms against insect herbivory. Direct methods involve the accumulation of insect repellents or toxic substances to minimize the level of damage. On the other hand, plants release signaling compounds as chemoattractants to signal predators, which may feed on the pest and minimize the plant damage [157]. Rotenone (isoflavonoid) is a very effective botanical insecticide used as a basis for insect repellents. Rotenone is a major component of insecticidal resins which may be extracted from roots of some legumes, including those in the genera *Lonchocarpus, Derris* and *Tephrosia.* Pyrethrum is another bioactive material against insect pests, also containing flavonoids as a major constituent. Pyrethrum can be extracted from flowers of *Chrysanthemum* species [158]. Likewise, the antifeeding effects of four isoflavonoids (genistein, formononetin, daidzein and biochanin A), isolated from two red clover cultivars, were investigated against clover root borer. The isoflavonoids decreased insect weight and activity. Genistein and formononetin had high anti-feeding activity against *Hylastinus obscurus* (clover root borer). The results could be utilized in controlling curculionid [159]. In addition, the insecticidal effects of phenolics of pea plants were tested against *Acyrthosiphon pisum.* The high concentration of phenolics and flavonoids in infested plants, as compared to controls, suggested the induced accumulation of antifeedants in response to insect pest presence. In addition, flavonoids (luteolin and genistein) added as supplements to artificial diets prolonged the stylet probing, onset of salivation and passive ingestion. Salivation and passive ingestion completely stopped at higher concentrations of flavonoids. such measures could be employed to induce resistance against certain invading pests [160]. Additionally, the role of phenolics and flavonoids as insect repellents is illustrated by several lines of research [161–163]

The antifeedant and toxic effects of four isolated flavonoids (isoglabratephrin (þ)-glabratephrin, tephroapollin-F and lanceolatin-A) from aerial parts of *Tephrosia apollinea L.* were determined. The flavonoids exhibited toxic effects against insect pests: *Sitophilus oryzae, Rhyzopertha dominica* and *Tribolium*, with mortality percentages of 78.6, 64.6 and 60.7%, respectively, at 3.5 mg mL^{-1}. [164]

6. Allelopathic/Phytotoxic Behavior of Flavonoids

Weeds are a very significant challenge to crop plants as they are constantly competing for light, nutrients and water, interfering with crop functioning and causing tremendous yield loss directly or indirectly. Reductions in crop yield are generally much greater due to weeds than other pests. It is estimated that about 34% of yield loss among the major crops is caused by weeds [165]. Some of the major crops affected by weeds are wheat, soybean, rice, maize, cotton and potato with yield reductions of 23, 37, 37, 40, 36 and 30% respectively [166]. Weed management strategies have always been a significant part of agricultural systems but have changed significantly based on the accessibility of tools and techniques, environmental and sustainability concerns, starting from ancient techniques such as pulling by hand and soil tilling with simple tools, to current use of herbicides and mechanized conventional tillage, which are the most recent and, so far, most effective techniques available [167]. However, despite being very effective, commercial herbicides are finding themselves eschewed by growers because of their toxic and residual effects which contravene the principles of sustainability, eventually contributing to climate change, which is a consequence of such unsustainable human activities. In addition, the continuous use of chemical herbicides induces herbicide resistance in weed populations, which is a crucial long-term consideration in weed management. In contrast, biopesticides are gaining significant popularity among crop scientists because of their environmentally friendly behavior, as they contain biochemicals with no, or minimal, residual effects. The concept of using plant-derived biochemicals as "weedicides" originated from the allelopathic effects mediated by certain plants by employing allelochemicals released into the environment.

The term allelopathy was first defined by Molisch [168], indicating that the effects that result (directly or indirectly) from exuded biochemicals transferred from one plant to another. This definition, suggested by Molisch, implies only plant activity. However, the term "allelopathy" was later refined to include microorganisms (bacteria, algae, fungi and viruses) in his definition, as a significant part of allelopathic processes [169]. Allelopathy is an interference process in which either plants or their dead parts exude phytotoxic chemicals which interfere with the physiology and growth of other plants [170]. The allelopathic behavior of certain entities (plants and microorganisms) has been demonstrated in the literature, however, given the extensive uncertainty, this area needs more exploration in order to understand allelochemical behavior, including the formation of allelopathic compounds and their chemical nature, viability, efficiency and mode of action in plant-plant and plant–microbe interactions, to improve their practical implementation in the field.

Several plants are known to have allelopathic natures through allelochemical exudation including wheat, rice, rye, barley, sorghum and sugarcane. These plants can be manipulated to suppress weeds through an allelopathic approach within crop rotations, intercropping, cover crops and mulch [171]. The phytotoxic effect of sunflower cultivars was evaluated against weed species in wheat either by growing with weeds or applied as residues over a wheat crop and weeds. Sunflower cultivars have shown strong allelopathic effects on weeds, however, variation among the cultivars was observed, indicating that allelochemical exudation or phytotoxic effects and weed suppression vary with cultivar/genotype.

In addition, the sunflower cultivars suppressed total weed density and biomass by 10–87% and 34–81%, respectively [172].

There is an array of biochemicals, produced as secondary metabolites in plants, or released during their decomposition by microbes, which act as active allelochemicals in plant ecosystems. These phytotoxic substances, based on their chemical nature, are classified into 14 categories including cinnamic acid and its derivatives: coumarin, flavonoids, tannins, terpenoids and steroids. Recent publications regarding flavonoids have evidenced their phytotoxicity and growth inhibitory effects, which could be a sustainable approach toward integrated weed management [173].

In early plants, bryophytes and ferns, some of the allelochemicals found are synthesized in the early stages of flavonoid biosynthesis, however, additional flavonoid classes accumulate in angiosperms and gymnosperms, reflecting the employment of genes beyond just those involved in flavonoid biosynthesis [137]. Flavonoids have been reported in the literature for over 50 years as allelochemicals in the rhizosphere [174]. They are either exuded from roots or released from decomposed plant tissues as leachates, persist for days in the soil and their activity (inhibitory or stimulatory) depends on their concentration and solubility. In addition, phytotoxic compounds can also accumulate in leaves and pollen, which eventually inhibit seed germination of other plants after falling onto the soil [12].

It has been observed that flavonoids are produced by many legumes; quercetin and kaempferol (aglycon and glycosylated) released from seeds and roots, possess phyto-inhibitory effects. If present in lower concentrations such compounds may stimulate seed germination while in higher concentrations they may inhibit seedling growth [175]. Flavonoids isolated from roots of *Stellera chamaejasme* L., a toxic and ecologically threatening weed, showed strong phytotoxic activity against *Arabidopsis thaliana*. The isolated flavonoids reduced seedling growth and root development. In addition, endogenous auxin distribution in *Arabidopsis thaliana* was also influenced, indicating a critical factor in phytotoxicity [176]. Spotted knapweed is one of the more noxious and economically devastating weeds of North America, destroying crops and other weeds by phytotoxicity. An allelochemical identified was flavan-3-ol (¬-)-catechin (flavonoid), was shown to be responsible for the invasive behavior and phytotoxicity of *Centaurea maculosa* (spotted knapweed) [103].

Recently, the herbicidal effects of 10 crude extracts obtained from Tunisian plants were assessed. Among the five phenolic compounds, three of the flavonoids had significant herbicidal effects on *Trifolium incarnatum*. Flavonoids inhibited seed germination and seedling growth and caused severe necrosis and chlorosis. Based on their efficiency, flavonoids were formulated into a natural herbicide and interestingly, the extracts showed the same herbicidal effects as an industrial biopesticide containing pelargonic acid [177]. A similar study was conducted, indicating the phytotoxicity of *Plantago major* extracts on germination and seedling growth of purslane (*Portulaca oleracea*). It was observed that the level of phytotoxicity or inhibition was directly proportional to extract concentration. Phytotoxicity of a higher extract dose (40 mg mL^{-1}) was greater than the lowest one evaluated (2.5 mg mL^{-1}) and these concentrations inhibited germination by 30.24 and 4.60%, respectively. In addition, the highest concentrations significantly inhibited radical and plumule growth. The biologically active organic compounds in plant extracts were analyzed and found to be phenolics, tannins, alkaloids, flavonoids and saponins [178]. However, the compounds were not tested alone in this study, leaving no evidence of individual phytotoxic intensity of biological compounds. The need for sustainable and eco-friendly approaches in agricultural systems fosters great interest in bioproducts and biological control agents. However, further studies are required to obtain a better understanding of the many phytotoxic flavonoids.

7. Conclusions

The indispensable role of flavonoids in stress mitigation and signaling behavior in plants is highlighted. More specifically, we reviewed the protective nature of flavonoids in plants against certain biotic and abiotic stress conditions. The polyphenolic structure and diverse chemical nature of flavonoids facilitate multiple mechanisms of action, favoring plant survival under a range of harsh conditions. Despite current knowledge of this matter, the use of flavonoids in agriculture is very limited. Soil microbiota are ecofriendly contributors in sustainable agriculture, and the iconic role of flavonoids in improving phyto-microbial associations is the "icing on the cake". Still, however, we know very little about the chemo-communications between plants and the vast number of microbial strains in the rhizosphere (phytomicrobiome members); much is left to be explored and elucidated. In addition, flavonoids play an indispensable role against plant biotic and abiotic stresses. Flavonoids could be employed as an ecofriendly and sustainable approach towards stress mitigation. The phytotoxic and pesticidal effects of flavonoids provide insight regarding how effective these biochemicals could be in the field if practically implemented. The use of bioflavonoids as natural herbicides is an area of growing interest in integrated weed management. Further research and investigations are required to understand the full range of activity of flavonoids produced naturally and/or applied artificially.

Author Contributions: Authors have contributed equally to this work. All authors have read and agreed to the published version of the manuscript.

Acknowledgements: The authors would like to acknowledge the support for this review paper was provided through the Biomass Canada Cluster (BMC), which is funded through Agriculture and Agri-Food Canada's AgriScience program and industry partners.

References

1. Observatory Earth. World of Change: Global Temperatures. Available online: https://earthobservatory.nasa.gov/world-of-change/global-temperatures (accessed on 24 March 2020).
2. Richardson, Y.; Blin, J.; Julbe, A. A short overview on purification and conditioning of syngas produced by biomass gasification: Catalytic strategies, process intensification and new concepts. *Prog. Energy Combust. Sci.* **2012**, *38*, 765–781. [CrossRef]
3. EPA. Global Emissions by Gas. Available online: https://www.epa.gov/ghgemissions/global-greenhouse-gas-emissions-data (accessed on 24 March 2020).
4. Scialabba, N.E.-H.; Müller-Lindenlauf, M. Organic agriculture and climate change. *Renew. Agric. Food Syst.* **2010**, *25*, 158–169. [CrossRef]
5. Chalker-Scott, L. Environmental significance of anthocyanins in plant stress responses. *Photochem. Photobiol.* **1999**, *70*, 1–9. [CrossRef]
6. Chen, S.; Wu, F.; Li, Y.; Qian, Y.; Pan, X.; Li, F.; Wang, Y.; Wu, Z.; Fu, C.; Lin, H.; et al. NtMYB4 and NtCHS1 are critical factors in the regulation of flavonoid biosynthesis and are involved in salinity responsiveness. *Front. Plant Sci.* **2019**, *10*. [CrossRef] [PubMed]
7. Shojaie, B.; Mostajeran, A.; Ghanadian, M. Flavonoid dynamic responses to different drought conditions: Amount, type, and localization of flavonols in roots and shoots of *Arabidopsis thaliana* L. *Turk. J. Biol.* **2016**, *40*, 612–622. [CrossRef]
8. Brunetti, C.; Di Ferdinando, M.; Fini, A.; Pollastri, S.; Tattini, M. Flavonoids as antioxidants and developmental regulators: Relative significance in plants and humans. *Int. J. Mol. Sci.* **2013**, *14*, 3540–3555. [CrossRef] [PubMed]
9. Cushnie, T.T.; Lamb, A.J. Antimicrobial activity of flavonoids. *Int. J. Antimicrob. Agents* **2005**, *26*, 343–356. [CrossRef]
10. Singla, P.; Garg, N. Plant flavonoids: Key players in signaling, establishment, and regulation of rhizobial and mycorrhizal endosymbioses. In *Mycorrhiza-Function, Diversity, State of the Art*; Springer: Berlin/Heidelberg, Germany, 2017; pp. 133–176.
11. Liu, C.W.; Murray, J.D. The role of flavonoids in nodulation host-range specificity: An update. *Plants (Basel)* **2016**, *5*, 33. [CrossRef] [PubMed]
12. Mierziak, J.; Kostyn, K.; Kulma, A. Flavonoids as important molecules of plant interactions with the environment. *Molecules* **2014**, *19*, 16240–16265. [CrossRef]

13. Dudek, B.; Warskulat, A.-C.; Schneider, B. The Occurrence of Flavonoids and Related Compounds in Flower Sections of Papaver nudicaule. *Plants* **2016**, *5*, 28. [CrossRef]
14. Cesco, S.; Mimmo, T.; Tonon, G.; Tomasi, N.; Pinton, R.; Terzano, R.; Neumann, G.; Weisskopf, L.; Renella, G.; Landi, L.; et al. Plant-borne flavonoids released into the rhizosphere: Impact on soil bio-activities related to plant nutrition. A review. *Biol. Fertil. Soils* **2012**, *48*, 123–149. [CrossRef]
15. Nabavi, S.M.; Samec, D.; Tomczyk, M.; Milella, L.; Russo, D.; Habtemariam, S.; Suntar, I.; Rastrelli, L.; Daglia, M.; Xiao, J.; et al. Flavonoid biosynthetic pathways in plants: Versatile targets for metabolic engineering. *Biotechnol. Adv.* **2020**, *38*, 107316. [CrossRef] [PubMed]
16. Wang, Y.; Chen, S.; Yu, O. Metabolic engineering of flavonoids in plants and microorganisms. *Appl. Microbiol. Biotechnol.* **2011**, *91*, 949–956. [CrossRef] [PubMed]
17. Aherne, S.A.; O'Brien, N.M. Dietary flavonols: Chemistry, food content, and metabolism. *Nutrition* **2002**, *18*, 75–81. [CrossRef]
18. Panche, A.N.; Diwan, A.D.; Chandra, S.R. Flavonoids: An overview. *J. Nutr. Sci.* **2016**, *5*, e47. [CrossRef]
19. Samanta, A.; Das, G.; Das, S.K. Roles of Flavonoids in Plants. *Int. J. Pharm. Sci. Technol.* **2011**, *6*, 12–35.
20. Sultana, B.; Anwar, F. Flavonols (kaempeferol, quercetin, myricetin) contents of selected fruits, vegetables and medicinal plants. *Food Chem.* **2008**, *108*, 879–884. [CrossRef]
21. Crozier, A.; Lean, M.E.; McDonald, M.S.; Black, C. Quantitative analysis of the flavonoid content of commercial tomatoes, onions, lettuce, and celery. *J. Agric. Food Chem.* **1997**, *45*, 590–595. [CrossRef]
22. López-Lázaro, M. Distribution and biological activities of the flavonoid luteolin. *Mini Rev. Med. Chem.* **2009**, *9*, 31–59. [CrossRef]
23. Heim, K.E.; Tagliaferro, A.R.; Bobilya, D.J. Flavonoid antioxidants: Chemistry, metabolism and structure-activity relationships. *J. Nutr. Biochem.* **2002**, *13*, 572–584. [CrossRef]
24. Del Valle, I.; Webster, T.M.; Cheng, H.-Y.; Thies, J.E.; Kessler, A.; Miller, M.K.; Ball, Z.T.; MacKenzie, K.R.; Masiello, C.A.; Silberg, J.J.; et al. Soil organic matter attenuates the efficacy of flavonoid-based plant-microbe communication. *Sci. Adv.* **2020**, *6*, eaax8254. [CrossRef] [PubMed]
25. Erlund, I. Review of the flavonoids quercetin, hesperetin, and naringenin. Dietary sources, bioactivities, bioavailability, and epidemiology. *Nutr. Res.* **2004**, *24*, 851–874. [CrossRef]
26. Andersen, O.M.; Markham, K.R. *Flavonoids: Chemistry, Biochemistry and Applications*; CRC Press: Boca Raton, FL, USA, 2005.
27. Arts, I.C.; van de Putte, B.; Hollman, P.C. Catechin contents of foods commonly consumed in The Netherlands. 1. Fruits, vegetables, staple foods, and processed foods. *J. Agric. Food Chem.* **2000**, *48*, 1746–1751. [CrossRef] [PubMed]
28. Tsao, R. Chemistry and biochemistry of dietary polyphenols. *Nutrients* **2010**, *2*, 1231–1246. [CrossRef]
29. Hillel, D.; Hatfield, J.L. *Encyclopedia of Soils in the Environment*; Elsevier: Amsterdam, The Netherlands, 2005; Volume 3.
30. Dykhuizen, D. Species numbers in bacteria. *Proc. Calif. Acad. Sci.* **2005**, *56*, 62.
31. Lal, R.; Francaviglia, R. *Sustainable Agriculture Reviews 29: Sustainable Soil Management: Preventive and Ameliorative Strategies*; Springer: Berlin/Heidelberg, Germany, 2019; Volume 29.
32. Mia, M.B.; Shamsuddin, Z. Rhizobium as a crop enhancer and biofertilizer for increased cereal production. *Afr. J. Biotechnol.* **2010**, *9*, 6001–6009.
33. Afzal, A.; Bano, A. Rhizobium and phosphate solubilizing bacteria improve the yield and phosphorus uptake in wheat (Triticum aestivum). *Int. J. Agric. Biol.* **2008**, *10*, 85–88.
34. Hijri, M. Analysis of a large dataset of mycorrhiza inoculation field trials on potato shows highly significant increases in yield. *Mycorrhiza* **2016**, *26*, 209–214. [CrossRef]
35. Winkel-Shirley, B. Biosynthesis of flavonoids and effects of stress. *Curr. Opin. Plant Biol.* **2002**, *5*, 218–223. [CrossRef]
36. Dixon, R.A.; Pasinetti, G.M. Flavonoids and isoflavonoids: From plant biology to agriculture and neuroscience. *Plant Physiol.* **2010**, *154*, 453–457. [CrossRef]
37. Cetinkaya, H.; Kulak, M.; Karaman, M.; Karaman, H.S.; Kocer, F. Flavonoid accumulation behavior in response to the abiotic stress: Can a uniform mechanism be illustrated for all plants. In *Flavonoids—From Biosynthesis to Human Health*; Intechopen: London UK, 2017. [CrossRef]
38. Trichopoulou, A.; Vasilopoulou, E.; Hollman, P.; Chamalides, C.; Foufa, E.; Kaloudis, T.; Kromhout, D.;

Miskaki, P.; Petrochilou, I.; Poulima, E.; et al. Nutritional composition and flavonoid content of edible wild greens and green pies: A potential rich source of antioxidant nutrients in the Mediterranean diet. *Food Chem.* **2000**, *70*, 319–323. [CrossRef]

39. Parvez, M.; Tomita-Yokotani, K.; Fujii, Y.; Konishi, T.; Iwashina, T. Effects of quercetin and its derivatives on the growth of Arabidopsis thaliana and Neurospora crassa. *Biochem. Syst. Ecol.* **2004**, *32*, 631–635. [CrossRef]
40. Škerget, M.; Kotnik, P.; Hadolin, M.; Hraš, A.R.; Simonič, M.; Knez, Ž. Phenols, proanthocyanidins, flavones and flavonols in some plant materials and their antioxidant activities. *Food Chem.* **2005**, *89*, 191–198. [CrossRef]
41. Díaz, M.; Rossini, C. Bioactive natural products from Sapindaceae deterrent and toxic metabolites against insects. In *Insecticides–Pest Engineering*; Perveen, F., Ed.; InTech: Rijeka, Croatia, 2012; pp. 287–308.
42. Tatsimo, S.J.N.; Tamokou, J.d.D.; Havyarimana, L.; Csupor, D.; Forgo, P.; Hohmann, J.; Kuiate, J.-R.; Tane, P. Antimicrobial and antioxidant activity of kaempferol rhamnoside derivatives from Bryophyllum pinnatum. *BMC Res. Notes* **2012**, *5*, 158. [CrossRef] [PubMed]
43. Mendki, P.; Salunke, B.; Kotkar, H.; Maheshwari, V.; Mahulikar, P.; Kothari, R. Antimicrobial and Insecticidal Activities of Flavonoid Glycosides from Calotropis procera L. for Post-harvest Preservation of Pulses. *Biopestic. Int.* **2005**, *1*, 193–200.
44. Manach, C.; Scalbert, A.; Morand, C.; Rémésy, C.; Jiménez, L. Polyphenols: Food sources and bioavailability. *Am. J. Clin. Nutr.* **2004**, *79*, 727–747. [CrossRef]
45. Wu, N.-L.; Fang, J.-Y.; Chen, M.; Wu, C.-J.; Huang, C.-C.; Hung, C.-F. Chrysin protects epidermal keratinocytes from UVA- and UVB-induced damage. *J. Agric. Food Chem.* **2011**, *59*, 8391–8400. [CrossRef]
46. Liggins, J.; Bluck, L.; Runswick, S.; Atkinson, C.; Coward, W.; Bingham, S. Daidzein and genistein contents of vegetables. *Br. J. Nutr.* **2000**, *84*, 717–725. [CrossRef]
47. Pei, Y.; Siemann, E.; Tian, B.; Ding, J. Root flavonoids are related to enhanced AMF colonization of an invasive tree. *AoB Plants* **2020**, *12*. [CrossRef]
48. Cavia-Saiz, M.; Busto, M.D.; Pilar-Izquierdo, M.C.; Ortega, N.; Perez-Mateos, M.; Muñiz, P. Antioxidant properties, radical scavenging activity and biomolecule protection capacity of flavonoid naringenin and its glycoside naringin: A comparative study. *J. Sci. Food Agric.* **2010**, *90*, 1238–1244. [CrossRef]
49. Wilmsen, P.K.; Spada, D.S.; Salvador, M. Antioxidant activity of the flavonoid hesperidin in chemical and biological systems. *J. Agric. Food Chem.* **2005**, *53*, 4757–4761. [CrossRef] [PubMed]
50. Tsai, S.M.; Phillips, D.A. Flavonoids released naturally from alfalfa promote development of symbiotic glomus spores in vitro. *Appl. Environ. Microbiol.* **1991**, *57*, 1485–1488. [CrossRef]
51. Harborne, J.B. The flavonoids: Advances in research since 1986 (Harborne, J.B.). *J. Chem. Educ.* **1995**, *72*, A73. [CrossRef]
52. Naeimi, A.F.; Alizadeh, M. Antioxidant properties of the flavonoid fisetin: An updated review of in vivo and in vitro studies. *Trends Food Sci. Technol.* **2017**, *70*, 34–44. [CrossRef]
53. Ong, K.C.; Khoo, H.-E. Biological effects of myricetin. *Gen. Pharmacol. Vasc. Syst.* **1997**, *29*, 121–126. [CrossRef]
54. Peters, N.; Frost, J.; Long, S. A plant flavone, luteolin, induces expression of Rhizobium meliloti nodulation genes. *Science* **1986**, *233*, 977–980. [CrossRef]
55. Garcia, K.; Delaux, P.M.; Cope, K.R.; Ane, J.M. Molecular signals required for the establishment and maintenance of ectomycorrhizal symbioses. *New Phytol.* **2015**, *208*, 79–87. [CrossRef]
56. Rice-evans, C.A.; Miller, N.J.; Bolwell, P.G.; Bramley, P.M.; Pridham, J.B. The relative antioxidant activities of plant-derived polyphenolic flavonoids. *Free Radic. Res.* **1995**, *22*, 375–383. [CrossRef]
57. Tsanova-Savova, S.; Ribarova, F.; Gerova, M. (+) Catechin and (−)-epicatechin in Bulgarian fruits. *J. Food Compos. Anal.* **2005**, *18*, 691–698. [CrossRef]
58. Pathan, S.I.; Ceccherini, M.T.; Sunseri, F.; Lupini, A. Rhizosphere as hotspot for plant-soil-microbe interaction. In *Carbon and Nitrogen Cycling in Soil*; Springer: Berlin/Heidelberg, Germany, 2020; pp. 17–43.
59. Mandal, S.M.; Chakraborty, D.; Dey, S. Phenolic acids act as signaling molecules in plant-microbe symbioses. *Plant Signal. Behav.* **2010**, *5*, 359–368. [CrossRef]
60. Sugiyama, A.; Shitan, N.; Yazaki, K. Involvement of a soybean ATP-binding cassette-type transporter in the secretion of genistein, a signal flavonoid in legume-Rhizobium symbiosis. *Plant Physiol.* **2007**, *144*, 2000–2008. [CrossRef]

61. Shaw, L.J.; Morris, P.; Hooker, J.E. Perception and modification of plant flavonoid signals by rhizosphere microorganisms. *Environ. Microbiol.* **2006**, *8*, 1867–1880. [CrossRef] [PubMed]
62. Kovács, E.; Keresztes, Á. Effect of gamma and UV-B/C radiation on plant cells. *Micron* **2002**, *33*, 199–210. [CrossRef]
63. Chaparro, J.M.; Sheflin, A.M.; Manter, D.K.; Vivanco, J.M. Manipulating the soil microbiome to increase soil health and plant fertility. *Biol. Fertil. Soils* **2012**, *48*, 489–499. [CrossRef]
64. Munoz Aguilar, J.M.; Ashby, A.M.; Richards, A.J.M.; Loake, G.J.; Watson, M.D.; Shaw, C.H. Chemotaxis of Rhizobium leguminosarum biovar phaseoli towards flavonoid inducers of the symbiotic nodulation genes. *Microbiology* **1988**, *134*, 2741–2746. [CrossRef]
65. Hassan, S.; Mathesius, U. The role of flavonoids in root-rhizosphere signalling: Opportunities and challenges for improving plant-microbe interactions. *J. Exp. Bot.* **2012**, *63*, 3429–3444. [CrossRef]
66. Bolaños-Vásquez, M.C.; Werner, D. Effects of Rhizobium tropici, R. etli, and R. leguminosarum bv. phaseoli on nod gene-inducing flavonoids in root exudates of Phaseolus vulgaris. *Mol. Plant Microbe Interact.* **1997**, *10*, 339–346. [CrossRef]
67. Stambulska, U.Y.; Bayliak, M.M. Legume-rhizobium symbiosis: Secondary metabolites, free radical processes, and effects of heavy metals. In *Bioactive Molecules in Food*; Springer: Berlin/Heidelberg, Germany, 2019; pp. 1–32. [CrossRef]
68. Davidson, E.A. The contribution of manure and fertilizer nitrogen to atmospheric nitrous oxide since 1860. *Nat. Geosci.* **2009**, *2*, 659–662. [CrossRef]
69. Hartwig, U.A.; Joseph, C.M.; Phillips, D.A. Flavonoids released naturally from alfalfa seeds enhance growth rate of Rhizobium meliloti. *Plant Physiol.* **1991**, *95*, 797–803. [CrossRef]
70. Maxwell, C.A.; Hartwig, U.A.; Joseph, C.M.; Phillips, D.A. A chalcone and two related flavonoids released from alfalfa roots induce *nod* genes of *Rhizobium meliloti. Plant Physiol.* **1989**, *91*, 842–847. [CrossRef]
71. Dolatabadian, A.; Sanavy, S.A.M.M.; Ghanati, F.; Gresshoff, P.M. Morphological and physiological response of soybean treated with the microsymbiont Bradyrhizobium japonicum pre-incubated with genistein. *S. Afr. J. Bot.* **2012**, *79*, 9–18. [CrossRef]
72. Subramanian, S.; Stacey, G.; Yu, O. Endogenous isoflavones are essential for the establishment of symbiosis between soybean and Bradyrhizobium japonicum. *Plant J.* **2006**, *48*, 261–273. [CrossRef] [PubMed]
73. Jones, F.R.; Tisdale, W. Effect of soil temperature upon the development of nodules on the roots of certain legumes. *J. Agric. Res.* **1921**, *22*, 17–37.
74. Lynch, D.; Smith, D. Soybean (Glycine max) modulation and N2-fixation as affected by exposure to a low root-zone temperature. *Physiol. Plant.* **1993**, *88*, 212–220. [CrossRef]
75. Pan, B.; Zhang, F.; Smith, D.L. Genistein addition to the rooting medium of soybean at the onset of nitrogen fixation increases nodulation. *J. Plant Nutr.* **1998**, *21*, 1631–1639. [CrossRef]
76. Gibson, A. Factors in the physical and biological environment affecting nodulation and nitrogen fixation by legumes. *Plant Soil* **1971**, *35*, 139–152. [CrossRef]
77. Layzell, D.; Rochman, P.; Canvin, D. Low root temperatures and nitrogenase activity in soybean. *Can. J. Bot.* **1984**, *62*, 965–971. [CrossRef]
78. Zhang, F.; Smith, D.L. Preincubation of bradyrhizobium japonicum with genistein accelerates nodule development of soybean at suboptimal root zone temperatures. *Plant Physiol.* **1995**, *108*, 961–968. [CrossRef]
79. Kasper, S.; Christoffersen, B.; Soti, P.; Racelis, A. Abiotic and biotic limitations to nodulation by leguminous cover crops in South Texas. *Agriculture* **2019**, *9*, 209. [CrossRef]
80. Hemida, M.; Issa, A.A.; Ohyam, T. Impact of harsh environmental conditions on nodule formation and dinitrogen fixation of legumes. In *Advances in Biology and Ecology of Nitrogen Fixation*; Intechopen: London, UK, 2014. [CrossRef]
81. Zhang, F.; Smith, D.L. Application of genistein to inocula and soil to overcome low spring soil temperature inhibition of soybean nodulation and nitrogen fixation. *Plant Soil* **1997**, *192*, 141–151. [CrossRef]
82. Belkheir, A.M.; Zhou, X.; Smith, D.L. Variability in yield and yield component responses to genistein pre-incubated Bradyrhizobium japonicum by soybean [Glycine max (L.) Merr] cultivars. *Plant soil* **2001**, *229*, 41–46. [CrossRef]
83. Belkheir, A.; Zhou, X.; Smith, D. Response of soybean [Glycine max (L.) Merr.] cultivars to

genistein-preincubated bradyrhizobium japonicum: Nodulation and dry matter accumulation under Canadian short-season conditions. *J. Agron. Crop Sci.* **2000**, *185*, 167–175. [CrossRef]

84. Morrison, M.; Cober, E.; Saleem, M.; McLaughlin, N.; Fregeau-Reid, J.; Ma, B.; Yan, W.; Woodrow, L. Changes in isoflavone concentration with 58 years of genetic improvement of short-season soybean cultivars in Canada. *Crop Sci.* **2008**, *48*, 2201–2208. [CrossRef]
85. Hinsinger, P. Bioavailability of soil inorganic P in the rhizosphere as affected by root-induced chemical changes: A review. *Plant soil* **2001**, *237*, 173–195. [CrossRef]
86. Mohammadi, K.; Khalesro, S.; Sohrabi, Y.; Heidari, G. A review: Beneficial effects of the mycorrhizal fungi for plant growth. *J. Appl. Environ. Biol. Sci.* **2011**, *1*, 310–319.
87. Parniske, M. Arbuscular mycorrhiza: The mother of plant root endosymbioses. *Nat. Rev. Microbiol.* **2008**, *6*, 763–775. [CrossRef]
88. Vierheilig, H.; Bago, B.; Albrecht, C.; Poulin, M.-J.; Piché, Y. Flavonoids and arbuscular-mycorrhizal fungi. In *Flavonoids in the Living System*; Manthey, J.A., Buslig, B.S., Eds.; Springer: Boston, MA, USA, 1998; pp. 9–33. [CrossRef]
89. Becard, G.; Taylor, L.P.; Douds, D.D.; Pfeffer, P.E.; Doner, L.W. Flavonoids are not necessary plant signal compounds in arbuscular mycorrhizal symbioses. *MPMI Mol. Plant Microbe Interact.* **1995**, *8*, 252–258. [CrossRef]
90. Siqueira, J.; Safir, G.; Nair, M. Stimulation of vesicular-arbuscular mycorrhiza formation and growth of white clover by flavonoid compounds. *New Phytol.* **1991**, *118*, 87–93. [CrossRef]
91. Davies, F.T.; Calderón, C.M.; Huaman, Z. Influence of arbuscular mycorrhizae indigenous to peru and a flavonoid on growth, yield, and leaf elemental concentration of 'yungay'potatoes. *HortScience* **2005**, *40*, 381–385. [CrossRef]
92. Davies, F.T., Jr.; Calderón, C.M.; Huaman, Z.; Gómez, R. Influence of a flavonoid (formononetin) on mycorrhizal activity and potato crop productivity in the highlands of Peru. *Sci. Hortic.* **2005**, *106*, 318–329. [CrossRef]
93. De Almeida Ribeiro, P.R.; dos SANTOS, J.V.; de Carvalho, T.S.; da Silva, J.S.; de Resende, P.M.; de Souza Moreira, F.M. Formononetin associated with phosphorus influences soybean symbiosis with mycorrhizal fungi and Bradyrhizobium. *Biosci. J.* **2016**, *32*.
94. Scervino, J.M.; Ponce, M.A.; Erra-Bassells, R.; Bompadre, J.; Vierheilig, H.; Ocampo, J.A.; Godeas, A. The effect of flavones and flavonols on colonization of tomato plants by arbuscular mycorrhizal fungi of the genera Gigaspora and Glomus. *Can. J. microbiol.* **2007**, *53*, 702–709. [CrossRef] [PubMed]
95. Nakabayashi, R.; Mori, T.; Saito, K. Alternation of flavonoid accumulation under drought stress in Arabidopsis thaliana. *Plant Signal. Behav.* **2014**, *9*, e29518. [CrossRef]
96. Dakora, F.D.; Phillips, D.A. Root exudates as mediators of mineral acquisition in low-nutrient environments. *Plant Soil* **2002**, *245*, 35–47. [CrossRef]
97. Gupta, R.; Chakrabarty, S. Gibberellic acid in plant: Still a mystery unresolved. *Plant Signal. Behav.* **2013**, *8*, e25504. [CrossRef]
98. Tomasi, N.; Weisskopf, L.; Renella, G.; Landi, L.; Pinton, R.; Varanini, Z.; Nannipieri, P.; Torrent, J.; Martinoia, E.; Cesco, S. Flavonoids of white lupin roots participate in phosphorus mobilization from soil. *Soil Biol. Biochem.* **2008**, *40*, 1971–1974. [CrossRef]
99. Cesco, S.; Neumann, G.; Tomasi, N.; Pinton, R.; Weisskopf, L. Release of plant-borne flavonoids into the rhizosphere and their role in plant nutrition. *Plant Soil* **2010**, *329*, 1–25. [CrossRef]
100. Cramer, G.R.; Urano, K.; Delrot, S.; Pezzotti, M.; Shinozaki, K. Effects of abiotic stress on plants: A systems biology perspective. *BMC Plant Biol.* **2011**, *11*, 163. [CrossRef]
101. Haak, D.C.; Fukao, T.; Grene, R.; Hua, Z.; Ivanov, R.; Perrella, G.; Li, S. Multilevel regulation of abiotic stress responses in plants. *Front. Plant Sci.* **2017**, *8*, 1564. [CrossRef]
102. Nihorimbere, V.; Ongena, M.; Smargiassi, M.; Thonart, P. Beneficial effect of the rhizosphere microbial community for plant growth and health. *Biotechnol. Agron. Soc. Environ.* **2011**, *15*, 327–337.
103. Walker, T.S.; Bais, H.P.; Grotewold, E.; Vivanco, J.M. Root exudation and rhizosphere biology. *Plant Physiol.* **2003**, *132*, 44–51. [CrossRef] [PubMed]
104. Chen, H.; Gao, H.; Fang, X.; Ye, L.; Zhou, Y.; Yang, H. Effects of allyl isothiocyanate treatment on postharvest

quality and the activities of antioxidant enzymes of mulberry fruit. *Postharvest Biol. Technol.* **2015**, *108*, 61–67. [CrossRef]

105. Solovchenko, A.; Schmitz-Eiberger, M. Significance of skin flavonoids for UV-B-protection in apple fruits. *J. Exp. Bot.* **2003**, *54*, 1977–1984. [CrossRef] [PubMed]
106. Mariz-Ponte, N.; Mendes, R.; Sario, S.; de Oliveira, J.F.; Melo, P.; Santos, C. Tomato plants use non-enzymatic antioxidant pathways to cope with moderate UV-A/B irradiation: A contribution to the use of UV-A/B in horticulture. *J. Plant Physiol.* **2018**, *221*, 32–42. [CrossRef]
107. Merzlyak, M.N.; Solovchenko, A.E.; Gitelson, A.A. Reflectance spectral features and non-destructive estimation of chlorophyll, carotenoid and anthocyanin content in apple fruit. *Postharvest Biol. Technol.* **2003**, *27*, 197–211. [CrossRef]
108. Schmitz-Hoerner, R.; Weissenböck, G. Contribution of phenolic compounds to the UV-B screening capacity of developing barley primary leaves in relation to DNA damage and repair under elevated UV-B levels. *Phytochemistry* **2003**, *64*, 243–255. [CrossRef]
109. Tossi, V.; Lombardo, C.; Cassia, R.; Lamattina, L. RETRACTED: Nitric oxide and flavonoids are systemically induced by UV-B in maize leaves. *Plant Sci.* **2012**, *193–194*, 103–109. [CrossRef]
110. Pathan, M.S.; Lee, J.-D.; Shannon, J.G.; Nguyen, H.T. Recent advances in breeding for drought and salt stress tolerance in soybean. In *Advances in Molecular Breeding Toward Drought and Salt Tolerant Crops*; Jenks, M.A., Hasegawa, P.M., Jain, S.M., Eds.; Springer: Dordrecht, The Netherlands, 2007; pp. 739–773. [CrossRef]
111. Ilangumaran, G.; Smith, D.L. Plant growth promoting rhizobacteria in amelioration of salinity stress: A systems biology perspective. *Front. Plant Sci.* **2017**, *8*, 1768. [CrossRef]
112. Greenway, H.; Munns, R. Mechanisms of salt tolerance in nonhalophytes. *Ann. Rev. Plant Physiol.* **1980**, *31*, 149–190. [CrossRef]
113. Chaves, M.M.; Flexas, J.; Pinheiro, C. Photosynthesis under drought and salt stress: Regulation mechanisms from whole plant to cell. *Ann. Bot.* **2009**, *103*, 551–560. [CrossRef]
114. Perri, S.; Entekhabi, D.; Molini, A. Plant osmoregulation as an emergent water-saving adaptation. *Water Resour. Res.* **2018**, *54*, 2781–2798. [CrossRef]
115. Bernstein, N. Plants and salt: Plant response and adaptations to salinity. In *Model Ecosystems in Extreme Environments*; Elsevier: Amsterdam, The Netherlands, 2019; pp. 101–112.
116. Jones, G.W.; Gorham, J. Intra-and Inter-Cellular Compartmentation of Ions. In *Salinity: Environment-Plants-Molecules*; Springer: Dordrecht, The Netherlands, 2002; pp. 159–180.
117. Flowers, T.J.; Colmer, T.D. Salinity tolerance in halophytes. *New Phytol.* **2008**, *179*, 945–963. [CrossRef] [PubMed]
118. Munns, R.; Tester, M. Mechanisms of salinity tolerance. *Ann. Rev. Plant Biol.* **2008**, *59*, 651–681. [CrossRef] [PubMed]
119. Koyro, H.-W. Effect of salinity on growth, photosynthesis, water relations and solute composition of the potential cash crop halophyte *Plantago coronopus* (L.). *Environ. Exp. Bot.* **2006**, *56*, 136–146. [CrossRef]
120. Rhodes, D.; Nadolska-Orczyk, A.; Rich, P. Salinity, osmolytes and compatible solutes. In *Salinity: Environment-Plants-Molecules*; Springer: Dordrecht, The Netherlands, 2002; pp. 181–204.
121. Al Hassan, M.; MartÍNez Fuertes, M.; Ramos SÁNchez, F.J.; Vicente, O.; Boscaiu, M. Effects of salt and water stress on plant growth and on accumulation of osmolytes and antioxidant compounds in cherry tomato. *Not. Bot. Horti Agrobot. Cluj Napoca* **2015**, *43*, 1–11. [CrossRef]
122. AbdElgawad, H.; Zinta, G.; Hegab, M.M.; Pandey, R.; Asard, H.; Abuelsoud, W. High salinity induces different oxidative stress and antioxidant responses in maize seedlings organs. *Front. Plant Sci.* **2016**, *7*, 276. [CrossRef]
123. Wang, H.-M.; Xiao, X.-R.; Yang, M.-Y.; Gao, Z.-L.; Zang, J.; Fu, X.-M.; Chen, Y.-H. Effects of salt stress on antioxidant defense system in the root of Kandelia candel. *Bot. Stud.* **2014**, *55*, 57. [CrossRef]
124. Abogadallah, G.M. Antioxidative defense under salt stress. *Plant Signal. Behav.* **2010**, *5*, 369–374. [CrossRef]
125. Engwa, G.A. Free radicals and the role of plant phytochemicals as antioxidants against oxidative stress-related diseases. In *Phytochemicals: Source of Antioxidants and Role in Disease Prevention. BoD–Books on Demand*; Intechopen: London, UK, 2018; pp. 49–74.
126. Halliwell, B.; Gutteridge, J.M.C. *Free Radicals in Biology and Medicine*, 5th ed.; Oxford University Press: New York, NY, USA, 2015.

127. Smith, A.M.; Ratcliffe, R.G.; Sweetlove, L.J. Activation and function of mitochondrial uncoupling protein in plants. *J. Biol. Chem.* **2004**, *279*, 51944–51952. [CrossRef]
128. Pietta, P.-G. Flavonoids as antioxidants. *J. Nat. Prod.* **2000**, *63*, 1035–1042. [CrossRef]
129. Lü, J.M.; Lin, P.H.; Yao, Q.; Chen, C. Chemical and molecular mechanisms of antioxidants: Experimental approaches and model systems. *J. Cell. Mol. Med.* **2010**, *14*, 840–860. [CrossRef] [PubMed]
130. Di Ferdinando, M.; Brunetti, C.; Fini, A.; Tattini, M. Flavonoids as antioxidants in plants under abiotic stresses. In *Abiotic Stress Responses in Plants*; Springer: New York, NY, USA, 2012; pp. 159–179. [CrossRef]
131. Kaliamoortiiy, S.; Rao, A. Effect of salinity on anthocyanin accumulation in the root of maize. *Science* **1994**, *248*, 1637–1638.
132. Sugiyama, A.; Yazaki, K. Flavonoids in plant rhizospheres: Secretion, fate and their effects on biological communication. *Plant Biotechnol.* **2014**, *31*, 431–443. [CrossRef]
133. Miransari, M.; Smith, D. Overcoming the stressful effects of salinity and acidity on soybean nodulation and yields using signal molecule genistein under field conditions. *J. Plant Nutr.* **2007**, *30*, 1967–1992. [CrossRef]
134. Miransari, M.; Smith, D. Alleviating salt stress on soybean (Glycine max (L.) Merr.)–Bradyrhizobium japonicum symbiosis, using signal molecule genistein. *Eur. J. Soil Biol.* **2009**, *45*, 146–152. [CrossRef]
135. Ma, D.; Sun, D.; Wang, C.; Li, Y.; Guo, T. Expression of flavonoid biosynthesis genes and accumulation of flavonoid in wheat leaves in response to drought stress. *Plant Physiol. Biochem.* **2014**, *80*, 60–66. [CrossRef]
136. Varela, M.C.; Arslan, I.; Reginato, M.A.; Cenzano, A.M.; Luna, M.V. Phenolic compounds as indicators of drought resistance in shrubs from Patagonian shrublands (Argentina). *Plant Physiol. Biochem.* **2016**, *104*, 81–91. [CrossRef]
137. Koes, R.E.; Quattrocchio, F.; Mol, J.N. The flavonoid biosynthetic pathway in plants: Function and evolution. *BioEssays* **1994**, *16*, 123–132. [CrossRef]
138. Maddox, C.E.; Laur, L.M.; Tian, L. Antibacterial activity of phenolic compounds against the phytopathogen Xylella fastidiosa. *Curr. Microbiol.* **2010**, *60*, 53. [CrossRef]
139. Rattan, R.S. Mechanism of action of insecticidal secondary metabolites of plant origin. *Crop Prot.* **2010**, *29*, 913–920. [CrossRef]
140. Kaplan, D.; Keen, N.; Thomason, I. Studies on the mode of action of glyceollin in soybean incompatibility to the root knot nematode, Meloidogyne incognita. *Physiol. Plant Pathol.* **1980**, *16*, 319–325. [CrossRef]
141. Chin, S.; Behm, C.A.; Mathesius, U. Functions of flavonoids in plant(-)nematode interactions. *Plants (Basel)* **2018**, *7*, 85. [CrossRef] [PubMed]
142. Huang, J.-S.; Barker, K.R. Glyceollin I in soybean-cyst nematode interactions: Spatial and temporal distribution in roots of resistant and susceptible soybeans. *Plant Physiol.* **1991**, *96*, 1302–1307. [CrossRef] [PubMed]
143. Soriano, I.; Asenstorfer, R.; Schmidt, O.; Riley, I. Inducible flavone in oats (Avena sativa) is a novel defense against plant-parasitic nematodes. *Phytopathology* **2004**, *94*, 1207–1214. [CrossRef] [PubMed]
144. Fawe, A.; Abou-Zaid, M.; Menzies, J.; Bélanger, R. Silicon-mediated accumulation of flavonoid phytoalexins in cucumber. *Phytopathology* **1998**, *88*, 396–401. [CrossRef]
145. del Río, J.A.; Gómez, P.; Baidez, A.G.; Arcas, M.C.; Botía, J.M.; Ortuño, A. Changes in the levels of polymethoxyflavones and flavanones as part of the defense mechanism of citrus sinensis (Cv. Valencia Late) fruits against phytophthora citrophthora. *J. Agric. Food Chem.* **2004**, *52*, 1913–1917. [CrossRef]
146. Ortuno, A.; Arcas, M.; Botia, J.; Fuster, M.; Del Río, J. Increasing resistance against Phytophthora citrophthora in tangelo Nova fruits by modulating polymethoxyflavones levels. *J. Agric. Food Chem.* **2002**, *50*, 2836–2839. [CrossRef]
147. Boué, S.M.; Carter, C.H.; Ehrlich, K.C.; Cleveland, T.E. Induction of the soybean phytoalexins coumestrol and glyceollin by Aspergillus. *J. Agric. Food Chem.* **2000**, *48*, 2167–2172. [CrossRef]
148. Cushnie, T.P.; Lamb, A.J. Recent advances in understanding the antibacterial properties of flavonoids. *Int. J. Antimicrob. Agents* **2011**, *38*, 99–107. [CrossRef]
149. Gnanamanickam, S.S.; Patil, S.S. Accumulation of antibacterial isoflavonoids in hypersensitively responding bean leaf tissues inoculated with Pseudomonas phaseolicola. *Physiol. Plant Pathol.* **1977**, *10*, 159–168. [CrossRef]
150. Lyon, F.M.; Wood, R. Production of phaseollin, coumestrol and related compounds in bean leaves inoculated with Pseudomonas spp. *Physiol. Plant Pathol.* **1975**, *6*, 117–124. [CrossRef]

151. Keen, N.; Kennedy, B. Hydroxyphaseollin and related isoflavanoids in the hypersensitive resistance reaction of soybeans to Pseudomonas glycinea. *Physiol. Plant Pathol.* **1974**, *4*, 173–185. [CrossRef]
152. Wyman, J.G. Antibacterial activity of selected isoflavonoids. *Phytopathology* **1978**, *68*, 583. [CrossRef]
153. Tanaka, H.; Sato, M.; Oh-Uchi, T.; Yamaguchi, R.; Etoh, H.; Shimizu, H.; Sako, M.; Takeuchi, H. Antibacterial properties of a new isoflavonoid from Erythrina poeppigiana against methicillin-resistant Staphylococcus aureus. *Phytomedicine* **2004**, *11*, 331–337. [CrossRef] [PubMed]
154. War, A.R.; Paulraj, M.G.; Ahmad, T.; Buhroo, A.A.; Hussain, B.; Ignacimuthu, S.; Sharma, H.C. Mechanisms of plant defense against insect herbivores. *Plant Signal. Behav.* **2012**, *7*, 1306–1320. [CrossRef] [PubMed]
155. Gao, Q.-M.; Zhu, S.; Kachroo, P.; Kachroo, A. Signal regulators of systemic acquired resistance. *Front. Plant Sci.* **2015**, *6*. [CrossRef]
156. Tridiptasari, A.; Brawijaya, U.; Leksono, A.S.; Siswanto, D. Antifeedant Effect of *Moringa oleifera* (L.) Leaf and Seed Extract on Growth and Feeding Activity of Spodoptera litura (Fab.) (Lepidoptera: Noctuidae). *J. Exp. Life Sci.* **2019**, *9*, 25–31. [CrossRef]
157. Aljbory, Z.; Chen, M.S. Indirect plant defense against insect herbivores: A review. *Insect Sci.* **2018**, *25*, 2–23. [CrossRef]
158. Rosell, G.; Quero, C.; Coll, J.; Guerrero, A. Biorational insecticides in pest management. *J. Pestic. Sci.* **2008**, *33*, 103–121. [CrossRef]
159. Quiroz, A.; Mendez, L.; Mutis, A.; Hormazabal, E.; Ortega, F.; Birkett, M.A.; Parra, L. Antifeedant activity of red clover root isoflavonoids on Hylastinus obscurus. *J. Soil Sci. Plant Nutr.* **2017**, *17*, 231–239. [CrossRef]
160. Goławska, S.; Lukasik, I. Antifeedant activity of luteolin and genistein against the pea aphid, Acyrthosiphon pisum. *J. Pest. Sci. (2004)* **2012**, *85*, 443–450. [CrossRef] [PubMed]
161. Johnson, E.T.; Dowd, P.F. Differentially enhanced insect resistance, at a cost, in Arabidopsis thaliana constitutively expressing a transcription factor of defensive metabolites. *J. Agric. Food Chem.* **2004**, *52*, 5135–5138. [CrossRef] [PubMed]
162. Simmonds, M.S. Importance of flavonoids in insect–plant interactions: Feeding and oviposition. *Phytochemistry* **2001**, *56*, 245–252. [CrossRef]
163. Lattanzio, V.; Lattanzio, V.M.; Cardinali, A. Role of phenolics in the resistance mechanisms of plants against fungal pathogens and insects. *Phytochem. Adv. Res.* **2006**, *661*, 23–67.
164. Nenaah, G.E. Toxic and antifeedant activities of prenylated flavonoids isolated from Tephrosia apollinea L. against three major coleopteran pests of stored grains with reference to their structure–activity relationship. *Nat. Prod. Res.* **2014**, *28*, 2245–2252. [CrossRef]
165. Weston, L.A.; Duke, S.O. Weed and crop allelopathy. *Crit. Rev. Plant Sci.* **2003**, *22*, 367–389. [CrossRef]
166. Oerke, E.C. Crop losses to pests. *J. Agric. Sci.* **2005**, *144*, 31–43. [CrossRef]
167. Young, S.L.; Pierce, F.J. *Automation: The Future of Weed Control*; Springer Science and Business Media: Berlin/Heidelberg, Germany, 2013.
168. Molisch, H. *Der Einfluss Einer Pflanze Auf die Andere, Allelopathie*; Fischer: Jena, Germany, 1937.
169. Rice, E. *Allelopathy*, 2nd ed.; Acad. Press Inc.: Orlando, FL, USA, 1984; p. 422.
170. Rob, M.M.; Hossen, K.; Iwasaki, A.; Suenaga, K.; Kato-Noguchi, H. Phytotoxic activity and identification of phytotoxic substances from schumannianthus dichotomus. *Plants* **2020**, *9*, 102. [CrossRef]
171. Jabran, K.; Mahajan, G.; Sardana, V.; Chauhan, B.S. Allelopathy for weed control in agricultural systems. *Crop Prot.* **2015**, *72*, 57–65. [CrossRef]
172. Alsaadawi, I.S.; Sarbout, A.K.; Al-Shamma, L.M. Differential allelopathic potential of sunflower (*Helianthus annuus* L.) genotypes on weeds and wheat (*Triticum aestivum* L.) crop. *Arch. Agron. Soil Sci.* **2012**, *58*, 1139–1148. [CrossRef]
173. Palma-Tenango, M.; Soto-Hernández, M.; Aguirre-Hernández, E. Flavonoids in agriculture. In *Flavonoids—From Biosynthesis to Human Health*; Intech: Rijeka, Croatia, 2017. [CrossRef]
174. Weston, L.A.; Mathesius, U. Flavonoids: Their structure, biosynthesis and role in the rhizosphere, including allelopathy. *J. Chem. Ecol.* **2013**, *39*, 283–297. [CrossRef] [PubMed]
175. Zhang, H.; Gao, J.-M.; Liu, W.-T.; Tang, J.-C.; Zhang, X.-C.; Jin, Z.-G.; Xu, Y.-P.; Shao, M.-A. Allelopathic substances from walnut (Juglans regia L.) leaves. *Allelopath. J.* **2008**, *21*, 425–431.
176. Yan, Z.; Guo, H.; Yang, J.; Liu, Q.; Jin, H.; Xu, R.; Cui, H.; Qin, B. Phytotoxic flavonoids from roots of Stellera chamaejasme L. (Thymelaeaceae). *Phytochemistry* **2014**, *106*, 61–68. [CrossRef] [PubMed]

177. Kaab, S.B.; Rebey, I.B.; Hanafi, M.; Hammi, K.M.; Smaoui, A.; Fauconnier, M.L.; De Clerck, C.; Jijakli, M.H.; Ksouri, R. Screening of Tunisian plant extracts for herbicidal activity and formulation of a bioherbicide based on Cynara cardunculus. *S. Afr. J. Bot.* **2020**, *128*, 67–76. [CrossRef]
178. Al-obaidi, A.F. Phytotoxicity of plantago major extracts on germination and seedling growth of purslane (portulaca oleracea). In *Seed Dormancy and Germination*; IntechOpen: London, UK, 2020.

12

Roles of Nitrogen Deep Placement on Grain Yield, Nitrogen use Efficiency and Antioxidant Enzyme Activities in Mechanical Pot-Seedling Transplanting Rice

Lin Li [1,2], Zheng Zhang [1,2], Hua Tian [1,2], Zhaowen Mo [1,2], Umair Ashraf [1,3], Meiyang Duan [1,2], Zaiman Wang [4], Shuli Wang [1,2], Xiangru Tang [1,2,*] and Shenggang Pan [1,2,*]

1 State Key Laboratory for Conservation and Utilization of Subtropical Agro-bioresources, College of Agriculture, South China Agricultural University, Guangzhou 510642, China; scaulilin@126.com (L.L.); scauzhangzheng@126.com (Z.Z.); tianhua@scau.edu.cn (H.T.); scaumozhw@126.com (Z.M.); umairashraf2056@gmail.com (U.A.); meiyang@scau.edu.cn (M.D.); wangshuli@scau.edu.cn (S.W.)

2 Scientific Observing and Experimental Station of Crop Cultivation in South China, Ministry of Agriculture PR China, Guangzhou 510642, China

3 Department of Botany, University of Education (Lahore), Faisalabad-Campus, Faisalabad 38000, Punjab, Pakistan

4 Key Laboratory of Key Technology for South Agricultural Machine and Equipment, Ministry of Education, Guangzhou 510642, China; wangzaiman@scau.edu.cn

* Correspondence: tangxr@scau.edu.cn (X.T.); panshenggang@scau.edu.cn (S.P.)

Abstract: Mechanical pot-seedling transplanting (PST) is an efficient transplanting method and deep nitrogen fertilization has the advantage of increasing nitrogen use efficiency. However, little information is available about the effect of PST when coupled with mechanized deep nitrogen (N) fertilization on grain yield, nitrogen use efficiency, and antioxidant enzyme activities in rice. A two-year field experiment was performed to evaluate the effect of PST coupled with deep N fertilization in both early seasons (March–July) of 2018 and 2019. All seedlings were transplanted by PST and three treatments were designed as follows. There was a mechanized deep placement of all fertilizer (MAF), broadcasting fertilizer (BF), no fertilizer (N0). MAF significantly increased grain yield by 52.7%. Total nitrogen accumulation (TNA) was enhanced by 27.7%, nitrogen partial factor productivity (NPFP) was enhanced by 51.4%. nitrogen recovery efficiency (NRE) by 123.7%, and nitrogen agronomic efficiency (NAE) was enhanced by 104.3%, compared with BF treatment. Moreover, MAF significantly improved peroxidase (POD), catalase (CAT), and notably reduced the malonic dialdehyde (MDA) content for both rice cultivars, compared to BF. Hence, the result shows that mechanical pot-seedling transplanting coupled with nitrogen deep placement is an efficient method with the increase of grain yield and nitrogen use efficiency in rice cultivation in South China.

Keywords: deep N fertilization; peroxidase activity; catalase activity; rice cultivation

---

1. Introduction

Rice (*Oryza sativa* L.) is one of the world's major crops and it provides food for over three billion people [1,2]. China is the main country of rice production, with rice planting area and yield in the forefront of the world [2,3]. Therefore, increasing rice production is essential for population growth in China and the world [4,5]. Transplanted rice is the most traditional planting method. Moreover, the traditional fertilization method is manually surface broadcast [6]. This method is not suitable for the stable improvement in Chinese agricultural systems because of some serious problems such as labor scarcity and low profits [7]. Therefore, improving mechanization is the main way to solve this problem.

The mechanical pot-seedling transplanting (PST) is an innovative technology for transplanting rice seedlings in the paddy field. It has the advantage of precise row and hill spacing without injury to the rice plants [8]. The technique not only reduces planting costs but improves the quality of transplanting rice seedlings [9]. Mechanical pot-seedling transplanting can transplant seedlings without root injury. It expects to reduce the transplanting shock and maintain the root activity, resulting in increased nutrient absorption and consequently a vigorous initial growth [10]. However, the application of nitrogen fertilizer under the flooding condition can cause nitrogen fertilizer to be lost through ammonia volatilization and runoff, reducing the utilization efficiency of nitrogen fertilizer [11,12]. Thus, the seedlings transplanted by mechanical pot-seedling transplanting were unable to access the N resource, restricting their performance irrespective of enhanced root activity [4]. To solve this problem, much application of chemical fertilizer is one of the approaches, but it can lead to problems of lower nitrogen use efficiency (NUE) and environmental pollution [11 13].

Alternative is the deep N fertilization in mechanical pot-seedling transplanting. Deep fertilizer application methods can maintain the nutrient and enhance nutrient use efficiency [14]. The nitrogen fertilization at about 5 cm depth notably improves the total above-ground biomass and grain yield, compared to the manual surface broadcast [15]. Moreover, deep fertilization could reduce the amount of fertilizer applications without reducing yield [16]. Deep fertilization is to bury the fertilizer near the rice root, which is beneficial for fertilizer absorption by rice root [17].

The appropriate fertilization depth could not only promote the growth of rice root but also improve the growth of rice plants in the early stage [18]. Some studies found that deep N fertilization could reduce CH_4 emissions by 40%and NO emissions by 54% [19,20]. Pan et al. [21] found that deep fertilization significantly increased peroxidase (POD) and catalase (CAT) in direct-seeded rice. Xu et al. [22] observed that ensuring the nutrient supply at the late stage of rice was conducive to improving antioxidant enzyme activities and photosynthetic performance of rice leaves. Moreover, Shu et al. [23] reported that mechanical deep fertilization could delay rice plan senescence by enhancing antioxidant enzyme activities and reducing the malonic dialdehyde (MDA) content in direct-seeded rice. Therefore, deep nitrogen fertilization is a feasible way to lessen environmental problems because of the excess fertilization in rice production.

Mechanical pot-seedling transplanting (PST) coupled with deep nitrogen fertilization is an emerging transplanting rice technology. However, little information is available about the effects of PST coupled with mechanized deep N fertilization on grain yield, nitrogen use efficiency, and antioxidant enzyme activities in rice. The aim of this study was to assess whether PST coupled with mechanized N deep fertilization could increase grain yield, nitrogen use efficiency, and antioxidant enzyme activities in rice.

2. Materials and Methods

2.1. *Mechanical Pot-Seedling Transplanting Machine*

A mechanical pot-seedling transplanting machine was developed by Changzhou YaMeiKe mechanical Co., Ltd. (Changzhou, China) (Figure 1). This method realized the synchronous operation of deep fertilization and transplanting seedlings and applied fertilizer quantitatively and fixed-point deep into the soil on the seedling side.

Figure 1. Pictorial view of mechanized transplanting rice machine coupled with N deep placement at the farm of South China Agricultural University, Guangzhou city, China.

2.2. Experimental Treatments and Design

Field experiments were conducted in early seasons of 2018 and 2019, respectively, at the Experimental Research Farm, College of Agriculture, South China Agricultural University, Guangzhou City, China (23°13′ N, 113°81′ E, altitude 11 m). The soil in the experimental field was sandy loam with 1010 mg kg^{-1} total N, 1080 mg kg^{-1} total *p*, 20,230 mg kg^{-1} total K, 73 mg kg^{-1} available *p*, 104 mg kg^{-1} available K, and 21,560 mg kg^{-1} organic C.

Two rice cultivars were *Yuxiangyouzhan* (*YXYZ*) and *Wufengyou615* (*WFY615*), which are inbred and hybrid rice, respectively, and widely grown in the local area. Moreover, the two rice cultivars have growth periods of 118 and 115 days for both early seasons, respectively. Both field experiments were used in a completely randomized design with three replicates with a plot area of 132 m^2 (8 m × 16.5 m). The YaraMila compound fertilizer (total nitrogen contents TN = 15%, N: P_2O_5: K_2O = 15%:15%:15%) was used in our experiment, which was manufactured by YaraMila Fertilizer Company, China. The application rate was 150 kg N ha^{-1} (pure N) for the fertilizer application treatment. All seedlings were transplanted by PST and three treatments were designed as follows. The mechanized deep N fertilization was a basal fertilizer in 10 cm soil depth (MAF) and fertilizers were broadcast manually on the soil surface two days before the transplanting as a basal fertilizer (BF). No fertilizer was applied during entire growth stage (N0). Water management strategies were adopted by local farmer's advice. Some chemical reagents such as herbicide, imidacloprid, tricyclazole, and carbendazim were adopted to prevent and control weeds, insects, and diseases.

2.3. Yield and Its Components

At maturity, rice grains were recorded from the harvested-area of 6 m^2. In total, 20 rice plants were collected randomly for each treatment and the averaged values were calculated for the number of productive panicles per hill. Six hills of rice plants were taken to investigate yield components and the yield components measurement were determined according to Pan et al. [21]. To divide the filled seeds, all spikelets were submerged in tap water, apart from the rachis (by manual threshing). To calculate the total number of spikelets, we counted the spikelets in three representative subsamples of 30 g. The average weight of half-filled spikelets was determined. Spikelets per panicle, grain-filling percentage, and 1000-grain-weight were also calculated from sampled plants and averaged.

2.4. Total Above-Ground Biomass (TAB) and Leaf Area Index (LAI) at Different Growth Stages

According to the average number of tillers in the plot, six plants were taken in the plot at all critical growth stages including mid-tillering (MT), panicle initiation (PI), heading stage (HS), and the maturity stage (MS). LAI and TAB were determined according to Pan et al. [21]. The soil on the rice plants were washed thoroughly. Then, leaf sheaths plus stems, leaves, and spikes were separated from the plant after the heading stage. The leaf area for all green leaf blades was measured with the Li-Cor area meter (Li Cor Model 3100, Lincoln, NE, USA) and the leaf area per m^2 (leaf area index, LAI) was then calculated. The separated part of the rice plants was oven-dried at 70 °C to constant weight and then the above-ground biomass was calculated.

2.5. Nitrogen Use Efficiency

Six plants were collected from each treatment in the physiological maturity stage. They were then divided into leaves, stems with leaf sheath, and grains. They were finally dried at 70 °C until constant weight, then stored to analyze the total N contents. Plant samples (0.2 g) were digested using the Kjeldhal method to analyze ammonia concentrations via an Alliance-Futura NP analyzer (Alliance Instruments, France) and then the N content was measured. The nitrogen use efficiency including nitrogen recovery efficiency (NRE), nitrogen agronomic use efficiency (NAE), nitrogen partial factor productivity (NPFP), nitrogen harvest index (NHI), and nitrogen grain production efficiency (NGPE) were evaluated by the formulae below:

1. NRE = (Nup − N0 up)/FN
2. NAE = (GY − GY0)/FN
3. NPFP = GY/FN
4. NGPE = GY/Nup
5. NHI = Ng/Nup

where N0 up and GY0 represented the total nitrogen uptake of above-ground plant parts and grain yields in the N0 plot, respectively. Nup and GY are the total nitrogen uptake of above-ground plant parts and grain yields in other N-fertilized plots, respectively. FN is the applied N fertilizer rate; Ng is the total nitrogen uptake in grain.

2.6. Determination of Antioxidant Enzyme Activities

About 25 leaves from each treatment were collected during the MT, PI, and HS stage. All samples were stored in −80 °C for enzyme activity determination i.e., peroxidase (POD), catalase (CAT), and malonic dialdehyde (MDA). POD and MDA were determined by the method established by Pan et al. [24]. Fresh leaf segments (<2 mm, 0.25 g) were homogenized in an ice bath in 5 mL of 50 mm borate buffer (pH 8.7) containing 5.0 mm sodium hydrogen sulfite and 0.1 g polyvinylpyrrolidone (PVP). The homogenate was centrifuged at 9000 ×g for 15 min at 4 °C. The supernatant was used as enzyme extract. POD activity was assayed by adding 0.1 mL of the enzyme extract to a substrate mixture containing acetate buffer (0.1 mol L^{-1}, pH 5.4), ortho-dianisidine (0.25% in ethyl alcohol) and 0.1 mL 0.8% H_2 O_2 was added to 0.1 mL of the enzyme extract. Absorbance change of the brown guaiacol at 460 nm was recorded for calculating POD activity. One POD unit of enzyme activity was defined as the absorbance increase because of guaiacol oxidation by 1-unit min^{-1} (U g^{-1} FW min^{-1}). Leaf samples (0.5 g) were homogenized in 5 mL of 5% trichloroacetic acid. The homogenate was centrifuged at 4000 ×g for 10 min at 25 °C and 3 mL of 2-thiobarbituric acid in 20% trichloroacetic acid was added to a 2 mL aliquot of the supernatant. The mixture was heated at 98 °C for 10 min and cooled rapidly in an ice bath. After centrifugation at 4000 ×g for 10 min, the absorbance was recorded at 532 nm. Measurements were corrected for non-specific turbidity by subtracting the absorbance at 600 nm. MDA concentration was determined by the extinction coefficient MDA (ε = 155 μm cm^{-1}). CAT activities were determined according to Dhindsa et al. [25]. The 3 mL reaction mixture contained 50 mM phosphate buffer, pH 7.0, 15 mM hydrogen peroxide, and 25 μL enzyme extract. The decrease in hydrogen peroxide was followed as a decline in A240 using a Perkin-Elmer double-beam spectrophotometer connected to a recorder. The activity was expressed in units where one unit of catalase converts one μmole of hydrogen peroxide per minute.

2.7. Data Analysis

The experimental data were analyzed using DPS3.11 (Data Processing System for Practical Statistics, Hangzhou, China). In the ANOVA model, the single effect of treatment, cultivar, year, and the interaction effect were fixed, while the replication effect in year was random. The differences amongst means of the experimental treatments were separated using the least significant difference (LSD) at 0.05 probability level (ANOVA). All figures were drawn with Origin 9.0.

3. Results

3.1. Grain Yield and Its Components

The grain yield and its components varied with different N fertilizer applications for both rice cultivars (Table 1). The highest grain yield for MAF treatment was found for the BF and N0 treatments in both years. Mean grain yields of both rice cultivars under MAF were 8.4 t ha^{-1}, which was 52.7% higher than the BF treatment. Regarding yield components, deep placement produced the highest number of productive panicles ha^{-1} and spikelet per panicle, which was 272.9 × 10^4 and 189.7, respectively. No significance was found between BF and N0 treatment in the 1000-grain-weight and grain filling. Significant differences were found in the number of productive panicles, spikelets per panicle, 1000-grain-weight, and grain yield between nitrogen treatments. Both rice cultivars differed significantly in 1000-grain-weight.

Table 1. Effects of mechanical deep placement of nitrogen fertilizer on average grain yield and its components for both rice cultivars in two-year (2018 and 2019).

Treatments	Productive Panicle (10^4 Ha^{-1})	Spikelet per Panicle	Grain Filling (%)	1000-Grain-Weight (G)	Harvested Yield (T Ha^{-1})
N0	151.4 b	150.3 b	80.9	20.1 b	3.9 c
MAF	272.9 a	189.7 a	82.7	22.2 a	8.4 a
BF	185.3 b	163.3 b	81.6	21.1 b	5.5 b
Anova					
Cultivar (C)	ns	ns	ns	**	ns
Nitrogen (N)	**	*	ns	**	**
N × C	ns	ns	ns	ns	ns

MAF: mechanized deep placement of all fertilizer; BF: broadcasting fertilizer; N0: no fertilizer. Average values followed by different letter represent LSD significant differences at $p < 0.05$. **: ($p < 0.01$); *: ($p < 0.05$); ns: not significant variance.

3.2. Nitrogen Use Efficiency

The TNA, NGPE, NHI, NAE, NPFP, and NRE varied with different N fertilizer application in both rice cultivars (Table 2). In two-years, the N fertilizer application (MAF and BF) treatments significantly increased TNA relative to N0 treatment. Moreover, MAF treatment significantly increased TNA compared to BF treatment. The TNA of both rice cultivars for MAF was 173.6 kg ha^{-1}, which was 27.7% higher than the BF treatment, respectively. NGPE was highest in MAF, followed by BF, and the lowest was the N0 treatment. MAF showed the maximum NRE and NAE, while the BF had the lowest value for NRE and NRE among all treatments, respectively. Furthermore, a significant difference was found between MAF and BF. The NPFP of both rice cultivars for MAF was 54.2 kg kg^{-1}, which was

51.4 % higher than that in the BF treatment, respectively. Moreover, no remarkable difference among all treatment was found in NHI. There were notable differences in TNA, NGPE, NPFP, NAE, and NRE between nitrogen treatments. Moreover, the N × C (Nitrogen × Cultivar) factor interactions also had an obvious impact on TNA and NHI (Table 2).

Table 2. Effects of mechanical deep placement of nitrogen fertilizer on average nitrogen use efficiency for both rice cultivars in two-year (2018 and 2019).

Treatments	TNA ($kg\ ha^{-1}$)	NGPE ($kg\ kg^{-1}$)	NPFP ($kg\ kg^{-1}$)	NHI	NRE (%)	NAE ($kg\ kg^{-1}$)
N0	100.6 c	39.2 b	26.8 c	53.2		
MAF	173.6 a	46.6 a	54.2 a	54.3	46.8 a	22.6 a
BF	135.9 b	41.4 b	35.8 b	50.8	22.9 b	10.1 b
Anova						
Cultivar(C)	**	ns	*	*	ns	*
Nitrogen(N)	**	*	**	ns	**	**
N × C	**	ns	ns	**	ns	ns

Average values followed by different letter represent LSD significant differences at $p < 0.05$. **: ($p < 0.01$); *: ($p < 0.05$); ns: not significant variance. TNA: Total nitrogen accumulation; NAE: N agronomic efficiency, NRE: N recovery efficiency, NGPE: N grain production efficiency; NHI: N harvest index, NPFP: nitrogen partial factor productivity.

3.3. Total Aboveground Biomass (TAB) and Leaf Area Index (LAI) at Different Growth Stage

The N fertilizer application remarkably affected the LAI for both rice cultivars (Figure 2). For example, during the MT stage, the LAI for BF and MAF treatments were significantly higher than N0, but no significant difference was found between MAF and BF. At the PI and HS stages, MAF was significantly larger for LAI, especially when compared to N0 and BF. The result manifested that deep placement of the N application could modulate a sustainable longer growth period than surface broadcasting.

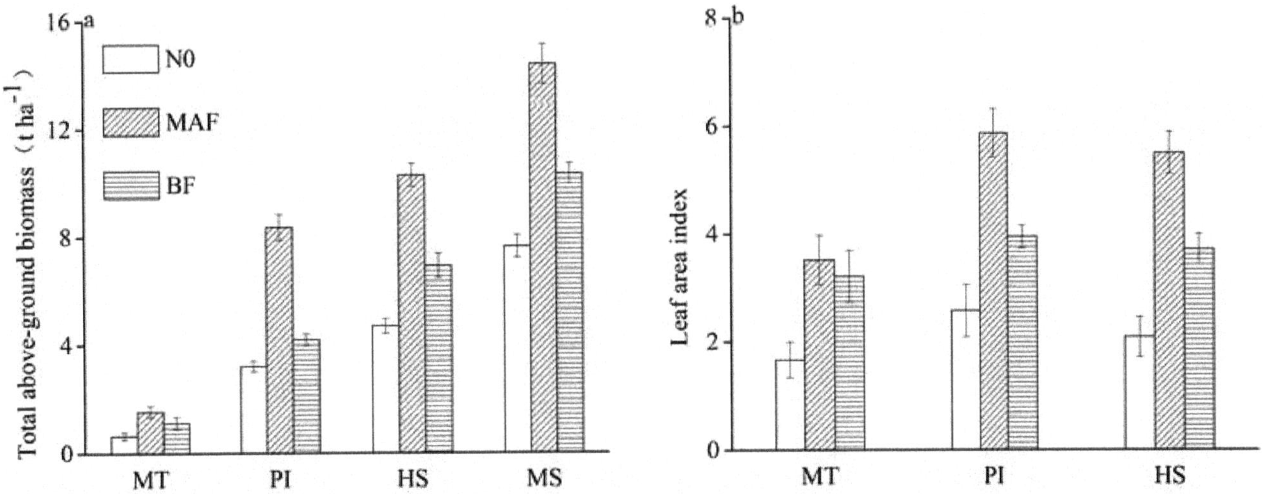

Figure 2. Effects of mechanical deep placement of nitrogen fertilizer on average for the total above-ground biomass and leaf area index for both rice cultivars (2018 and 2019). (**a**): total above-ground biomass, (**b**): leaf area index. MT: Mid-tillering stage; PI: Panicle initiation stage; HS: Heading stage; MS: Maturity stage.

The N fertilizer application remarkably affected the TAB for both rice cultivars (Figure 2). For example, at the MT stage, the TAB for BF and MAF treatments were significantly higher than N0. There were no significant differences between MAF and BF, and the TAB for MAF was higher than BF. At the HS and MS stages, N fertilizer application (MAF and BF) treatments remarkably increased TBA, especially when compared to the N0 treatment. Moreover, there were remarkable differences between

MAF and BF treatments. In the whole growth period, a similar trend for TAB of both rice cultivars was observed.

3.4. *Determination of Antioxidant Enzymatic Activity*

3.4.1. POD Activity

The POD activity in the leaves at all critical growth stages including the MT, PI, and HS stages were shown in Figure 3. At the MT stage, the highest POD activities were observed in the MAF treatment. However, the POD activity of the BF treatment was higher than the N0 treatment while lower than the MAF treatment. At the PI stage, MAF treatment significantly improved POD activity, especially when compared to the BF treatment. However, the POD activity did not differ significantly between BF and N0 treatment. A significant difference was found among all treatments at the HS stage. In the whole growth period, the similar trend for POD activity of both rice cultivars was observed.

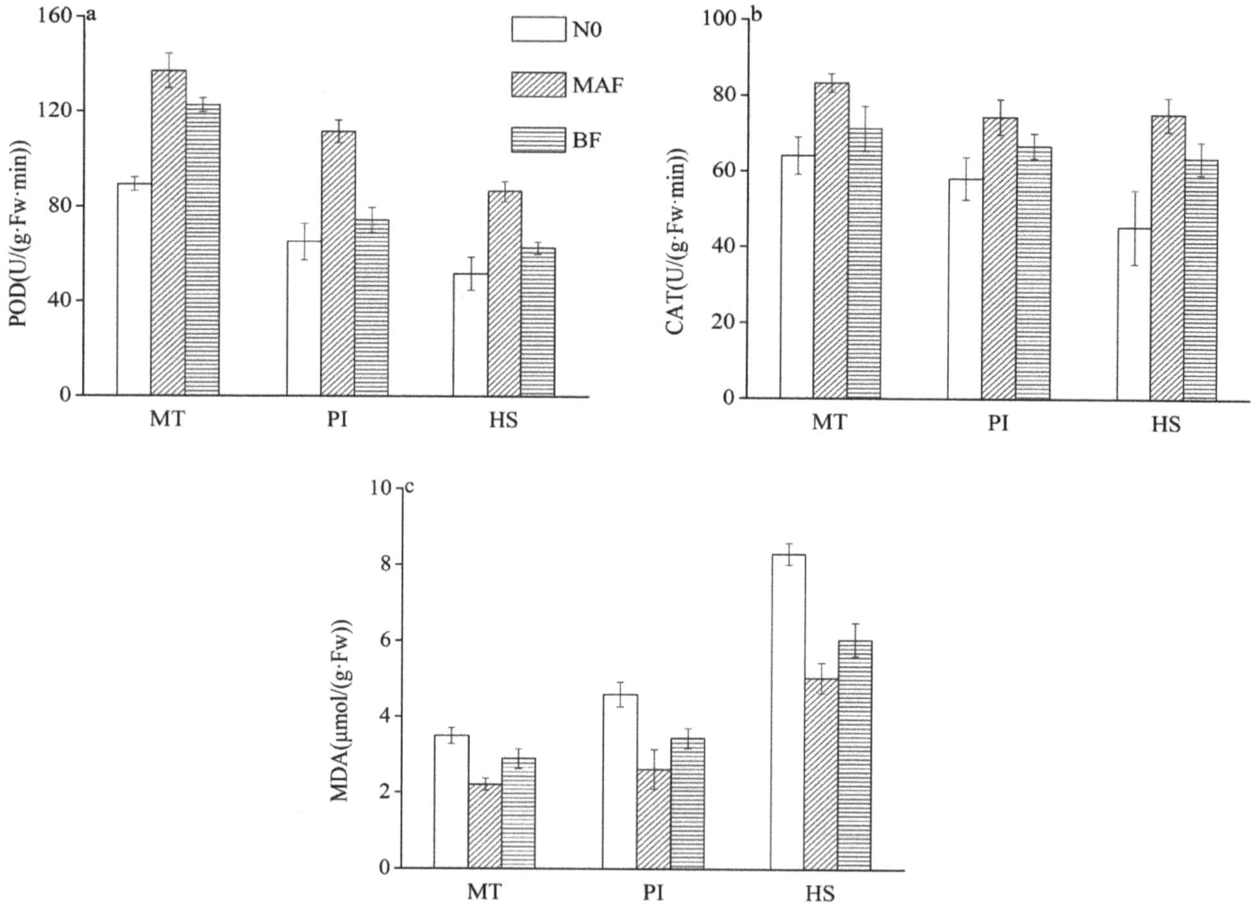

Figure 3. Effects of mechanical deep placement of nitrogen fertilizer on average for POD, CAT activity, and MDA content for both rice cultivars (2018 and 2019). (**a**): POD, (**b**): CAT, (**c**): MDA.

3.4.2. CAT Activity

The CAT activity in the leaves at all critical growth stages including the MT, PI, and HS stages were shown in Figure 3. There were significant differences in CAT activity found between the MAF and BF treatments at the MT stage. However, the BF and N0 treatments did not differ significantly. A significant difference of CAT activity was found between MAF and N0 treatment at the PI stage. At the HS stage, the highest CAT activities were observed in the MAF treatment over two years. However, CAT activity was higher in the BF treatment than the N0 treatment, while remaining lower than MAF treatment. Furthermore, a significant difference was found among all treatments.

3.4.3. MDA Content

The MDA content in the leaves at all critical growth stages including the MT, PI, and HS stages were shown in Figure 3. At the MT stage, the lowest MDA contents for both rice cultivars over two years were observed in the MAF treatment. However, the MDA content was higher in the BF treatment than the MAF treatment, while remaining lower than the N0 treatment. At the PI stage, compared to the N0 treatment, the BF and MAF treatments could significantly decrease MDA content. Marginal differences in MDA content was found between the MAF and BF treatments. Significant differences of MDA content were found among all treatments at the HS stage.

3.5. Correlation Analysis

The relationship among grain yield and its components TAB, TNA, LAI, and antioxidant enzyme activities for both rice cultivars over two years (Table 3). The variation of N treatment was focused on in the correlation analysis. Rice yield was significantly and positively correlated with the number of productive panicles ha^{-1}, spikelet per panicle, TAB, LAI, and antioxidant enzyme activities, including POD and CAT activities. The spikelet per panicle also significantly correlated with LAI, TAB, and MDA content. Furthermore, TAB at the maturity stage and LAI at the HS stage significantly correlated with antioxidant enzyme activities and TNA.

Table 3. Relationship among grain yield, and its components, TAB, TNA, LAI, and antioxidant enzyme activities for both rice cultivars (2018 and 2019).

Parameter	Productive Panicle	Spikelet per Panicle	Grain Filling	1000-Grain -Weigh (G)	LAI	TAB	POD	CAT	MDA	TNA
Productive panicle										
Spikelet per panicle	0.28									
Grain filling	−0.03	−0.18								
1000-grain-weight(g)	0.37 *	0.18	0.33 *							
LAI	0.53 **	0.47 **	0.20	0.52 **						
TAB	0.68 **	0.44 **	−0.06	0.51 **	0.85 **					
POD activity	0.55 **	0.27	0.46 **	0.45 **	0.76 **	0.60 **				
CAT activity	0.49 **	0.51 **	−0.58 **	0.03	0.47 **	0.60 **	0.28			
MDA content	−0.48 **	−0.44 **	0.54 **	−0.43 **	−0.52 **	−0.70 **	−0.18	−0.74 **		
TNA	0.65 **	0.62	−0.03	0.41 *	0.74 **	0.71 **	0.62 **	0.52 **	−0.56 **	
Yield	0.73 **	0.39	−0.06	0.44 **	0.75 **	0.79 **	0.68 **	0.62 **	−0.61 **	0.79 **

LAI: Leaf area index at HS stage; TAB: Total above-ground biomass at MS stage; POD: Peroxidase; CAT: Catalase, MDA: malonic dialdehyde at HS stage, TNA: Total nitrogen accumulation at MS stage. The same as below. **: ($p < 0.01$); *: ($p < 0.05$)

4. Discussion

4.1. Grain Yield and Its Components

Compared with broadcasting fertilizer treatment (BF), mechanized deep placement of all fertilizers at once (MAF) significantly enhanced the grain yields in mechanical pot-seedling transplanting (PST). The highest yields of MAF treatment for both rice cultivars were mainly due to the number of productive panicles ha^{-1} and spikelet per panicle, which was in agreement with Bandaogo et al. [26]. Moreover, yield increase via a mechanized deep placement in PST was 52.7% for both rice cultivars. The increase was far larger than in previous studies on mechanized deep placement in mechanical carpet-seedling transplanting (6.3–11.6%) [15], suggesting that the mechanized deep placement method could be more effective under PST conditions than non-PST conditions. Moreover, Pan et al. [21] found that deep fertilization remarkably improved spikelet number per panicle in direct-seeded rice compared to manual surface broadcast fertilizers. Tracing it to the cause, the PST was used in our experiment, which opened a fassula and then uniformly placed the N fertilizer at 10 cm depth. Finally, the fassula was covered by this applicator immediately. This method provided a continuous nitrogen supply for rice growth and the NH_4^+ absorbed by the root system remained in the soil for a longer period of time, thus promoting the growth of rice plants throughout the growth period and thereby increasing the nitrogen absorption and grain yield [27,28]. Deep placement fertilizer also maintained higher antioxidant enzyme activities at the heading stage, which was one of the reasons for the higher grain yield.

Compared with BF treatment, MAF significantly enhanced the leaf area index (LAI) and total above-ground biomass (TAB) at the PI and HS stages in PST. The main reason for this was that MAF treatment promoted rice growth in the middle and late stages by reducing nutrition loss. Broadcasting fertilizers did not meet this demand and led to insufficient nutrient supply. Moreover, LAI and TAB increased by mechanized deep placement in PST, reaching 58.5% and 40.2% for both rice cultivars, respectively. The increase was larger than in previous studies on mechanized deep placement in mechanical carpet-seedling transplanting (36.1–38.9% and 8.7–10.6%) [15,29]. In addition, a larger TAB was beneficial to the transportation of dry matter to the panicle, leading to more spikelet per panicle and 1000-grain-weight of rice, so as to improve the yield of rice. Our results showed that TAB at the MS stage remarkably correlated with LAI at the HS stage, and both of them were positively related to grain yield. The result indicated that MAF had some superiority in larger LAI and TAB, which thus resulted in a higher grain yield.

4.2. *Nitrogen Use Efficiency and Antioxidant Enzyme Activities*

We discovered that MAF led to a substantial increase in total N accumulation (TNA), N recovery efficiency (NRE), and agronomic efficiency (NAE) compared to BF treatment. Moreover, NAE and NRE increase by mechanized deep placement in PST was 104.3% and 123.7% for both rice cultivars, respectively. The increase was far larger than in previous studies on mechanized deep placement in mechanical carpet-seedling transplanting (17.9–43.1% and 19.6–37.4%), suggesting that the mechanized deep placement method could be more effective to improve NRE and NAE under PST conditions than non-PST conditions [29,30]. Some researchers have showed that deep fertilization could improve NUE by reducing nitrogen loss and prolonging the duration of fertilizer, which was compared with surface broadcasting [31,32]. Previous results also found that the deep N fertilization could reduce urease activity by increasing NH_4^+ concentration in soil depth, thereby reducing NH_4^+ concentration in flood water [33,34]. The reason was that the deep nitrogen fertilizer in the anaerobic soil layer caused the NH_4^+ to move slowly from the depth to soil surface, thereby the NH_4^+ content in the flood was low. Moreover, the decrease of N concentration in the flood reduced the loss of N through runoff, ammonia volatilization, and denitrification, thus improving the NUE. Deep placement of fertilizers was a concentrated application of fertilizer near the roots of rice, which was conducive to the absorption of roots and improved the NUE [18,35]. We also found that TNA content was significantly related to grain yield and antioxidant enzyme activities in a positive correlation, because the higher antioxidant enzyme activity at the full heading stage was decisive to the transfer of nutrients to grains, and increased the accumulation of nitrogen in grains, thus increasing the NAE and NRE of rice plants.

Our results showed that MAF notably increased antioxidant enzyme activities including POD and CAT and reduced the MDA content of both rice cultivars. Some reports have showed that the appropriate application of nitrogen fertilizer could maintain high antioxidant enzyme activities at the MS stage, which would be beneficial when delaying the senescence of functional leaves of rice [36,37]. The reason was that the deep application of fertilizer could fulfill nutrient requirements of rice growth and development in time, providing sufficient energy for the antioxidant enzyme activities of the plant, enhancing the activity of the plant leaf protection enzyme system, accelerating the scavenging of free radicals, and reducing membrane lipid peroxidation and the MDA content in the rice plant [38]. Moreover, the rice plant had a stronger root system and longer green leaf duration under deep N fertilization [4]. We also found a significant positive correlation between POD and CAT activities and grain yield. N-fertilized plants need such effective antioxidant machinery to cope with excessive reactive oxygen species production. In this way, it can delay the senescence of leaves in the late growth stage [23,38]. Therefore, maintaining high POD and CAT activities at the HS stage was conducive to a higher rice yield.

5. Conclusions

Compared to broadcasting fertilizer treatment, mechanized deep N fertilization enhanced rice growth under mechanical pot-seedling transplanting (PST) in terms of leaf area index and total above-ground biomass. Mechanized deep N fertilization also improved the grain yield, nitrogen use efficiency, and antioxidant enzyme activities in PST. Therefore, mechanized deep fertilization in mechanical pot-seedling transplanting should be recommended in actual production.

Author Contributions: Conceptualization, S.P. and X.T.; Methodology, S.P. and X.T.; Data curation, H.T., Z.W., M.D., and S.W.; Formal analysis, L.L., H.T., Z.Z., and M.D.; Project administration, S.W., Z.W., and Z.Z.; Supervision, L.L., H.T., S.P., and X.T.; Validation, L.L., H.T., Z.M., M.D, and U.A.; Writing—original draft, L.L., H.T., Z.M., and M.D.; Writing—review and editing, L.L., H.T., Z.M., M.D., U.A., Z.W., S.P., and X.T. All authors have read and agreed to the published version of the manuscript.

Acknowledgments: We thank Leilei Kong who provided the rice cultivars. The research was supported by Key-Area Research and Development Program of Guangdong Province (No. 2019 B020221003) and National Natural Science Foundation of China (31471442).

References

1. Liu, K.; Deng, J.; Lu, J.; Wang, X.; Lu, B.; Tian, X.; Zhang, Y. High Nitrogen Levels Alleviate Yield Loss of Super Hybrid Rice Caused by High Temperatures During the Flowering Stage. *Front. Plant Sci.* **2019**, *10*, 357. [CrossRef] [PubMed]
2. He, A.; Wang, W.; Jiang, G.; Sun, H.; Jiang, M.; Man, J.; Cui, K.; Huang, J.; Peng, S.; Nie, L. Source-sink regulation and its effects on the regeneration ability of ratoon rice. *Field Crop. Res.* **2019**, *236*, 155–164. [CrossRef]
3. Kong, L.; Ashraf, U.; Cheng, S.; Rao, G.; Mo, Z.; Tian, H.; Pan, S.; Tang, X. Short-term water management at early filling stage improves early-season rice performance under high temperature stress in South China. *Eur. J. Agron.* **2017**, *90*, 117–126. [CrossRef]
4. Chen, Y.; Fan, P.; Mo, Z.; Kong, L.; Tian, H.; Duan, M.; Li, L.; Wu, L.; Wang, Z.; Tang, X.; et al. Deep Placement of Nitrogen Fertilizer Affects Grain Yield, Nitrogen Recovery Efficiency, and Root Characteristics in Direct-Seeded Rice in South China. *J. Plant Growth Regul.* **2020**, *2*, 1–9. [CrossRef]
5. Ashraf, U.; Hussain, S.; Akbar, N.; Anjum, S.A.; Hassan, W.; Tang, X. Water management regimes alter Pb uptake and translocation in fragrant rice. *Ecotoxicol. Environ. Saf.* **2018**, *149*, 128–134. [CrossRef]
6. Tao, Y.; Chen, Q.; Peng, S.; Wang, W.; Nie, L. Lower global warming potential and higher yield of wet direct-seeded rice in Central China. *Agron. Sustain. Dev.* **2016**, *36*, 24. [CrossRef]
7. Paman, U.; Inaba, S.; Uchida, S. The mechanization of small-scale rice farming: Labor requirements and costs. *Eng. Agric. Environ. Food* **2014**, *7*, 122–126. [CrossRef]
8. Hu, Y.J.; Xing, Z.P.; Gong, J.L.; Liu, G.T.; Zhang, H.C.; Dai, Q.G.; Huo, Z.Y.; Xu, K.; Wei, H.Y.; Guo, B.W.; et al. Study on population characteristics and formation mechanisms for high yield of pot-seedling mechanical transplanting rice. *Sci. Agri. Sin.* **2014**, *47*, 865–879.
9. Jiang, X.; Li, X.; Chi, Z.; Wang, S.; Yang, F.; Zheng, J. Research on potted-tray grown rice seedling transplanting by machine. *Agri. Sci. Tech.* **2014**, *15*, 1923–1927.
10. Zhu, C.; Yang, H.J.; Guan, Y.X.; Chen, Z. Research progress on green and high efficiency cultivation technique of rice pot seedling mechanical transplanting in Jiangsu province. *China Rice* **2019**, *25*, 37–41.
11. Gaihre, Y.K.; Singh, U.; Islam, S.M.; Huda, A.; Satter, M.A.; Sanabria, J.; Islam, R.; Shah, A. Impacts of urea deep placement on nitrous oxide and nitric oxide emissions from rice fields in Bangladesh. *Geoderma* **2015**, *260*, 370–379. [CrossRef]
12. Watanabe, T.; Son, T.T.; Hung, N.N.; Van Truong, N.; Giau, T.Q.; Hayashi, K.; Ito, O. Measurement of ammonia volatilization from flooded paddy fields in Vietnam. *Soil Sci. Plant Nutr.* **2009**, *55*, 793–799. [CrossRef]

13. Yao, Y.; Zhang, M.; Tian, Y.; Zhao, M.; Zhang, B.; Zhao, M.; Zeng, K.; Yin, B. Urea deep placement for minimizing NH3 loss in an intensive rice cropping system. *Field Crop. Res.* **2018**, *218*, 254–266. [CrossRef]
14. Du, B.; Luo, H.W.; He, L.X.; Zheng, A.X.; Chen, Y.L.; Zhang, T.T.; Wang, Z.M.; Hu, L.; Tang, X.R. DEEP FERTILIZER PLACEMENT IMPROVES RICE GROWTH AND YIELD IN ZERO TILLAGE. *Appl. Ecol. Environ. Res.* **2018**, *16*, 8045–8054. [CrossRef]
15. Zhu, C.; Xiang, J.; Zhang, Y.; Zhang, Y.; Defeng, Z.; Chen, H. Mechanized transplanting with side deep fertilization increases yield and nitrogen use efficiency of rice in Eastern China. *Sci. Rep.* **2019**, *9*, 5653. [CrossRef]
16. Guo, L.; Ning, T.; Nie, L.; Li, Z.; Lal, R. Interaction of deep placed controlled-release urea and water retention agent on nitrogen and water use and maize yield. *Eur. J. Agron.* **2016**, *75*, 118–129. [CrossRef]
17. Li, X.; Chen, X.; Wang, H.; Lu, D.; Zhou, J.; Chen, Z.; Zhu, D. Effects and principle of root-zone one-time N fertilization on enhancing rice (Oryza sativa L.) N use efficiency. *Soils* **2017**, *49*, 868–875.
18. Wu, M.; Liu, M.; Liu, J.; Li, W.; Jiang, C.-Y.; Li, Z.-P. Optimize nitrogen fertilization location in root-growing zone to increase grain yield and nitrogen use efficiency of transplanted rice in subtropical China. *J. Integr. Agric.* **2017**, *16*, 2073–2081. [CrossRef]
19. Linquist, B.A.; Adviento-Borbe, M.A.; Pittelkow, C.M.; Van Kessel, C.; Van Groenigen, K.J. Fertilizer management practices and greenhouse gas emissions from rice systems: A quantitative review and analysis. *Field Crop. Res.* **2012**, *135*, 10–21. [CrossRef]
20. Yao, Z.; Zheng, X.; Zhang, Y.; Liu, C.; Wang, R.; Lin, S.; Zuo, Q.; Butterbach-Bahl, K. Urea deep placement reduces yield-scaled greenhouse gas (CH4 and N2O) and NO emissions from a ground cover rice production system. *Sci. Rep.* **2017**, *7*, 11415. [CrossRef]
21. Pan, S.; Wen, X.; Wang, Z.; Ashraf, U.; Tian, H.; Duan, M.; Mo, Z.; Fan, P.; Tang, X. Benefits of mechanized deep placement of nitrogen fertilizer in direct-seeded rice in South China. *Field Crop. Res.* **2017**, *203*, 139–149. [CrossRef]
22. Xu, Y.L.; Fu, A.B.; Liu, T.X. Effects of different nitrogen application patterns on leaf physiological characteristics and grain yield of double cropping rice. *Acta Agric. Bor. Sin.* **2019**, *34*, 155–163.
23. Shu, S.; Tang, X.R.; Luo, X.W.; Li, G.X.; Wang, Z.M.; Zheng, T.X.; Jia, X.N. Effects of deep mechanized application of slow-release fertilizers on physiological characteristics of precision hill-direct-seeding super rice. *Trans. Chin. Soc. Agric. Eng.* **2011**, *27*, 89–92.
24. Pan, S.; Rasul, F.; Li, W.; Tian, H.; Mo, Z.; Duan, M.; Tang, X. Roles of plant growth regulators on yield, grain qualities and antioxidant enzyme activities in super hybrid rice (Oryza sativa L.). *Rice* **2013**, *6*, 9. [CrossRef]
25. Plumb-Dhindsa, P.; Dhindsa, R.S.; Thorpe, T.A. Leaf Senescence: Correlated with Increased Levels of Membrane Permeability and Lipid Peroxidation, and Decreased Levels of Superoxide Dismutase and Catalase. *J. Exp. Bot.* **1981**, *32*, 93–101. [CrossRef]
26. Bandaogo, A.; Fofana, B.; Youl, S.; Safo, E.; Abaidoo, R.C.; Andrews, O. Effect of fertilizer deep placement with urea supergranule on nitrogen use efficiency of irrigated rice in Sourou Valley (Burkina Faso). *Nutr. Cycl. Agroecosystems* **2014**, *102*, 79–89. [CrossRef]
27. Kargbo, M.; Pan, S.; Mo, Z.; Wang, Z.; Luo, X.; Tian, H.; Hossain, F.; Ashraf, U.; Tang, X. Physiological Basis of Improved Performance of Super Rice (Oryza sativa) to Deep Placed Fertilizer with Precision Hill-drilling Machine. *Int. J. Agric. Boil.* **2016**, *18*, 797–804. [CrossRef]
28. Kapoor, V.; Singh, U.; Patil, S.K.; Magre, H.; Shrivastava, L.K.; Mishra, V.N.; Das, R.O.; Samadhiya, V.K.; Sanabria, J.; Diamond, R. Rice Growth, Grain Yield, and Floodwater Nutrient Dynamics as Affected by Nutrient Placement Method and Rate. *Agron. J.* **2008**, *100*, 526–536. [CrossRef]
29. Zhu, C.H.; Zhang, Y.P.; Xiang, J.; Zhang, Y.K.; Wu, H.; Wang, Y.L.; Zhu, D.F.; Chen, H.Z. Effects of side deep fertilization on yield formation and nitrogen utilization of mechanized transplanting rice. *Sci. Agric. Sin.* **2019**, *52*, 4228–4239.
30. Zhong, X.M.; Huang, T.P.; Peng, J.W.; Lu, W.L.; Kang, X.R.; Song, S.M.; Zhou, X. Effects of machine-transplanting synchronized with one-time precision fertilization on nutrient uptake and use efficiency of double cropping rice. *Chin. J. Rice Sci.* **2019**, *33*, 436–446.
31. Huda, A.; Gaihre, Y.K.; Islam, M.R.; Singh, U.; Islam, R.; Sanabria, J.; Satter, M.A.; Afroz, H.; Halder, A.; Jahiruddin, M. Floodwater ammonium, nitrogen use efficiency and rice yields with fertilizer deep placement and alternate wetting and drying under triple rice cropping systems. *Nutr. Cycl. Agroecosystems* **2016**, *104*, 53–66. [CrossRef]

32. Zhang, M.; Yao, Y.; Zhao, M.; Zhang, B.; Tian, Y.; Yin, B.; Zhu, Z. Integration of urea deep placement and organic addition for improving yield and soil properties and decreasing N loss in paddy field. *Agric. Ecosyst. Environ.* **2017**, *247*, 236–245. [CrossRef]

33. Danying, W.; Chang, Y.; Chunmei, X.; Zaiman, W.; Song, C.; Guang, C.; Zhang, X. Soil Nitrogen Distribution and Plant Nitrogen Utilization in Direct-Seeded Rice in Response to Deep Placement of Basal Fertilizer-Nitrogen. *Rice Sci.* **2019**, *26*, 404–415. [CrossRef]

34. Yao, Y.; Zhang, M.; Tian, Y.; Zhao, M.; Zhang, B.; Zeng, K.; Zhao, M.; Yin, B. Urea deep placement in combination with Azolla for reducing nitrogen loss and improving fertilizer nitrogen recovery in rice field. *Field Crop. Res.* **2018**, *218*, 141–149. [CrossRef]

35. Liu, T.; Fan, D.; Zhang, X.; Chen, J.; Li, C.; Cao, C. Deep placement of nitrogen fertilizers reduces ammonia volatilization and increases nitrogen utilization efficiency in no-tillage paddy fields in central China. *Field Crop. Res.* **2015**, *184*, 80–90. [CrossRef]

36. Zhou, J.; Meng, G.Y.; Long, J.R.; Ma, G.H.; Wan, Y.Z. Effects of application amount of slow-release nitrogen fertilizer on physiological characteristics and grain yield of early super hybrid rice zhuliangyou 02. *Hybrid Rice* **2011**, *26*, 62–68.

37. Wang, Y.; Lin, A.; Loake, G.J.; Chu, C. H2O2?induced Leaf Cell Death and the Crosstalk of Reactive Nitric/Oxygen SpeciesF. *J. Integr. Plant Boil.* **2013**, *55*, 202–208. [CrossRef]

38. Tang, H.; Xiao, X.P.; Li, C.; Tang, W.G.; Guo, L.J.; Wang, K.; Cheng, K.; Pan, X.C.; Sun, G. Effects of different long-term fertilization managements on the physiological characteristics of leaves and yield of rice in double cropping paddy field. *J. China Agric. Univ.* **2018**, *11*, 60–71.

13

Role of *Trichoderma aggressivum* f. *europaeum* as Plant-Growth Promoter in Horticulture

Brenda Sánchez-Montesinos [1], Fernando Diánez [1,*], Alejandro Moreno-Gavíra [1], Francisco J. Gea [2] and Mila Santos [1]

[1] Departamento de Agronomía, Escuela Superior de Ingeniería, Universidad de Almería, 04120 Almería, Spain; brensam@hotmail.com (B.S.-M.); alejanmoga@gmail.com (A.M.-G.); msantos@ual.es (M.S.)

[2] Centro de Investigación, Experimentación y Servicios del Champiñón (CIES), Quintanar del Rey, 16220 Cuenca, Spain; fjgea.cies@dipucuenca.es

* Correspondence: fdianez@ual.es.

Abstract: The main objective of this study was to determine the capacity of *Trichoderma aggressivum* f. *europaeum* to promote pepper and tomato seedling growth compared to that of *T. saturnisporum*, a species recently characterised as a biostimulant. Consequently, in vitro seed germination and seedling growth tests were performed under commercial plant nursery conditions. Additionally, the effects of different doses and a mixture of both species on seedling growth under plant nursery and subsequently under greenhouse conditions were determined. Furthermore, mass production of spores was determined in different substrates, and their siderophore and indole acetic acid production and phosphate (P) solubilisation capacity were also determined. Direct application of *Trichoderma aggressivum* f. *europaeum* to seeds in vitro neither increases the percentage of pepper and tomato seed germination nor improves their vigour index. However, substrate irrigation using different doses under commercial plant nursery conditions increases the quality of tomato and pepper seedlings. Tomato roots increased by 66.66% at doses of 10^6 spores per plant. Applying *T. aggressivum* f. *europaeum* or *T. saturnisporum* under plant nursery conditions added value to seedlings because their growth-promoting effect is maintained under greenhouse conditions up to three months after transplantation. The combined application of the two species had no beneficial effect in relation to that of the control. The present study demonstrates the biostimulant capacity of *T. aggressivum* f. *europaeum* in pepper and tomato plants under commercial plant nursery and greenhouse conditions.

Keywords: *Trichoderma*; plant growth promotion; tomato; pepper; biostimulant

---

1. Introduction

The success of applying *Trichoderma* in agriculture results from the multiple benefits that it generates in plants. Thus, the genus *Trichoderma* is characterised by its strong competitive and reproductive potential, presenting high survival rates under unfavourable or abiotic stress conditions, such as salinity [1], water stress [2], or the presence of various toxic chemicals, including fungicides [3], among others. Similarly, *Trichoderma* exhibits high efficiency in the promotion of nutrient uptake [4], the capacity to modify the rhizosphere and root structure in which the fungus is established [5,6], high aggressiveness against plant pathogenic fungi, efficiency in the promotion of plant growth [7–12], and the ability to induce plant defence mechanisms, among many additional benefits [8,9,13]. The properties of *Trichoderma* have generated considerable research interest in these fungi for use in agriculture, and a large number of commercial products have been developed using different *Trichoderma* species [10,14]. Many formulations contain mixtures of different species that provide a wider range of direct and indirect beneficial effects for the plants. Numerous studies have reported the benefits of *Trichoderma* application for plant growth and even increased production yield. Thus, applying *Trichoderma* species, to both soil and seeds, allows the multiplication of the fungus in conjunction with the developing root system [15]. Its ability to colonise plant roots from the appressorium-like structure directly enhances seed vigour [16] and germination and promotes seedling growth [10,11,17]; thereby, suggesting that these fungi should be applied from the plant nursery stage in the case of horticultural, ornamental, or forest species, which would allow the early colonisation of the roots by *Trichoderma*, before transplanting the seedlings in the field.

It has been reported that plant growth is enhanced in association with *Trichoderma* species similar to that of other plant-growth-promoting microorganisms (PGPMs), but the effects are greater with *Trichoderma* when plants are under biotic, abiotic, or physiological stress conditions [9,18–21]. Recently, *T. aggressivum* f. *europaeum* has been described as a melon seedling growth promoter under saline stress conditions, in addition to its capacity to control *Pythium ultimum*, decreasing the severity of the disease in seedlings [1]. *Trichoderma aggressivum* Samuels & W. Gams is the causal agent of the green mould disease, which causes economic losses in the cultivation of white button mushrooms (*Agaricus bisporus* (J.E. Lange) Imbach) worldwide. There are two subspecies, *T. aggressivum* f. *aggressivum* and *T. aggressivum* f. *europaeum* found in North America and Europe, respectively [22]. *Trichoderma aggressivum*, a fast-growing filamentous fungus, colonises compost and casings used as growth substrates in mushroom cultivation and produces dense white mycelial colonies that change colour to green after sporulation [23]. This aggressive competitor is known to produce metabolites that are toxic to *A. bisporus* [24,25]. In areas colonised by *T. aggressivum*, fruit body formation is retarded, and fruit bodies may be of poor quality because of damage or discolouration [23]. Numerous *Trichoderma* species have been isolated from *Agaricus* compost and *Pleurotus* substrates, such as *T. harzianum*, *T. longibrachiatum*, *Trichoderma ghanense*, *T. asperellu*, and *T. atroviride*, although its aggressiveness has not been determined [26]. Sánchez-Montesinos et al. [1] demonstrated its high mycelial growth and sporulation on roots. Thus, *T. aggressivum* f. *europaeum* is a potential biofertilizer for different crops. In our study, the growth-promoting capacity of this species has been analysed in comparison to that of *T. saturnisporum* Ca1606, which was recently characterised as a biocontrol agent and a seedling growth promoter for different horticultural plants [11,16,27]. Since the effectiveness of microorganisms as growth promoters will depend on the crop, dose and application method, among many other factors, further studies on *T. aggressivum* f. *europaeum* are needed to determine its efficacy.

Consequently, in the present study, *T. aggressivum* f. *europaeum* Tae52481 and *T. saturnisporum* Ca1606, were tested to evaluate: (a) the effects of direct application to seeds of a fungus suspension on root colonisation of tomatoes and peppers and subsequent plant vigour; (b) the promotion of growth and quality of pepper and tomato seedlings under a conventional production system; and (c) the effects of applying different doses and the synergistic effect of both isolates on tomato seedlings and on their subsequent transplantation under greenhouse conditions.

2. Materials and Methods

2.1. Fungal Isolates

Trichoderma saturnisporum Ca1606 (TS), already known for their plant growth promotion properties, were extracted from suppressive soils. TS was cultivated on potato dextrose agar (PDA) for 7 days at 25 °C in lightless conditions. The growth results measured were used to establish a comparison value.

For this study *Trichoderma aggressivum* f. *europaeum* Tae52481 (TA) were isolated from samples of substrate used for *Agaricus bisporus* cultivation at mushroom farms. These fungal spore samples were similarly cultivated on potato dextrose agar (PDA) for 7 days at 25 °C in dark conditions. The corresponding growth results were recorded. The spore suspensions for both samples were prepared using sterile distilled water. A concentration of 1×10^7 spores/mL was achieved with a Neubauer haemocytometer.

2.2. Analysis of Plant Growth-Promoting Attributes

In accordance with the method of Louden et al. [28], by the transference of fungal mycelial discs (5 mm) of active culture onto Chrome-Azurol S (CAS) agar medium, siderophore production was determined. At 24, 48 and 72 h the diameter of the siderophore colony indicative orange halos on blue were measured.

Indole-3-acetic acid (IAA) production was estimated according to the procedure described by Diánez et al. [16]. Five independent replicates of TA and TS were analysed. This process is described as follows. A glucose peptone broth (GPB) of 50 mL, amended with or without L-tryptophan (Sigma-Aldrich) at a concentration of 100 mg L^{-1} was prepared. Flasks containing this broth inoculated with TA and TS were incubated on an orbital shaker at 150 rpm in dark conditions for 7 days at 25 °C. Subsequently the supernatants from each flask, having first being centrifuged for 30 min at 12,000× *g* and filtered through sterile Millipore membranes (pore size 0.22 μm), were collected into sterile test tubes. In order to determine the quantity of IAA, optical density tests were carried out and compared to a standard IAA curve. For both the TA and TS, 3 mL of the culture supernatant and 2 mL (0.5 mol L^{-1} $FeCl_3$ + 98 mL of 35% $HClO_4$) Salkowski reagent were combined and left for 30 min. The intensity of the resulting red pigmentation density was measured at 530 nm using a scanning spectrophotometer for each of the samples.

To determine the quantitative estimation of phosphate solubilisation, a modified version of the Lima–Rivera procedure [29] was followed. Then, 250 mL capacity flasks containing 50 mL National Botanical Research Institute's phosphate (NBRIP) broth, inoculated with two 5 mm pure *Trichoderma* isolates agar disks were agitated at 100 rpm and incubated at 26 °C for 3, 5, 7, 10 and 15 days. As a control the procedure was carried out on uninoculated flasks containing the same NBRIP broth. The experiments were conducted in triplicate.

Using the Fiske and Subbarow method [30] phosphate concentrations in culture supernatants were estimated as equivalent phosphate (μg mL^{-1}), mean values expressed and pH analysed. The total P (phosphate) in the flasks was 10 mg mL^{-1}.

2.3. Mass Production of TA and TS on Solid Substrates

A mixture of two kinds of substrates, one containing buckwheat husk (BH) and oat (O), the other containing BH and rice (R) were tested for the mass multiplication of TA and TS [31]. Different proportions of BH-O (90–10%, 80–20% and 70–30% *v/v*) and BH-R (90–10%, 80–20% and 70–30% *v/v*) were submerged in 30% *v/v* of water for 24 h. Each mixture was sterilised for 1 h at 125 °C twice on consecutive days. Each mixture was placed on a tray and aseptically inoculated by spraying with 5 mL of spore suspension containing 1×10^7 spores mL^{-1} of each isolate. The trays were kept at 25 °C in the dark for 15 days. In total, three samples (2 g) of the fungus-colonised substrate were removed from the trays in each treatment. The samples were successively diluted in sterile distilled water + 0.01% Tween 20® and the number of conidia g^{-1} of the solid substrate was quantified for each replicate using a Neubauer haemocytometer. There were three replications per treatment. The collected spores were used in the different experiments conducted in this study.

2.4. Analysis of Effects of TA and TS on Seed Germination under Laboratory Conditions

Three treatments (control, TA and TS) and four repetitions following a random block experimental design were implemented in this study. For each repetition of the three treatments 50 seeds of tomato (*Solanum lycopersicum* 'Red Cherry') and pepper (*Capsicum annuum* 'Largo de Reus') were germinated on two sheets of sterile distilled water moistened Whatman No. 1 filter paper in (150 mm) Petri dishes. These seeds were first surface sterilized for 5 min with 1.5% sodium hypochlorite (NaOCl), rinsed twice with sterile distilled water and dried under laminar airflow on sterile paper [16]. Germination was achieved by treating the seeds with 50 μL of spore suspension (1×10^5 spores mL^{-1}) of TA, TS or 50 μL of sterile water (control). The trays were placed in a lightless incubator at 25 ± 1 °C, 7 days for tomato and 10 days for pepper seeds. For each Petri dish treated with one of the three solutions (control, TA and TS), percent germination, root length and shoot length of tomato and pepper seeds were recorded. A Seed Vigour Index (SVI) was calculated as follows: SVI (length) = seed germination% (mean root length + mean shoot length) [32].

2.5. *Analysis of Promoter Effects of TA and TS on Pepper and Tomato Seedlings: Experiment 1*

The following experiment was conducted in autumn using a completely randomised design at a commercial nursery (Almería, Spain). Pepper (*Capsicum annuum* 'Largo de Reus') and tomato (*Solanum lycopersicum* 'Red Cherry') seeds were sown in 96-cell commercial peat mix filled nursery polystyrene planting trays (70 mL volume) and covered with vermiculite. Trays were relocated to a greenhouse and rinsed with sterile distilled water (control), or a 5 mL (TA or TS) spore suspension per cell at 10^5 spores per plant, after a 2 day (tomato) or 4 day (pepper) period in a germination room (relative humidity (RH) = 95%; 25 °C). Four trays of seedlings for each treatment were cultivated under standard nursery culture conditions (18–28 °C; 75.4 ± 6.7% RH). Then, 20 plants per treatment and control were randomly selected from the four replications at 45 days after sowing across the four replications. Different growth parameters: number of leaves, stem length, stem base diameter, total leaf area and root dry weights, as well as leaf area using the WINDIAS 3.1 of the plants, were measured. The formula: DQI = TDW/((LS/D) + SDW/RDW)) where TDW is the total dry weight (g), LS is stem length (cm), D is stem diameter (mm), SDW and RDW are stem and root dry weight (g), respectively; they were employed to determine the Dickson Quality Index (DQI) [33].

2.6. *Analysis of Effects of Applying Different Doses of TA and TS to Tomatoes: Experiment 2*

The experimental procedure followed for experiment 2 was similar to that described for experiment 1, although conducted in winter. Again, propagated in substrate appropriately irrigated according to climate and crop necessity under commercial plant nursery conditions and supplemented with a commercial complex nutrient fertiliser, 96 tomato seedlings per replicate of four were treated with three solutions of spore suspension, each with 5 mL of TA, TS conidia and TA + TS (M) (TA D1, TS D1 and M D1: 10^5 spores mL^{-1}; TA D2, TS D2 and M D2: 10^6 spores mL^{-1}; and TA D3, TS D3 and M D3: 10^7 spores mL^{-1}). After 30 days of sowing, twenty plants from each of the three treatment batches and control were randomly selected for harvest. The plants were measured, and data were recorded for the same parameters described in experiment 1. In mid-February a further 25 plants were transplanted into a sandy soil and analysed in mid-May.

In all tests, roots inoculated with *Trichoderma* isolates were collected at the end of the tests. Roots were surface sterilized in 0.1% sodium hypochlorite and washed with sterilised water. Root fragments were placed in PDA medium to determine root colonisation by the fungal isolate.

2.7. *Statistical Analysis*

The experimental results are presented as the means and standard error (± SE) for the different replicates. Mean separation was carried out using Fisher's Least Significant Difference (LSD) test. The data were tested by one-way analysis of variance (ANOVA) or Student's *t*-test with significance defined as *p*-values less than 0.05 ($p < 0.05$). Statgraphics Centurion 18 Software was utilised for statistical analysis.

3. Results

3.1. *Mass Production of Trichoderma Isolates on Solid Substrates*

The results are outlined in Table 1. Both isolates grew and sporulated well in all mixtures tested. The proportion of 70 + 30% for buckwheat husk and oats (Figure 1), respectively, and 80 + 20% for buckwheat husk and rice, resulted in significantly higher spore production for both species, followed by

90 + 10% and 70 + 30% of BH + 10% R (Table 1). The lowest spore production rate was observed for 80 BH + 20% O.

Table 1. Mass production of spores on solid substrates (CFU g^{-1}).

Treatments	*T. aggressivum* f. *europaeum*	*T. saturnisporum*
90% BH + 10% O	$6.65\cdot10^{8} \pm 3.04\cdot10^{7}$ c	$6.48\cdot10^{8} \pm 2.84\cdot10^{7}$ c
80% BH + 20% O	$5.63\cdot10^{8} \pm 3.20\cdot10^{7}$ d	$5.17\cdot10^{8} \pm 6.60\cdot10^{7}$ d
70% BH + 30% O	$1.04\cdot10^{9} \pm 1.44\cdot10^{7}$ a	$9.98\cdot10^{8} \pm 5.69\cdot10^{7}$ a
90% BH + 10% R	$8.32\cdot10^{8} \pm 1.61\cdot10^{7}$ b	$7.88\cdot10^{8} \pm 6.45\cdot10^{7}$ b
80% BH + 20% R	$1.04\cdot10^{9} \pm 1.04\cdot10^{7}$ a	$1.02\cdot10^{8} \pm 6.26\cdot10^{7}$ a
70% BH + 30% R	$8.00\cdot10^{8} \pm 5.00\cdot10^{7}$ b	$7.12\cdot10^{8} \pm 4.25\cdot10^{7}$ bc
p-value	0.0000	0.0000

BH: buckwheat husk; O: oat; R: rice; CFU: colony forming unit. Data were analysed by ANOVA and treatment means were compared according to Fisher's Least Significant Difference (LSD) statistical procedure (*F*-test at $p < 0.05$). Different letters indicate significant differences according to the one-way ANOVA test ($p = 0.05$).

Figure 1. Mass production of (**A**) *Trichoderma aggressivum* f. *europaeum* and (**B**) *Trichoderma saturnisporum* on 70 + 30% for buckwheat husk and oats.

3.2. Siderophore Production, IAA and P Solubilisation

TA and TS siderophore production was observed in the formation of an orange-coloured zone around the fungal colonies at 24 and 48 h, and the production of TA was higher, in both cases. No increase in the diameter of the halo (mm) was detected at 72 h in any isolate (Table 2).

Table 2. Siderophores and IAA production by *Trichoderma* isolates.

Radius of Siderophores Production (mm)				IAA (mg mL^{-1})	
Treatment	**24 h**	**48 h**	**72 h**	**+Trp**	**−Trp**
p-value	0.0000	0.0000	-	0.0068	0.0304
T. aggressivum	9.73 ± 0.89	18.50 ± 1.70	-	0.145 ± 0.011	0.085 ± 0.009
T. saturnisporum	5.45 ± 0.31	9.82 ± 0.56	-	0.199 ± 0.014	0.129 ± 0.021

Values are average of five replications; values after ± represent standard deviation. IAA: indole-3-acetic acid; +Trp: with L-tryptophan; −Trp: without L-tryptophan.

Although both *Trichoderma* strains exhibited an in vitro ability to produce IAA in medium supplemented with and without 100 mg L^{-1} tryptophan during a 7-day period, the production of *T. saturnisporum* was higher under both conditions tested (Table 2). In both TA and TS, IAA production increased in the medium supplemented with tryptophan.

The effects of TA and TS on the soluble phosphate concentration are shown in Figure 2. The initial concentration of P in the medium was used to quantify the concentration of P solubilised by both isolates.

As shown in Figure 2, P solubilisation was significant from the fifth day of incubation, with no significant differences between the two isolates. Furthermore, no change in the pH of the medium was detected, which remained at approximately 6.5–7.

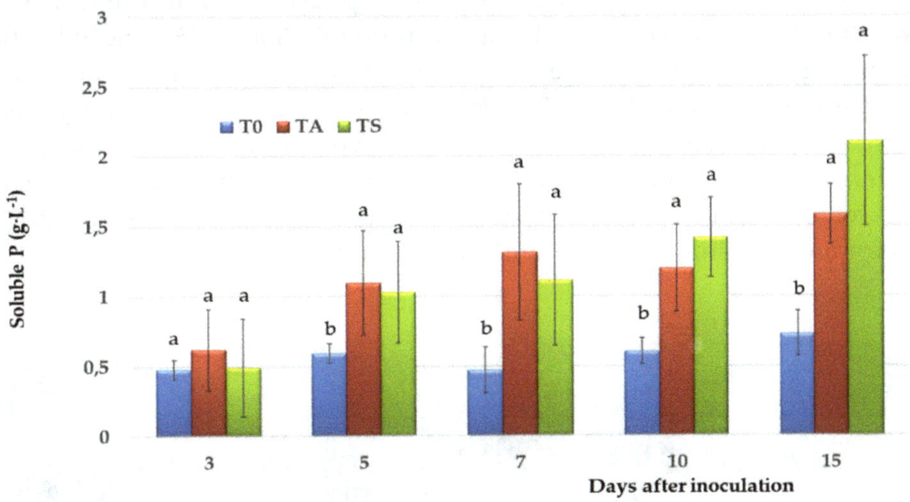

Figure 2. Effects of *Trichoderma aggressivum* f. *europaeum* (TA) and *Trichoderma saturnisporum* (TS) on phosphate solubilisation in National Botanical Research Institute's phosphate (NBRIP) broth containing tribasic calcium phosphate (10 g). T0: NBRIP broth without *Trichoderma* isolates. The results are shown as the average of the three replicates, in g L^{-1}. Mean standard deviation is expressed in the error bar ($n = 3$). For each isolate, columns marked with different letters indicate a significant difference at $p < 0.05$.

3.3. Effects of TA and TS Treatment on Germination and Vigour Index

The results from the in vitro application of TA and TS spores to tomato and pepper seeds are outlined in Table 3. No significant effects on pepper and tomato seed germination percentages were observed in either treatment. However, the application of both *Trichoderma* isolates led to a decrease in the radicle and hypocotyl length (growth) parameters and significantly decreased the SVI in peppers. The tomato seed vigour index was not affected by TA or TS treatment ($p = 0.1918$).

Table 3. Effects of *T. aggressivum* f. *europaeum* and *T. saturnisporum* on tomato and pepper seed germination 7 and 10 days after treatment, respectively.

Treatment	% Germination	Root Length (cm)	Shoot Length (cm)	Seed Vigour Index
		Pepper		
p-value	0.5420	0.0126	0.0010	0.0030
T. aggressivum	83 ± 6.83a	0.64 ± 0.18b	1.66 ± 0.55b	138.14 ± 35.48b
T. saturnisporum	80 ± 3.26a	0.91 ± 0.11b	1.29 ± 0.17b	176.50 ± 18.14b
Control	78 ± 4.61a	2.16 ± 1.01a	1.98 ± 0.56a	320.32 ± 83.36a
		Tomato		
p-value	0.5268	0.0020	0.3154	0.1918
T. aggressivum	89 ± 6.83a	4.53 ± 0.31a	2.97 ± 0.47a	671.80 ± 112.91a
T. saturnisporum	92 ± 7.30a	3.25 ± 0.37b	2.55 ± 0.27a	536.29 ± 76.26a
Control	85 ± 8.32a	3.95 ± 0.28b	2.81 ± 0.29a	580.13 ± 96.37a

Different letters indicate significant differences according to the one-way ANOVA test ($p = 0.05$).

3.4. *Effects of Trichoderma Inoculation on Tomato and Pepper Seedlings*

The effects of TA and TS application on morphological parameters and DQI are shown in Table 4. Unlike the results from the direct application of both *Trichoderma* to the seeds, the application to the substrate increased the study parameters compared to that of the control, and the results were better in peppers than in tomatoes, with better quality seedlings, according to the DQI values (Figure 3). There were no significant differences after the application of TA and TS in both horticultural plants. The increased percentage assessed in pepper seedlings for each species (TA/TS) was 8%/8.5% for stem length, 12.32/~0.01 for stem diameter, 7.77/5.5 for leaf number, 22.22/25 for shoot dry weight, 36.36/63.63 for root dry weight and 13.83/13.74 for leaf area, respectively. In tomato seedlings, the percentages were 9/6 for stem length, 0.5/1.5 for stem diameter, 6/8.8 for leaf number, 12.5/5.3 for shoot dry weight, 0/–6.6 for root dry weight and 8/9.2 for leaf area. No significant differences in DQI were found in tomato seedlings for any treatment applied with respect to that of the control.

Figure 3. (**A**) Differential growth of pepper seedlings with *T. aggressivum* f. *europaeum* (TA), compared to control. (**B**) Tomato plants grown under field transplantation conditions (60 days).

3.5. *Effects of Dose of Application of T. aggressivum f. europaeum and T. saturnisporum*

Since no significant plant growth-promoting results were found in tomato seedlings, the effects of applying three doses of both species separately, as well as jointly, were determined. The results are outlined in Table 5, wherein values significantly higher than that of the control are highlighted in green, and negative values in red, for better visualisation.

The increase in the dose of both species improved seedling quality, increasing all study parameters in TA D2 and TS D3 treatments, with respect to that of the control. In treatment TA D2, stem length increased 14.37%, plant diameter 9.4%, leaf number 21.58%, shoot dry weight 16.66% and root dry weight 66.66%. In treatment TS D3, stem length increased 39.05%, plant diameter 15.22%, leaf number 11.55%, shoot dry weight 12.5% and root dry weight 33.33%. Although most treatments favoured the development of seedling shoots, no favourable results were found in roots; therefore, the seedling quality was not improved. The combination or mixture of the two species for the three doses tested did not improve the results compared to that of their separate application.

The results of the study parameters after transplantation of the seedlings into the soil are outlined in Table 5 (Figure 3). Three treatments, TA D1, TA D2 and TS D2, led to a good relationship between tomato shoots and roots, with significantly higher plant quality, compared to that of the control, without a new application of *Trichoderma*. Thus, shoot dry weight increased 43.20%, 22.84% and 29.58% and root dry weight increased 29.94%, 39.32% and 31.51% after the TA D1, TA D2 and TS D2 treatment, respectively. The establishment of the endophytic fungus at the root (Figure 4) enabled its effects to persist after transplantation.

Table 4. Morphological parameters and quality index of pepper and tomato seedlings treated with *T. aggressivum* f. *europaeum* and *T. saturnisporum* at 45 days after sowing.

Treatment	Length of Stem (cm)	Diameter (mm)	Number of Leaves	Aerial Dry Weight (g)	Root Dry Weight (g)	Leaf Area (mm^2)	DQI
Pepper							
p-value	0.0000	0.0000	0.0092	0.0000	0.0000	0.0001	0.0000
T. aggressivum	29.18 ± 1.36a	4.10 ± 0.24a	7.22 ± 0.76a	0.44 ± 0.06a	0.15 ± 0.30b	86.38 ± 12.58a	0.06 ± 0.01a
T. saturnisporum	29.31 ± 1.75a	3.68 ± 0.22b	7.07 ± 0.82a	0.45 ± 0.06a	0.18 ± 0.09a	86.31 ± 13.22a	0.06 ± 0.01a
Control	27.01 ± 2.07b	3.65 ± 0.23b	6.70 ± 0.72b	0.36 ± 0.04b	0.11 ± 0.03c	75.88 ± 11.15b	0.04 ± 0.01b
Tomato							
p-value	0.0000	0.4387	0.0031	0.0232	0.0245	0.0295	0.0793
T. aggressivum	27.93 ± 1.99a	3.86 ± 1.84a	4.45 ± 0.50a	0.63 ± 0.09a	0.15 ± 0.02ab	74.63 ± 10.99a	0.06 ± 0.01a
T. saturnisporum	27.17 ± 1.62a	3.90 ± 0.24a	4.57 ± 0.50a	0.59 ± 0.10ab	0.14 ± 0.02b	75,52 ± 13.40a	0.06 ± 0.01a
Control	25.61 ± 2.07b	3.84 ± 0.26a	4.20 ± 0.46b	0.56 ± 0.10b	0.15 ± 0.02a	69.12 ± 6.96b	0.06 ± 0.00a

Different letters indicate significant differences according to the one-way ANOVA test ($p = 0.05$). DQI: Dickson Quality Index.

Table 5. Morphological parameters and DQI of tomato seedlings and plants treated with different doses (10^5, 10^6 and 10^7 spores per plant; D1, D2 and D3, respectively) of *T. aggressivum* f. *europaeum* (TA), *T. saturnisporum* (TS) and mix (M) of two species.

Tomato Seedling						
Treatment	**Length of Stem (mm)**	**Diameter (mm)**	**N° Leaves**	**Aerial Dry Weight (g)**	**Root Dry Weight (g)**	**DQI**
p-value	0.0000	0.0000	0.0000	0.0000	0.0000	0.0000
TA D1	135.21 ± 28.79de	3.91 ± 0.46ab	4 ± 0.65b	0.27 ± 0.09a	0.03 ± 0.02cd	0.025 ± 0.01de
TA D2	143.80 ± 32.58d	3.81 ± 0.34b	4 ± 0.51b	0.28 ± 0.07a	0.05 ± 0.018a	0.034 ± 0.01a
TA D3	127.02 ± 25.41e	3.76 ± 0.46b	3.98 ± 0.64b	0.27 ± 0.10a	0.04 ± 0.02ab	0.030 ± 0.01ab
TS D1	141.25 ± 19.83d	3.55 ± 0.49c	3.65 ± 0.60cd	0.23 ± 0.10b	0.03 ± 0.01cd	0.024 ± 0.01de
TS D2	188.44 ± 30.29a	4.07 ± 0.33a	4.25 ± 0.73a	0.28 ± 0.07a	0.03 ± 0.01d	0.021 ± 0.01e
TS D3	174.83 ± 27.78b	4.01 ± 0.45a	3.67 ± 0.63cd	0.27 ± 0.09a	0.04 ± 0.02a	0.030 ± 0.01ab
M D1	157.25 ± 29.14c	3.5 ± 0.39c	3.56 ± 0.62de	0.27 ± 0.061a	0.04 ± 0.01bc	0.026 ± 0.00bcd
M D2	99.08 ± 21.02b	3.13 ± 0.23b	3.35 ± 0.39bc	0.28 ± 0.08a	0.04 ± 0.01bcd	0.029 ± 0.00bcd
M D3	98.54 ± 21.37f	3.11 ± 0.45d	3.33 ± 0.63ef	0.18 ± 0.08c	0.04 ± 0.01ab	0.029 ± 0.00abc
Control	125.73 ± 22.3e	3.48 ± 0.31c	3.29 ± 0.62f	0.24 ± 0.07b	0.03 ± 0.01cd	0.025 ± 0.01cde
Tomato Plants						
Treatment	**Length of Stem (cm)**	**Diameter (mm)**	**Internodes**	**Aerial Dry Weight (g)**	**Root Dry Weight (g)**	**DQI**
p-value	0.0000	0.5379	0.0159	0.0015	0.2769	0.5373
TA D1	101.40 ± 15.06ab	12.11 ± 0.90a	15.00 ± 1.15ab	61.19 ± 11.27a	4.99 ± 0.68ab	3.23 ± 0.24a
TA D2	102.80 ± 9.47a	12.11 ± 1.24a	14.10 ± 1.29ab	52.49 ± 11.26	5.35 ± 0.57a	3.20 ± 0.46a
TA D3	92.00 ± 9.82bc	11.26 ± 1.17ab	14.40 ± 0.84abc	54.74 ± 13.09abc	4.70 ± 1.10abc	3.01 ± 0.58ab
TS D1	77.55 ± 7.76e	11.74 ± 1.18ab	13.90 ± 1.37bc	49.26 ± 14.78bcde	4.40 ± 1.04bc	2.91 ± 0.56ab
TS D2	88.75 ± 10.63cd	11.92 ± 1.57ab	15.20 ± 1.03a	55.37 ± 8.24ab	5.05 ± 1.23ab	3.19 ± 0.42a
TS D3	81.00 ± 12.39de	11.58 ± 1.37ab	13.20 ± 2.57c	51.71 ± 7.97abcd	4.34 ± 1.17bc	3.01 ± 0.94ab
M D1	79.05 ± 13.20e	10.87 ± 1.38b	13.90 ± 1.29bc	47.00 ± 11.38bcde	4.14 ± 1.11bc	2.73 ± 0.64ab
M D2	76.18 ± 11.91e	11.61 ± 0.95ab	13.20 ± 1.23c	45.30 ± 9.79cde	4.23 ± 0.99bc	2.85 ± 0.59ab
M D3	78.45 ± 8.08e	11.22 ± 1.18ab	14.30 ± 1.16abc	40.91 ± 11.56e	4.28 ± 1.37bc	2.72 ± 0.86ab
Control	77.96 ± 7.29e	11.52 ± 2.03ab	14.25 ± 1.16abc	42.73 ± 6.46de	3.84 ± 0.62c	2.60 ± 0.43b

Different letters indicate significant differences according to the one-way ANOVA test ($p = 0.05$). Green: favourable; Red: unfavourable; Orange: no effect compared to control. DQI: Dickson Quality Index.

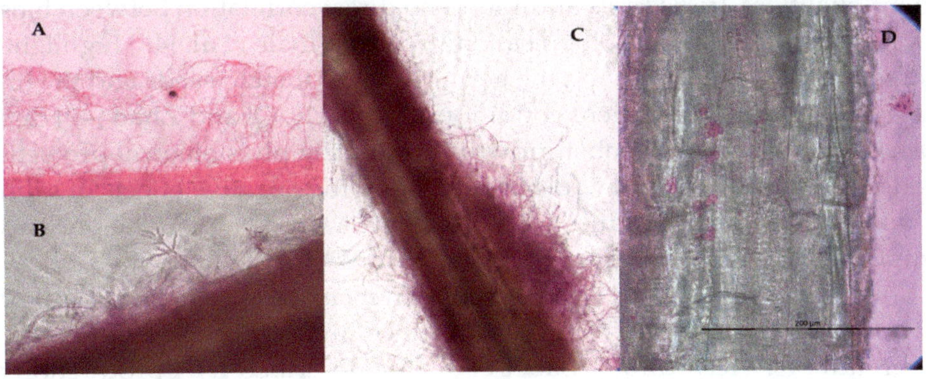

Figure 4. (**A**) Colonization of pepper and tomato roots by *T. aggressivum* f. *europaeum*. (**C**) Mycelium in pepper root (100×). (**B**) Conidiophores and mycelium in tomato root (100×). (**D**) Chlamydospores in pepper root (200×).

4. Discussion

Numerous *Trichoderma* species have been described as plant-growth promoters, including *T. harzianum*, *T. longipile*, *T. tomentosum*, *T. viride*, *T. koningii*, *T. asperellum*, *T. aureoviride* and *T. saturnisporum*, among others [34]. This ability to promote growth depends on several factors, including the existence of isolates of the same species that may or may not promote plant growth, or for example, the crop and/or variety to which the species is applied [34]. Similarly, the use of a mixture of species has been extensively studied and commercialised to increase this activity [10]. In this study, the plant growth-promoting capacity of a new species, *T. aggressivum* f. *europaeum*, which is characterised by its rapid growth and sporulation, was analysed and compared to that of *T. saturnisporum*, a species characterised as a plant-growth promoter by Diánez et al. [16,18]. Although Allaga et al. [35] recommend not using species that produce green mould disease, these species do not create any problems in horticultural crops or pose any danger to mushroom crops, as long as they are applied in different geographical areas. Additionally, mushrooms are produced in closed locations and under completely different conditions. Furthermore, plant remains under horticultural production are not used to prepare substrates for mushroom cultivation, as shown in many commercial species; neither are plant remains that have been studied with plant-growth promoters, which may also cause green mould disease, such as *T. harzianum* [36] or *T. longibrachiatum* [37].

The first objective was to obtain viable spores with high yield on low-cost substrates. This product was used for additional tests, which demonstrated that the nutritional composition of the substrates used did not affect the biostimulant capacity of either *Trichoderma* species. Lane [38] determined that the nutrients provided in the medium could affect the biocontrol or biostimulant capacity of the agent. Different substrates have been used for *Trichoderma* spore production, including barley straw [39], wheat, rice, corn kernels [40] or a mixture of substrates, such as wheat straw, bran, cassava, potato starch and sugar beets [41,42], among others. In general, in our study, high yields, expressed as colony forming unit (CFU) g^{-1}, were assessed in all substrate mixtures tested; the yields increased both in 80% buckwheat husk + 20% rice and 70% buckwheat husks + 30% oats. Although in laboratory tests, extraction could be performed without a problem in all mixtures, in the extractor tank, mixtures containing rice adhered to the walls and pipes, complicating the subsequent extraction and filtration processes. For this reason, to develop low-cost production methods for industrial scale-up, rice was rejected as a constituent of the production substrate for TA and TS. A high siderophore and IAA production and P solubilisation by TA and TS compared to other *Trichoderma* species or isolates were demonstrated in our study. These three components play key roles in plant biostimulation by increasing nutrient availability to plants, such as for hormone production [43,44]. However, the direct relationship between IAA production and plant-growth promotion is not yet clear because numerous species can produce IAA, but they do not promote plant growth [45]. Hoyos et al. [45] concluded that IAA production is not a species-dependent quality of *Trichoderma* and found no direct correlation between biostimulation and IAA and siderophore production or P solubilisation. In turn, Vinale et al. [46] highlighted the effects of siderophore (harzianic acid) production on the germination of tomato seeds and the improved growth of the seedlings even under iron-deficient conditions. Similarly, Qi and Zhao [47] demonstrated that applying *T. asperellum* enhanced cucumber growth by inducing physiological protection under saline stress, and its siderophores played a key role in mitigating the negative effects of salinity.

Many *Trichoderma* species can produce IAA, and high IAA secretion in the presence of tryptophan indicates the importance of tryptophan as a precursor for IAA production [48,49]. Gravel et al. [50] reported that IAA production induced by L-tryptophan increased the fresh weight of tomato shoots and roots. Our results indicate that TA and TS produce much higher amounts of IAA than those assessed by other authors. Accordingly, Saber et al. [48] described IAA production of *T. harzianum* isolates that were 10 times lower than that of *T. aggressivum* f. *europaeum* and *T. saturnisporum* assessed in this study. Bader et al. [51] reported that IAA production ranged from 13.38 to 21.14 $\mu g\ mL^{-1}$ in *T. brevicompactum*, *T. gamsii* and *T. harzianum*. Diánez et al. [16] described a highly similar IAA production for *T. saturnisporum*; therefore, the in vitro production capacity of IAA was preserved despite maintaining the isolate in the laboratory for 10 years. Similarly, phosphate solubilisation by *Trichoderma* species has been described both in vitro and in vivo [52–54]. Recently, Tandon et al. [55] evaluated P solubilisation of different *Trichoderma koningiopsis* isolates under abiotic stress conditions and determined a range from 1.6 to 71 $\mu g\ mL^{-1}$. Bononi et al. [12] found that *Trichoderma* isolated from soils of the Amazon rainforest demonstrated a high potential for phosphate solubilisation, which ranged from 51.7 to 90.3% 10 days after inoculation. Despite their high P solubilisation capacity, some of these isolates inhibited the germination of soybean seeds. In our study, the P solubilisation range of both isolates was lowest on the tenth day of incubation, at 5.9% and 6.16% for TA and TS, respectively.

Applying PGPMs to seeds makes it possible to use a lower concentration of spores while ensuring that the PGPMs are readily accessible at germination and during early developmental plant stages, stimulating healthy and rapid establishment, and consequently, maximising crop production [43]. However, the direct application of different *Trichoderma* isolates or species to seeds (biopriming) has not always had beneficial effects. In this study, the seed germination rate was not affected by *T. aggressivum* f. *europaeum* or *T. saturnisporum* application. Similar results were found by Azarmi et al. [18] after applying *T. harzianum* isolates to tomato seeds. Hajieghrari et al. [56] demonstrated that direct exposure of corn seeds to *Trichoderma* spores decreased the percentage of seed germination, as well as radicle and shoot length. However, You et al. [57] demonstrated that *T. harzianum* and *T. koningiopsis* isolates significantly enhanced the tomato seed vigour index when they were used to treat tomato seeds. Our results demonstrated that direct *T. aggressivum* f. *europaeum* and *T. saturnisporum* application decreased seed vigour, significantly so in peppers but not in tomatoes. However, the application of either species under commercial plant nursery conditions, via substrate irrigation, similarly enhanced pepper seedling quality significantly, albeit again non-significantly for tomatoes. Optimising the application dose for each species is a factor that should be considered, among other factors, to enable companies and producers to adopt this technology with higher security [58]. Increasing the dose of *T. aggressivum* f. *europaeum* and *T. saturnisporum* applied to tomato seedlings increased most of the study parameters, as well as the DQI value in treatments TA D2, TA D3 and TS D3. The endophytic establishment of *Trichoderma* in plant nurseries may ensure its colonisation once transplanted. As such, in the TA D2 treatment, tomato plants continued to show better quality, without any additional application of *Trichoderma*, and plant quality improved in other treatments with *Trichoderma* applied separately. The poorest results were obtained for mixtures of both species, with no improvement in study parameters for any dose tested, and even a reduction of 21.62%, 10.63% and 25% in stem length, diameter and shoot dry weight of tomato seedlings in the MD3 treatment, respectively. Similar results were found by Liu et al. [59], who reported that the combination of three species, *T. afroharzianum*, *T. pseudoharzianum* and *T. asperelloides*, decreased the biocontrol and growth-promoting effects in comparison to the application of each species separately.

Although major reductions in the use of chemical fertilisers without production losses is currently difficult in many farming systems, their gradual decrease accompanied by the use of biostimulants or biofertilizers is a tool that can optimise the use of chemical inputs while reducing environmental pollution and food crop contamination.

5. Conclusions

The present study demonstrated, for the first time, the biostimulant capacity of *T. aggressivum* f. *europaeum* in pepper and tomato plants under commercial plant nursery and greenhouse conditions, with similar results to those of *T. saturnisporum*.

6. Patents

This isolate was patented with a Spanish patent number ES2706099: New strain of *T. aggressivum* f. *europaeum*, compositions and applications.

Author Contributions: F.D. and M.S. conceived and designed the experiments; B.S.-M., F.J.G., and A.M.-G. performed the experiments; M.S. analysed the data; F.D. wrote the paper. All authors have read and agreed to the published version of the manuscript.

Acknowledgments: The present work benefited from the input of the project RTC-2017-6486-2 and was supported by the Spanish Ministry of Science, Innovation and Universities.

References

1. Sánchez-Montesinos, B.; Diánez, F.; Moreno-Gavíra, A.; Gea, F.J.; Santos, M. Plant growth promotion and biocontrol of *Pythium ultimum* by saline tolerant *Trichoderma* isolates under salinity stress. *Int. J. Environ. Res. Public Health* **2019**, *16*, 2053. [CrossRef] [PubMed]
2. Khoshmanzar, E.; Aliasgharzad, N.; Neyshabouri, M.R.; Khoshru, B.; Arzanlou, M.; Asgari Lajayer, B. Effects of *Trichoderma* isolates on tomato growth and inducing its tolerance to water-deficit stress. *Int. J. Environ. Sci. Technol.* **2020**, *17*, 869–878. [CrossRef]
3. Adnan, M.; Islam, W.; Shabbir, A.; Khan, K.A.; Ghramh, H.A.; Huang, Z.; Chen, H.Y.H.; Lu, G. Plant defense against fungal pathogens by antagonistic fungi with *Trichoderma* in focus. *Microb. Pathog.* **2019**, *129*, 7–18. [CrossRef] [PubMed]
4. Halifu, S.; Deng, X.; Song, X.; Song, R. Efects of Two *Trichoderma* Strains on Plant Growth, Rhizosphere Soil Nutrients, and Fungal Community of *Pinus sylvestris* var. mongolica Annual Seedlings. *Forests* **2019**, *10*, 758.
5. Vargas, W.A.; Crutcher, F.K.; Kenerley, C.M. Functional characterization of a plant-like sucrose transporter + from the beneficial fungus *Trichoderma virens*. Regulation of the symbiotic association with plants by sucrose metabolism inside the fungal cells. *New Phytol.* **2010**, *189*, 777–789. [CrossRef]
6. Garnica-Vergara, A.; Barrera-Ortiz, S.; Muñoz-Parra, E.; Raya-Gonzalez, J.; Mendez-Bravo, A.; Macias-Rodriguez, L.; Ruiz-Herrera, L.F.; Lopez-Bucio, J. The volatile 6-pentyl-2H-pyran-2-one from *Trichoderma* atroviride regulates *Arabidopsis thaliana* root morphogenesis via auxin signaling and ethylene insensitive 2 functioning. *New Phytol.* **2016**, *209*, 1496–1512. [CrossRef]
7. Harman, G.E.; Howell, C.R.; Viterbo, A.; Chet, I.; Lorito, M. *Trichoderma* species -opportunistic, avirulent plant symbionts. *Nat. Rev. Microbiol.* **2004**, *2*, 43–56. [CrossRef]
8. Shoresh, M.; Harman, G.E.; Mastouri, F. Induced systemic resistance and plant responses to fungal biocontrol agents. *Annu. Rev. Phytopathol.* **2010**, *48*, 21–43. [CrossRef]
9. Hermosa, R.; Viterbo, A.; Chet, I.; Monte, E. Plant-beneficial effects of *Trichoderma* and of its genes. *Microbiology* **2012**, *158*, 17–25. [CrossRef]
10. López-Bucio, J.; Pelagio-Flores, R.; Herrera-Estrella, A. *Trichoderma* as biostimulant: Exploiting the multilevel properties of a plant beneficial fungus. *Sci. Hortic.* **2015**, *196*, 109–123. [CrossRef]
11. Diánez, F.; Santos, M.; Carretero, F.; Marín, F. Biostimulant activity of *Trichoderma saturnisporum* in melon (Cucumis melo). *Hortscience* **2018**, *53*, 810–815.

12. Bononi, L.; Chiaramonte, J.B.; Pansa, C.C.; Moitinho, M.A.; Melo, I.S. Phosphorus-solubilizing *Trichoderma* pp. from Amazon soils improve soybean plant growth. *Sci. Rep.* **2020**, *10*, 2858. [CrossRef] [PubMed]
13. Yuan, M.; Huang, Y.; Ge, W.; Jia, Z.; Song, S.; Zhang, L.; Huang, Y. Involvement of jasmonic acid, ethylene and salicylic acid signaling pathways behind the systemic resistance induced by *Trichoderma longibrachiatum* H9 in cucumber. *BMC Genom.* **2019**, *20*, 144–157. [CrossRef] [PubMed]
14. Sanjeev, K.; Manibhushan, T.; Archana, R. *Trichoderma*: Mass production, formulation, quality control, delivery and its scope in commercialization in India for the management of plant diseases. *Afr. J. Agric. Res.* **2014**, *9*, 3838–3852.
15. Khan, M.R.; Mohiddin, F.A. Trichoderma: Its Multifarious Utility in Crop Improvement. In *Crop Improvement through Microbial Biotechnology. New and Future Developments in Microbial Biotechnology and Bioengineering*; Aligarh Muslim University: Aligarh, India; Srinagar, India, 2018; Volume 13, pp. 263–291.
16. Diánez, F.; Santos, M.; Carretero, F.; Marín, F. *Trichoderma saturnisporum*, a new biological control agent. *J. Sci. Food Agric.* **2016**, *96*, 1934–1944. [CrossRef]
17. Rocha, I.; Ma, Y.; Souza-Alonso, P.; Vosátka, M.; Freitas, H.; Oliveira, R.S. Seed coating: A tool for delivering beneficial microbes to agricultural crops. *Front. Plant Sci.* **2019**, *10*, 1357. [CrossRef]
18. Azarmi, R.; Hajieghrari, B.; Giglou, A. Effect of *Trichoderma* isolates on tomato seedling growth response and nutrient uptake. *Afr. J. Biotechnol.* **2011**, *10*, 5850–5855.
19. Druzhinina, I.S.; Seidl-Seiboth, V.; Herrera-Estrella, A.; Horwitz, B.A.; Kenerley, C.M.; Monte, E.; Mukherjee, P.K.; Zeilinger, S.; Grigoriev, I.V.; Kubicek, C.P. *Trichoderma: The* genomics of opportunistic success. *Nat. Rev. Microbiol.* **2011**, *9*, 749–759. [CrossRef]
20. Bimenya, G.S.; Kaviri, D.; Mbona, N.; Byarugaba, W. Monitoring the severity of iodine deficiency disorders in Uganda. *Afr. Health Sci.* **2002**, *2*, 63–68.
21. Berg, G. Plant–microbe interactions promoting plant growth and health: Perspectives for controlled use of microorganisms in agriculture. *Appl. Microbiol. Biot.* **2009**, *84*, 11–18. [CrossRef]
22. Samuels, G.J.; Dodd, S.L.; Gams, W.; Castlebury, L.A.; Petrini, O. *Trichoderma* species associated with the green mold epidemic of commercially grown *Agaricus bisporus*. *Mycologia* **2002**, *94*, 147–170. [CrossRef]
23. Largeteau, M.L.; Savoie, J.M. Microbially induced diseases of *Agaricus bisporus*: Biochemical mechanisms and impact on commercial mushroom production. *App. Microbiol. Biotechnol.* **2010**, *86*, 63–73. [CrossRef]
24. Krupke, O.A.; Castle, A.J.; Rinker, D.L. The North American mushroom competitor, *Trichoderma aggressivum* f. *aggressivum*, produces antifungal compounds in mushroom compost that inhibit mycelial growth of the commercial mushroom *Agaricus bisporus*. *Mycol. Res.* **2003**, *102*, 1467–1475.
25. Guthrie, J.L.; Castle, A.J. Chitinase production during interaction of *Trichoderma aggressivum* and *Agaricus bisporus*. *Can. J. Microbiol.* **2006**, *96*, 961–967. [CrossRef]
26. Hatvani, L.; Antal, Z.; Manczinger, L.; Szekeres, A.; Druzhinina, I.S.; Kubicek, C.P.; Nagy, A.; Nagy, E.; Vagvolgyi, C.; Kredics, L. Green mold diseases of *Agaricus* and *Pleurotus* spp. are caused by related but phylogenetically different *Trichoderma* species. *Phytopathology* **2007**, *97*, 532–537. [CrossRef]
27. Sharma, V.; Shanmugam, V. Unraveling the multilevel aspects of least explored plant beneficial *Trichoderma saturnisporum* isolate GITX-Panog. *Eur. J. Plant Pathol.* **2018**, *152*, 169–183. [CrossRef]
28. Louden, B.C.; Harmann, D.; Lynne, A.M. Use of Blue Agar CAS Assay for Siderophore Detection. *J. Microbiol. Biol. Educ.* **2011**, *12*, 51–53. [CrossRef] [PubMed]
29. Lima-Rivera, D.L.; Lopez-Lima, D.; Desgarennes, D.; Velázquez-Rodríguez, A.S.; Carrion, G. Phosphate solubilization by fungi with nematicidal potential. *J. Soil Sci. Plant Nutr.* **2016**, *16*, 507–524. [CrossRef]
30. Fiske, C.H.; Subbarow, Y. The colorimetry determination of phosphorous. *J. Biol. Chem.* **1925**, *66*, 375–400.
31. Moreno-Gavíra, A.; Diánez, F.; Sánchez-Montesinos, B.; Santos, M. *Paecilomyces variotii* as a plant-growth promoter in horticulture. *Agronomy* **2020**, *10*, 597. [CrossRef]
32. Murali, M.; Amruthesh, K.N.; Sudisha, J.S.; Niranjana, R.; Shetty, H.S. Screening for plant growth promoting fungi and their ability for growth promotion and induction of resistance in pearl millet against downy mildew disease. *J. Phytol.* **2012**, *4*, 30–36.
33. Dickson, A.; Leaf, A.L.; Hosner, J.F. Quality appraisal of white spruce and white pine seedling stock in nurseries. *Forest Chron.* **1960**, *36*, 10–13. [CrossRef]
34. Stewart, A.; Hill, R. Applications of Trichoderma in Plant growth promotion. In *Biology and Applications of Trichoderma*; Mukherjee, P.K., Horwitz, B.A., Singh, U.S., Mukherjee, M., Schmoll, M., Eds.; CABI: Wallingford, CT, USA, 2013; pp. 415–425.

35. Allaga, H.; Bóka, B.; Poór, P.; Nagy, V.D.; Szuzs, A.; Stankovics, I.; Takó, M.; Manczinger, L.; Vágvölgyi, C.; Kredics, L.; et al. Composite Bioinoculant Based on the Combined Application of Beneficial Bacteria and Fungi. *Agronomy* **2020**, *10*, 220. [CrossRef]
36. Wang, G.; Cao, X.; Ma, X.; Guo, M.; Liu, C.; Yan, L.; Bian, Y. Diversity and effect of *Trichoderma* spp. associated with green mold disease on *Lentinula edodes* in China. *Microbiologyopen* **2016**, *5*, 709–718. [CrossRef] [PubMed]
37. Zhang, T.; Lu, M.Z.; Zhang, C.L.; Xu, J.Z. First Report of *Trichoderma longibrachiatum* Causing Green Mold Disease on *Ganoderma lingzhi*. *Plant Dis.* **2019**, *103*, 156. [CrossRef]
38. Lane, B.S.; Trinci, A.P.J.; Gillespie, A.T. Influence of culture conditions on the virulence of conidia and blastospores of *Beauveria bassiana* to the green leafhopper, *Nephotettix virescens*. *Mycol. Res.* **1991**, *95*, 829–833. [CrossRef]
39. Serna-Díaz, M.G.; Mercado-Flores, Y.; Jiménez-González, A.; Anducho-Reyes, M.A.; Medina-Marín, J.; Seck Tuoh-Mora, J.C.; Téllez-Jurado, A. Use of barley straw as a support for the production of conidiospores of *Trichoderma harzianum*. *Biotechnol. Rep.* **2020**, *26*, e00445. [CrossRef]
40. Flodman, H.R.; Noureddini, H. Effects of intermittent mechanical mixing on solid-state fermentation of wet corn distiller grain with *Trichoderma reesei*. *Biochem. Eng. J.* **2013**, *81*, 24–28. [CrossRef]
41. Deschamps, F.; Giuliano, C.; Asther, M.; Huet, M.C.; Roussos, S. Cellulase production by *Trichoderma harzianum* in static and mixed solid-state fermentation reactors under nonaseptic conditions. *Biotechnol. Bioeng.* **1985**, *27*, 1385–1388. [CrossRef]
42. Michel-Aceves, A.C.; Otero-Sánchez, M.A.; Martínez-Rojero, R.D.; Rodríguez-Morán, N.L.; Ariza-Flores, R.; Barrios-Ayala, A. Producción masiva de *Trichoderma harzianum* Rifai en diferentes sustratos orgánicos. *Rev. Chapingo Ser. Hortic.* **2008**, *14*, 185–191. [CrossRef]
43. Zhang, F.G.; Yuan, J.; Yang, X.M.; Cui, Y.Q.; Chen, L.H.; Ran, W.; Shen, Q.R. Putative *Trichoderma harzianum* mutant promotes cucumber growth by enhanced production of indoleacetic acid and plant colonization. *Plant Soil* **2013**, *368*, 433–444. [CrossRef]
44. Contreras-Cornejo, H.A.; Macías-Rodríguez, L.; Cortés-Penagos, C.; López-Bucio, J. *Trichoderma virens*, a plant beneficial fungus, enhances biomass production and promotes lateral root growth through an auxin-dependent mechanism in Arabidopsis. *Plant Phys.* **2009**, *149*, 1579–1592. [CrossRef] [PubMed]
45. Hoyos-Carvajal, L.; Orduz, S.; Bissett, J. Growth stimulation in bean (*Phaseolus vulgaris* L.) by *Trichoderma*. *Biol. Control.* **2009**, *51*, 409 416. [CrossRef]
46. Vinale, V.; Nigro, M.; Sivasithamparam, K.; Flematti, G.; Ghisalberti, E.L.; Ruocco, M.; Varlese, R.; Marra, R.; Lanzuise, S.; Eid, A.; et al. Harzianic acid: A novel siderophore from *Trichoderma harzianum*. *FEMS Microbiol. Lett.* **2013**, *347*, 123–129. [CrossRef] [PubMed]
47. Qi, W.; Zhao, L. Study of the siderophore-producing *Trichoderma asperellum* Q1 on cucumber growth promotion under salt stress. *J. Basic Microbiol.* **2013**, *53*, 355–364. [CrossRef]
48. Saber, W.I.A.; Ghoneem, K.; Rashad, Y.M.; AlAskar, A.A. *Trichoderma Harzianum* WKY1: An indole acetic acid producer for growth improvement and anthracnose disease control in sorghum. *Biocontrol. Sci. Technol.* **2017**, *27*, 654–676. [CrossRef]
49. Vinale, F.; Sivasithamparam, K.; Ghisalberti, E.L.; Woo, S.L.; Nigro, M.; Marra, R.; Lombardi, N.; Pascale, A.; Ruocco, M.; Lanzuise, S.; et al. Trichoderma secondary metabolites active on plants and fungal pathogens. *Open Mycol. J.* **2014**, *8*, 127–139. [CrossRef]
50. Gravel, V.; Antoun, V.; Tweddell, R.J. Growth stimulation and fruit yield improvement of greenhouse tomato plants by inoculation with *Pseudomonas putida* or *Trichoderma atroviride*: Possible role ofindoleacetic acid (IAA). *Soil Biol. Biochem.* **2007**, *39*, 1968–1977. [CrossRef]
51. Bader, A.N.; Salerno, G.L.; Covacevich, F.; Consolo, V.F. Native *Trichoderma harzianum* strains from Argentina produce indole 3 acetic acid and phosphorus solubilization, promote growth and control wilt disease on tomato (*Solanum lycopersicum* L.). *J. King Saud Univ.-Sci.* **2020**, *32*, 867–873. [CrossRef]
52. Altomare, C.; Norvell, W.A.; Bjorkman, T.G.; Harman, E. Solubilization of phosphates and micronutrients by the plant-growth-promoting and biocontrol fungus *Trichoderma harzianum* Rifai. *Appl. Environ. Microbiol.* **1999**, *65*, 2926–2933. [CrossRef]
53. Kapri, A.; Tiwari, L. Phosphate solubilization potential and phosphatase activity of rhizospheric *Trichoderma* spp. *Braz. J. Microbiol.* **2010**, *41*, 787–795. [CrossRef]
54. Tandon, A.; Fatima, T.; Gautam, A.; Yadav, U.; Srivastava, S.; Singh, P.C. Effect of *Trichoderma koningiopsis* on chickpea rhizosphere activities under different fertilization regimes. *Open J. Soil Sci.* **2018**, *8*, 261–275. [CrossRef]

55. Tandon, A.; Fatima, T.; Shukla, D.; Tripathi, P.; Srivastava, S.; Singh, P.C. Phosphate solubilization by *Trichoderma koningiopsis* (NBRI-PR5) under abiotic stress conditions. *J. King Saud Univ.-Sci.* **2020**, *32*, 791–798. [CrossRef]
56. Hajieghrari, B. Effects of some Iranian *Trichoderma* isolates on maize seed germination and seedling vigor. *Afr. J. Biotechnol.* **2010**, *9*, 4342–4347.
57. You, J.; Zhang, J.; Wua, M.; Yang, L.; Chen, W.; Li, G. Multiple criteria-based screening of *Trichoderma* isolates for biological control of *Botrytis cinerea* on tomato. *Biol. Control* **2016**, *101*, 31–38. [CrossRef]
58. Diniz, K.A.; Silva, P.A.; Oliveira, J.A.; Evangelista, J.R.E. Sweet pepper seed responses to inoculation with microorganisms and coating with micronutrients, aminoacids and plant growth regulators. *Sci. Agric.* **2009**, *66*, 293–297. [CrossRef]
59. Liu, B.; Jib, S.; Zhang, H.; Wang, Y.; Liu, Z. Isolation of *Trichoderma* in the rhizosphere soil of *Syringa oblata* from Harbin and their biocontrol and growth promotion function. *Microbiol. Res.* **2020**, *235*, 126445. [CrossRef]

Permissions

All chapters in this book were first published in MDPI; hereby published with permission under the Creative Commons Attribution License or equivalent. Every chapter published in this book has been scrutinized by our experts. Their significance has been extensively debated. The topics covered herein carry significant findings which will fuel the growth of the discipline. They may even be implemented as practical applications or may be referred to as a beginning point for another development.

The contributors of this book come from diverse backgrounds, making this book a truly international effort. This book will bring forth new frontiers with its revolutionizing research information and detailed analysis of the nascent developments around the world.

We would like to thank all the contributing authors for lending their expertise to make the book truly unique. They have played a crucial role in the development of this book. Without their invaluable contributions this book wouldn't have been possible. They have made vital efforts to compile up to date information on the varied aspects of this subject to make this book a valuable addition to the collection of many professionals and students.

This book was conceptualized with the vision of imparting up-to-date information and advanced data in this field. To ensure the same, a matchless editorial board was set up. Every individual on the board went through rigorous rounds of assessment to prove their worth. After which they invested a large part of their time researching and compiling the most relevant data for our readers.

The editorial board has been involved in producing this book since its inception. They have spent rigorous hours researching and exploring the diverse topics which have resulted in the successful publishing of this book. They have passed on their knowledge of decades through this book. To expedite this challenging task, the publisher supported the team at every step. A small team of assistant editors was also appointed to further simplify the editing procedure and attain best results for the readers.

Apart from the editorial board, the designing team has also invested a significant amount of their time in understanding the subject and creating the most relevant covers. They scrutinized every image to scout for the most suitable representation of the subject and create an appropriate cover for the book.

The publishing team has been an ardent support to the editorial, designing and production team. Their endless efforts to recruit the best for this project, has resulted in the accomplishment of this book. They are a veteran in the field of academics and their pool of knowledge is as vast as their experience in printing. Their expertise and guidance has proved useful at every step. Their uncompromising quality standards have made this book an exceptional effort. Their encouragement from time to time has been an inspiration for everyone.

The publisher and the editorial board hope that this book will prove to be a valuable piece of knowledge for researchers, students, practitioners and scholars across the globe.

List of Contributors

Ruijun Qin and Scott Lukas
Hermiston Agricultural Research and Extension Center, Oregon State University, Hermiston, OR 97838, USA

Christos Noulas
Institute of Industrial and Forage Crops, Hellenic Agricultural Organization-"Demeter", 41335 Larissa, Greece

Don Wysocki
Columbia Basin Agricultural Research Center, Oregon State University, Adams, OR 97810, USA

Xi Liang
Aberdeen Research and Extension Center, University of Idaho, Aberdeen, ID 83210, USA

Guojie Wang
Eastern Oregon Agriculture and Natural Resources Program, La Grande, OR 97850, USA

Radu-Liviu Sumalan, Isidora Radulov and Renata-Maria Sumalan
Faculty of Horticulture and Forestry, Banat's University of Agriculture Science and Veterinary Medicine "King Michael Ist of Romania" from Timisoara, Calea Aradului nr 119, 300645, Timisoara, Romania

Mihaela Petric and Manuela Crisan
"Coriolan Dragulescu" Institute of Chemistry, 24 Mihai Viteazul Blvd., 300223, Timisoara, Romania

Lilia Croitor
"Coriolan Dragulescu" Institute of Chemistry, 24 Mihai Viteazul Blvd., 300223, Timisoara, Romania
Institute of Applied Physics, Academiei Street 5, MD2028, Chisinau, Moldova

Paulina Bourosh
Institute of Applied Physics, Academiei Street 5, MD2028, Chisinau, Moldova

Motsim Billah
Department of Soil Science, University of Haripur, Haripur 21120, Pakistan

Matiullah Khan
Pakistan Agriculture Research Council, NARC, Islamabad 44000, Pakistan

Asghari Bano
Department of Biosciences, University of Wah, Wah Cantt. 47000, Pakistan

Sobia Nisa
Department of Microbiology, University of Haripur, Haripur 21120, Pakistan

Ahmad Hussain
Department of Forestry, University of Haripur, Haripur 21120, Pakistan

Khadim Muhammad Dawar
Department of Soil and Environmental Science, The University of Agriculture, Peshawar 25000, Pakistan

Asia Munir
Soil and Water Testing Laboratory, Rawalpindi 46000, Pakistan

Alejandro Moreno-Gavíra and Brenda Sánchez-Montesinos
Departamento de Agronomía, Escuela Superior de Ingeniería, Universidad de Almería, 04120 Almería, Spain

Yaser M. Hafez
Plant Pathology and Biotechnology Lab., Faculty of Agriculture, Kafrelsheikh University, Kafr El Sheikh 33516, Egypt

Ahmed M. Eid
Botany and Microbiology Department, Faculty of Science, Al-Azhar University, Nasr City 11884, Egypt

Mohamed A. M. Ali
Department of Horticulture, Faculty of Agriculture, New Valley University, El-Kharga, New Valley 72511, Egypt

El Chami Daniel and Galli Fabio
Timac Agro Italia S.p.A., S.P.13—Località Ca' Nova, I-26010 Ripalta Arpina (CR), Italy

Muhammad Adnan Shahid
Department of Agriculture, Nutrition and Food Systems, University of New Hampshire, Durham, NH 03824, USA

Ali Sarkhosh
Horticultural Sciences Department, IFAS-University of Florida, Gainesville, FL 32611, USA

Rashad Mukhtar Balal
Department of Horticulture, College of Agriculture, University of Sargodha, Sargodha 40100, Pakistan

Shahid Ali
Plant Epigenetic and Development, Northeast Forestry University Harbin, Heilongjiang 150040, China

Lorenzo Rossi
Horticultural Sciences Department, University of Florida, IFAS-Indian River Research and Education Center, Fort Pierce, FL 34945, USA

Celina Gómez
Environmental Horticulture Department, IFAS-University of Florida, Gainesville, FL 32611, USA

Neil Mattson
Horticulture Section, School of Integrative Plant Science, Cornell University, Ithaca, NY 14853, USA

Wajid Nasim
Department of Agronomy, Faculty of Agriculture and Environmental Sciences, The Islamia University of Bahawalpur (IUB), Punjab 63100, Pakistan

Francisco Garcia-Sanchez
CEBAS-CSIC, Departamento De Nutrición Vegetal, Campus Universitario de Espinardo, Espinardo, 30100 Murcia, Spain

Israel Benítez García, Ana Karen Ducñas Ledezma, Jesús Aarón Salazar Leyva and Idalia Osuna Ruiz
Maestría en Ciencias Aplicadas, Unidad Académica de Ingeniería en Biotecnología, Universidad Politécnica de Sinaloa (UPSIN), Carretera Municipal Libre Mazatlán Higueras Km 3., Mazatlán, SIN 82199, Mexico

Emmanuel Martínez Montaño
Maestría en Ciencias Aplicadas, Unidad Académica de Ingeniería en Biotecnología, Universidad Politécnica de Sinaloa (UPSIN), Carretera Municipal Libre Mazatlán Higueras Km 3, Mazatlán, SIN 82199, Mexico

Esther Carrera
Instituto de Biología Molecular y Celular de Plantas (UPV-CSIC), Instituto de Biología Molecular y Celular de Plantas (IBMCP), Universidad Politécnica de Valencia-Consejo Superior de Investigaciones Científicas (CSIC), 46022 Valencia, Spain

Fanrong Zeng
School of Agriculture, Yangtze University, 88 Jingmi Road, Jingzhou 434025, China

Munazza Zahoor, Muhammad Rizwan and Tahira Yasmeen
Department of Environmental Sciences and Engineering, Government College University, Faisalabad 38000, Pakistan

Muhammad Waseem
Department of Microbiology, Government College University, Faisalabad 38000, Pakistan

Alia Anayat
Soil & Water Testing Lab., Ayub Agricultural Research Institute, Jhang Road, Faisalabad 38000, Pakistan

Awais Ahmad
Department of Chemistry, the University of Lahore, Lahore 54590, Pakistan

Shafaqat Ali
Department of Environmental Sciences and Engineering, Government College University, Faisalabad 38000, Pakistan
Department of Biological Sciences and Technology, China Medical University, Taichung 40402, Taiwan

Mohamed A. El-Sheikh, Mohammed Nasser Alyemeni and Leonard Wijaya
Department of Botany and Microbiology, College of Science, King Saud University, Riyadh 11451, Saudi Arabia

Muneera D. F. ALKahtani
Biology Department, College of Science, Princess Nourah Bint Abdulrahman University, Riyadh POX 102275-11675, Saudi Arabia

Amr Fouda, Ahmed M. Eid, Emad El-Din Ewais and Saad El-Din Hassan
Botany and Microbiology Department, Faculty of Science, AL-Azhar University, Nasr City, Cairo 11884, Egypt

Kotb A. Attia
Center of Excellence in Biotechnology Research, King Saud University, Riyadh POX 2455-11451, Saudi Arabia

Fahad Al-Otaibi
Biodiversity Centre, Institut de Recherche en Biologie Végétale, Université de Montréal and Jardin botanique de Montréal, Montréal, QC 22001, Canada
Department of Soil Science, King Saud University, Riyadh 11564, Saudi Arabia

Mohamed Hijri
Biodiversity Centre, Institut de Recherche en Biologie Végétale, Université de Montréal and Jardin botanique de Montréal, Montréal, QC 22001, Canada
AgroBioSciences, Mohammed VI Polytechnic University (UM6P), 43150 Ben Guerir, Morocco

Marc St-Arnaud
Biodiversity Centre, Institut de Recherche en Biologie Végétale, Université de Montréal and Jardin botanique de Montréal, Montréal, QC 22001, Canada

Naeem Khan
Department of Agronomy, Institute of Food and Agricultural Sciences, University of Florida, Gainesville, FL 32611, USA

Khaled A. A. Abdelaal
Excellence Center (EPCRS), Plant Pathology and Biotechnology Laboratory, Faculty of Agriculture, Kafrelsheikh University, Kafr El Sheikh 33516, Egypt

Ateeq Shah and Donald L. Smith
Department of Plant Science, Macdonald Campus, McGill University, Sainte-Anne-de-Bellevue, QC H9X 3V9, Canada

Lin Li, Zheng Zhang, Hua Tian, Zhaowen Mo, Meiyang Duan, Shuli Wang, Xiangru Tang and Shenggang Pan
State Key Laboratory for Conservation and Utilization of Subtropical Agro-bioresources, College of Agriculture, South China Agricultural University, Guangzhou 510642 China
Scientific Observing and Experimental Station of Crop Cultivation in South China, Ministry of Agriculture PR China, Guangzhou 510642, China

Umair Ashraf
State Key Laboratory for Conservation and Utilization of Subtropical Agro-bioresources, College of Agriculture, South China Agricultural University, Guangzhou 510642 China
Department of Botany, University of Education (Lahore), Faisalabad-Campus, Faisalabad 38000, Punjab, Pakistan

Zaiman Wang
Key Laboratory of Key Technology for South Agricultural Machine and Equipment, Ministry of Education, Guangzhou 510642, China

Brenda Sánchez-Montesinos, Fernando Diánez, Alejandro Moreno-Gavíra and Mila Santos
Departamento de Agronomía, Escuela Superior de Ingeniería, Universidad de Almería, 04120 Almería, Spain

Francisco J. Gea
Centro de Investigación, Experimentación y Servicios del Champiñón (CIES), Quintanar del Rey, 16220 Cuenca, Spain

Index

R

S

T

W

Printed in the USA
CPSIA information can be obtained
at www.ICGtesting.com
LVHW080055211023
761596LV00010BA/2062

9 781641 168151